Charles Seale-Hayne Library

University of Plymouth

(01752) 588 588

LibraryandITenquiries@plymouth.ac.uk

PROGRESS IN GEOMORPHOLOGY

papers in honour of David L. Linton

INSTITUTE OF BRITISH GEOGRAPHERS
SPECIAL PUBLICATION

No. 7 June 1974

Compiled and edited for the British Geomorphological Research Group by
E. H. BROWN AND R. S. WATERS

LONDON
INSTITUTE OF BRITISH GEOGRAPHERS

1 KENSINGTON GORE, SW7 2AR

1974

0121377806 ✓
0113648

89481

551·4
INS

Printed in Great Britain by Alden & Mowbray Ltd
at the Alden Press, Oxford

Contents

PREFACE

THE purpose of this volume of essays in honour of the late Professor David L. Linton, founder and first chairman of the British Geomorphological Research Group, is to record developments in those aspects of geomorphology in which his own contributions were influential and stimulating. The fifteen essays have been contributed by members of the research group and colleagues in Europe, Australia and New Zealand. They are arranged in five thematic groups, in the introductions to which reference is made to the relevant publications of Professor Linton. A bibliography of his published work is given in the Transactions of the Institute of British Geographers, no. 55, March 1972.

E. H. BROWN
R. S. WATERS

Geomorphology in the United Kingdom since the first world war

E. H. BROWN

Professor of Geography, University College London

AND

R. S. WATERS

Professor of Geography, University of Sheffield

Revised MS received 1 August 1973

ABSTRACT. Before the first world war, geomorphologists in Britain were mainly geologists, concerned with the chronological significance of upland plateaux, the origins of rivers, and the impact of glaciation. Between the wars, geomorphology began to be studied in university departments of geography, and the continued interest in denudation chronology was much influenced by the ideas of W. M. Davis and H. Baulig. Since the second world war, a great expansion in the number of geomorphologists, again largely amongst geographers in the universities, has been matched by an explosion of geomorphological research. In the early post-war years, regional studies of denudation chronology following the model devised for south-east England by Wooldridge and Linton dominated. Later, studies of slope form and evolution developed from a late appreciation of the contribution of W. Penck to geomorphology. Climatic geomorphology found expression in studies of periglacial phenomena and landforms and processes in arid and semi-arid regions. Only recently has there been much interest in the humid tropics. A more quantitative approach to geomorphology, together with a greater concern for the study of the fluvial process, spread from the United States. But the glacial theme continues to be well supported, and approximately 36 per cent of all active geomorphological research relates to glacial and periglacial phenomena. Meltwater channels and raised beaches have received particular attention. There is an increasing application of geomorphological research to environmental problems, especially soils, slope stability and terrain evaluation.

IN the United Kingdom up to and including the first world war, most interest in geomorphology was to be found, not unnaturally, among geologists. The geomorphological work was, for the most part, concerned with the evolution of landforms within a stratigraphical framework. Geologists assumed that structure and lithology were the primary controls of the detailed shape of the ground and paid little attention to the physical processes acting upon geological materials to produce the landforms on whose development they passed comment. Although W. M. Davis published his classic paper on the geographical cycle in London in 1899, and had made geomorphological excursions to the United Kingdom, as a consequence of which he had written papers on aspects of the geomorphology of a number of areas in Britain (Davis, 1909 and 1912), nevertheless, the impact of his ideas seems to have been minimal. Interest amongst British geomorphologists was at this time focused principally upon the origins of river courses and upland plateaux. The latter in particular, ever since Ramsay first described the 'great inclined plane' in central Wales in 1846, had a special fascination for British geologists. But above all, geological geomorphologists in Britain were interested in the erosional work of the Pleistocene ice sheets and associated depositional forms. Much controversy revolved around the relative efficacy of ice as an agent of erosion. Major contributions were made from many parts of Britain and summarized by W. B. Wright in 1914, for forty years the only comprehensive review. Particularly distinctive was P. F. Kendall's study of glacial meltwater channels in the North York Moors (Kendall, 1902), a theme which was to figure prominently in later research. The

3

coast of Britain also excited interest but because geologists were mainly concerned with cliff exposures of bed rock, and not with the shape of cliffs or forms of coastal deposition, publications in the field of coastal geomorphology before 1920 were somewhat limited. A notable exception was E. N. A. Arber's book on the coast scenery of north Devon (1911). The reports and minutes of evidence of the Royal Commission on Coast Erosion (1907–11), a founding document in studies of coastal geomorphology, arose out of a need to protect the coast from erosion, and was primarily the work of engineers.

BETWEEN THE WARS

The three-fold concern of British geomorphology with denudation chronology, glaciation and coastal forms, was continued between the first and second world wars. Whereas before 1914 geomorphological work had been mostly a by-product of geological research, from 1920 homes for geomorphologists were created in newly established departments of geography in British universities. Many such departments started life as sub-departments in departments of geology, and not unnaturally, some teachers of geology were commissioned to teach the required courses in geography. In this way, a number of people, trained initially in geology, were to embark upon geographical careers with a particular interest in research and teaching in geomorphology. Foremost amongst these were S. W. Wooldridge, for over thirty years a dominant personality at King's College London, and A. Austin Miller at the University of Reading. But geologists in the universities continued to research and publish in geomorphology, notably A. E. Trueman at Swansea, O. T. Jones at Cambridge, and J. K. Charlesworth in Belfast. In the Geological Survey, many officers such as S. E. Hollingworth pursued matters of geomorphological interest and published their findings. Geomorphology had a particular appeal for amateur geologists who made valuable contributions in the field of denudation chronology, with particular reference to the nature and origins of river terraces. In south-east England the members of the Weald Research Committee of the Geologists' Association studied not only its solid geology but also the terraces and long profiles of its rivers. J. F. N. Green, one of their number, gave his presidential address to the Geological Society of London on the subject of river terraces in southern England (1936).

In the fullness of time the new departments of geography produced their own research geomorphologists. In Cambridge, J. A. Steers and W. V. Lewis became the stimuli of an influential school of geomorphological research concentrating upon coastal and glacial geomorphology respectively, whilst in London D. L. Linton was amongst the first graduates in geography to take up geomorphological research under the supervision of S. W. Wooldridge before moving to Edinburgh.

The predominant theme in geomorphological research between the two wars remained denudation chronology, with particular emphasis upon the twin themes of drainage evolution and planation surfaces. This school of geomorphology found its principal adherents in London but others contributed from Reading and Scotland. In Cambridge, interests were focused upon glaciation, especially the process of glacial erosion and coasts, with particular reference to processes involved in coastal deposition. In 1935, Henri Baulig visited London from Strasbourg to give a series of lectures published under the title of *The Changing Sea Level*. These appear to have triggered off a prodigious flow of studies of pre-glacial and glacial sea levels in Britain which lasted for 25 years. Little or no interest was shown in studies of geomorphological processes other than glacial and coastal, with the exception of the solitary work of R. A. Bagnold which was little known until the publication of his book on the *Physics of Blown Sand and Desert Dunes* in 1941, by which time the second world war was demanding the interest of most geomorphologists.

POST SECOND WORLD WAR

The years 1946–50 saw a sudden increase in the number of research workers in geomorphology as the universities began to expand to accommodate the flood of returning ex-servicemen and the increased number of school leavers going to university. Many geography graduates of the time developed a particular interest in geomorphology and chose to follow the traditional theme of denudation chronology in their research, inspired particularly by Wooldridge and Linton's *Structure, surface and drainage in South-east England*. Its impact when first published in 1939 was lost during the second world war but the geomorphological research explosion of the late 1940s and 1950s was based upon attempts to explore the themes of drainage evolution and land surface development in other parts of Britain using south-eastern England as the model (E. H. Brown, 1960).

The writings of Walter Penck, on the relationships between slope form and processes, published as early as 1924 in German, had an even longer delayed impact in Britain. Because of the linguistic ignorance of most British geomorphologists, they had remained largely unknown until after the second world war. They then became known to British students in a variety of ways. First and foremost through the report of the symposium on Penck's contribution to geomorphology held in Chicago in 1939 (O. D. Von Engeln, 1940) which did not become widely available in Britain until after the war. Secondly through new American text books (Von Engeln, 1942) and thirdly only in 1953 from the translation of Penck by Czech and Boswell. One of the earliest manifestations in Britain of Penck's views on the parallel nature of slope retreat was Alan Wood's paper on slope evolution published in wartime in 1942 and not fully appreciated until the post-war era. Discussions on the relative efficacy of slope down-wasting, as proposed by W. M. Davis, and parallel retreat, following Penck, were extremely active in the 1950s, and led eventually to the beginnings of the serious study of slope evolution in Britain (A. F. Pitty 1969; C. A. Carson and M. J. Kirkby, 1972). Nowhere was interest greater than in the University of Sheffield where D. L. Linton was instrumental in encouraging the work of R. A. G. Savigear (1953) and A. Young (1972).

A symposium in the U.S.A. in honour of the 100th anniversary of the birth of William Morris Davis (K. Bryan, 1950) had a major impact also upon British geomorphological research, especially the contribution of L. C. Peltier (1950) on the geographic cycle in periglacial regions as it is related to climatic morphology. It was principally through Peltier's paper that the concept of climatic geomorphology came to Britain even though it was of some antiquity in Europe. It also renewed interest in earlier British publications dealing with evidence of periglacial fashioning of landforms, notably the important paper on terminology by Dines (1940). The British contribution to climatic geomorphology has been made particularly in semi-arid environments (R. F. Peel, 1964; R. U. Cooke and A. Warren, 1973), largely because in the 1950s a considerable number of British geomorphologists found employment in the new universities of what was rapidly ceasing to be the colonial empire. Their work was to a considerable extent focused on the nature and origins of pediments and they were much stimulated by contemporary work in California and by the publications of L. C. King in South Africa. A number of visits by King to British universities helped to maintain a vigorous interest in all forms of slope development as well as in pediment formation (L. C. King, 1953). The periglacial theme has continued to develop in the past 25 years and it has become an integral part of studies of Quaternary landform evolution in Britain.

At the time when studies of denudation chronology were at their zenith in the early 1950s in Britain, a new approach to geomorphology was being developed by A. N. Strahler (1950) in New York. Critical of the fact that W. M. Davis' approach to geomorphology was entirely qualitative, what was needed he wrote was a much more rigorous measurement of landform shapes and a more penetrating analysis of erosional processes. A reflection of this quantitative approach to geomorphology was the development in Britain of a system of geomorphological mapping of slopes inspired by David Linton and developed by R. A. G. Savigear and R. S. Waters (1958) at Sheffield. But the full impact of Strahler's views was not to be felt in Britain until they were introduced in a vigorous manner in the writings of R. J. Chorley (1958), a graduate of Oxford, who had done postgraduate work in Columbia University under Strahler. Amongst the many facets of this dynamic approach to geomorphology the measurement of the morphometric properties of landforms, especially from published maps, has had a particular appeal in Britain.

Research into fluvial processes in the United States after 1945 stimulated especially by hydraulic engineers such as Horton (1945) and studies such as those of Schumm (1956) at Perth Amboy, led in 1964 to the publication of L. B. Leopold, M. G. Wolman and J. P. Miller's trend-setting book, *Fluvial Processes in Geomorphology*. Its emphasis upon process created particular interest amongst younger geomorphologists in Britain and its impact was perhaps the greater because the year of its publication was also the occasion of the 20th International Geographical Congress in London, and it formed one of the major items of conversation amongst geomorphological delegates at the Congress. Many of the ideas it contains have been followed up by younger research workers and it undoubtedly led to an increased interest in the processes operating in small drainage catchments (K. J. Gregory and D. Walling, 1974) and to increased links with hydrologists, a situation which is still developing. The establishment of the Hydraulics Research Station in 1946 and the Institute of Hydrology in 1960 were additional stimuli. Process studies in coastal geomorphology have also increased, notable amongst them are those of C. A. M. King (1959). Research with an engineering approach to coastal problems at the Hydraulic Research Station has led to considerable developments in geomorphological knowledge, especially concerning the longshore transport of materials and their accumulation as depositional features both above and below mean tide level.

One of the notable characteristics of scientific research in Britain since the second world war has been an increasing concern with the application of the results obtained from theoretical studies to practical problems. This is no less so in geomorphology than in any other science and in the past thirty years there have been a number of significant applications of geomorphological knowledge to the solution of environmental problems. Early among these was the application of geomorphological knowledge to the study of soil genesis. The appreciation that the geomorphological development of the land surface is an important clue to the origin of the nature of soils on that surface, has formed the background to much of the work of the Soil Survey of Britain (B. W. Avery, 1964). In return soil surveyors have made significant contributions to geomorphological knowledge, they have been especially concerned with the nature and origins of superficial deposits, and have pointed to characteristic weathering forms and products on ancient land surfaces (J. Loveday, 1962). They have also brought to light the presence of an aeolian contribution in many soils in England. Problems of slope stability, especially of those slopes developed in clays, have been studied by both engineers and geomorphologists. The importance of a knowledge of their Quaternary history has been amply demonstrated (A. W. Skempton, 1953). National concern for the conservation of coastal amenities prompted J. A. Steers' (1946) pioneer study of the coastline of England and Wales carried out on behalf of the then Ministry of Town and Country Planning. This was followed by the establishment of the Coastal Physiography Unit of the Nature Conservancy which carried out valuable research especially into the origins of depositional landforms (C. Kidson, 1963). Practical offshoots of the geomorphologist's concern with slope angles and the morphometry of landforms have been various systems of terrain evaluation used to assess the suitability of land surfaces for particular economic activities, especially agricultural and road development in the underdeveloped tropics (Beckett *et al.*, 1972).

Since the mid-1960s, study of the evolution of the pre-glacial land surface has virtually ceased and the emphasis in chronological studies has been on the development of landforms and especially deposits during the Quaternary (C. Embleton and C. A. M. King, 1968). As far as glacial erosion is concerned there has been a continued interest in the nature and origin of cirques, and a renewed study of glacial troughs and ice moulded rock surfaces (D. L. Linton, 1963) with particular reference to their morphometry and spatial distribution. J. B. Sissons (1960, 1961) initiated a wholesale reappraisal of the nature and origins of glacial meltwater channels from which it has been concluded that most were not related to overflow from proglacial lakes but were located marginally or sub-marginally to the ice. Contemporary proglacial deposition outside Britain has also been studied (R. J. Price, 1969). Geomorphologists have also made a major contribution to work on the chronology of glacial events in Britain (A. Straw, 1961; D. Q. Bowen, 1973). They have been particularly concerned with raised beaches (K. Walton, 1966).

There has developed since the second world war a numerically small but very active geomorphological interest in karst landforms (M. Sweeting, 1972). These studies have been influenced by the trends in geomorphological research in other fields. An early interest in denudation chronology has been followed by a concern with the measurement of processes and especially the load carried in solution by streams emerging from karst systems from which rates of solution of limestone are calculated. Interest has continued in the problems posed by the recurrence of dry valleys, especially in the Cretaceous chalk (G. T. Warwick, 1964; R. J. Small, 1964). The question of their possible origins has been pursued through analyses of their morphometry and the nature and age of the deposits which frequently mask their floors (M. P. Kerney *et al.*, 1964).

The post-war explosion in geomorphological publications had by 1960 reached such proportions that it was no longer possible for an individual to keep abreast of developments. Geomorphologists throughout the world were greatly helped at this juncture by the publication of Geomorphological Abstracts inspired and edited by K. M. Clayton.

The British Geomorphological Research Group was founded in January 1961 to encourage research in geomorphology, to undertake large-scale projects of research or compilation and to hold field-meetings and symposia. It was conceived by Professor David L. Linton and held its first meetings at the University of Sheffield and in December 1963 began to publish a register of current research in geomorphology. The first contained 163 entries, the latest (1973) has 896 relating to 467 individuals. Glacial and periglacial geomorphology dominates the scene with some 36 per cent of the entries, fluvial 25 per cent, weathering and slopes 18 per cent, coastal and regional studies each account for 6 per cent, and karst and arid land interests 4 per cent each.

David L. Linton was one of the first British geomorphologists to be born, academcially speaking, into the subject. He practised his craft through its infant pre-war years and had a major influence upon the form it took in its post-war maturity. Although he did not pursue a statistical approach to the subject he was alive to its possibilities and encouraged those who would. His personal prowess lay in the quality and style of his written argument and the lucidity of his field sketches. The denudation chronology of fluvial landscapes, the delimitation of morphological regions, the use of air photographs, problems of Scottish scenery, the geomorphology of the Sheffield region and Lincolnshire, the problem of tors, the forms of glacial erosion, the geomorphology of the moon and the assessment of scenery as a resource, all were illumined by his appraisal. The contributors to this volume seek to pay homage to the intellectual stimulus received from him by reporting the results of their own work on those aspects of geomorphology for which he would have been an obvious referee.

REFERENCES

ARBER, E. N. A. (1911) *The coast scenery of north Devon*

AVERY, B. W. (1964) 'The soils and land use of the district around Aylesbury and Hemel Hempstead', *Mem. Soil Surv. Gt. Br.*

BAGNOLD, R. A. (1941) *Physics of blown sand and desert dunes*

BAULIG, H. (1935) 'The changing sea level', *Trans. Inst. Br. Geogr.* 3, 1–46

BECKETT, P. H. T., R. WEBSTER, G. M. McNEIL and C. W. MITCHELL (1972) 'Terrain evaluation by means of a data bank', *Geogrl J.* 138, 430–56

BOWEN, D. Q (1973) 'The Pleistocene history of Wales and the borderland', *Geol. J.* 8, 207–24

BRITISH GEOMORPHOLOGICAL RESEARCH GROUP (1973) *Current research in geomorphology* (Norwich)

BROWN, E. H. (1960) *Relief and drainage of Wales* (Cardiff)

BRYAN, K. (1950) 'Symposium on geomorphology in honor of the 100th anniversary of the birth of William Morris Davis', *Ann. Ass. Am. Geogr.* 40, 172–236

CARSON, C. A. and M. J. KIRKBY (1972) *Hillslope form and process*

CHORLEY, R. J. (1958) 'Aspects of the morphometry of a "poly-cyclic" drainage basin', *Geogrl J.* 124, 370–4

CLAYTON, K. M. Ed. (1960) *Geomorphological Abstracts* (1966) *Geographical Abstracts A Geomorphology*, (1972) *Landforms and the Quaternary* (Norwich)

COOKE, R. U. and A. WARREN (1973) *Geomorphology in deserts*

DAVIS, W. M. (1899) 'The geographical cycle', *Geogrl J.* 14, 481–504

DAVIS, W. M. (1909) 'Glacial erosion in North Wales', *Q. J. geol. Soc. Lond.* 65, 281–350

DAVIS, W. M. (1912) 'A geographical pilgrimage from Ireland to Italy', *Ann. Ass. Am. Geogr.* 2, 73–100

DINES, H. G., S. E. HOLLINGWORTH, W. EDWARDS, S. BUCHAN and F. B. A. WELCH (1940) 'The mapping of head deposits', *Geol. Mag.* 77, 198–226

EMBLETON, C. and C. A. M. KING (1968) *Glacial and periglacial geomorphology*

GREEN, J. F. N. (1936) 'The terraces of southernmost England', *Q. J. geol. Soc. Lond.* 92, lviii–lxxxviii

GREGORY, K. J. and D. E. WALLING (1974) 'Fluvial processes in instrumented watersheds: studies of small watersheds in the British Isles, *Special Publication No. 6 Inst. Br. Geogr.*

HORTON, R. E. (1945) 'Erosional development of streams and their drainage basins: hydrophysical approach to quantitative geomorphology', *Bull. geol. Soc. Am.* 56, 275–370

KENDALL, P. F. (1902) 'A system of glacier lakes in the Cleveland hills', *Q. J. geol. Soc. Lond.* 58, 471–571

KERNEY, M. P., E. H. BROWN and T. J. CHANDLER (1964) 'The Late-glacial and Post-glacial history of the chalk escarpment near Brook, Kent', *Phil. Trans. R. Soc.* B, 248, 135–204

KIDSON, C. (1963) 'The growth of sand and shingle spits across estuaries', *Z. Geomorph.* 7, 1–22

KING, C. A. M. (1959) *Beaches and Coasts*

KING, L. C. (1953) 'Canons of landscape evolution', *Bull. geol. Soc. Am.* 64, 721–52

LEOPOLD, L. B., M. G. WOLMAN, and J. P. MILLER (1964) *Fluvial processes in geomorphology*

LINTON, D. L. (1963) 'The forms of glacial erosion', *Trans. Inst. Br. Geogr.* 33, 1–27

LOVEDAY, J. (1962) 'Plateau deposits of the southern Chiltern hills', *Proc. Geol. Ass.* 73, 83–102

PEEL, R. F. (1960) 'Some aspects of desert geomorphology', *Geography* 45, 241–62

PENCK, W. (1953) *Morphological analysis of land forms* (translated by H. Czech and K. C. Boswell)

PELTIER, L. C. (1950) 'The geographic cycle in periglacial regions as it is related to climatic morphology', *Ann. Ass. Am. Geogr.* 40, 214–36

PITTY, A. F. (1969) 'A scheme for hillslope analysis, 1. Initial considerations and calculations', *Univ. Hull. Occ. Pap. in Geogr.* 9

PRICE, R. J. (1969) 'Moraines, sandar, kames and eskers near Breidermerkujokull, Iceland', *Trans. Inst. Br. Geogr.* 46, 17–37

RAMSAY, A. C. (1846) 'The denudation of South Wales and adjacent English counties', *Mem. geol. Surv. Gt. Br.*

ROYAL COMMISSION ON COAST EROSION (AND AFFORESTATION) (1907–11) Coast erosion, the reclamation of tidal lands and afforestation in the United Kingdom *H.M. Stationery Office* 1 (1), cd 3683, and 1 (2) cd 3684, 1907: 2 (1), cd 4460 and 2 (2), cd 4461, 1909: 3 (1) cd 5708 and 3 (2), cd 5709, 1911

SAVIGEAR, R. A. G. (1953) 'Some observations on slope development in South Wales', *Trans. Inst. Br. Geogr.* 18, 31–51

SCHUMM, S. A. (1956) 'Evolution of drainage systems and slopes in badlands at Perth Amboy, New Jersey', *Bull. geol. Soc. Am.* 67, 597–646

SISSONS, J. B. (1960, 1961) 'Some aspects of glacial drainage channels in Britain', *Scott. Geogr. Mag.* 76, 131–46, 77, 15–36

SKEMPTON, A. W. (1953) 'Soil mechanics in relation to geology', *Proc. Yorks. Geol. Soc.* 29, 33–62

SMALL, R. J. (1964) 'The escarpment dry valleys of the Wiltshire Chalk', *Trans. Inst. Br. Geogr.* 34, 33–52

STEERS, J. A. (1946) *The coastline of England and Wales*

STRAHLER, A. (1950) 'Equilibrium theory of erosional slopes approached by frequency distribution analysis', *Am. J. Sci.* 248, 673–96

STRAW, A. (1961) 'Drifts, meltwater channels and ice margins in the Lincolnshire Wolds, *Trans. Inst. Br. Geogr.* 29, 115–28

SWEETING, M. (1972) *Karst Landforms*

VON ENGELN, O. D. (1942) *Geomorphology systematic and regional*

VON ENGELN, O. D. (1940) 'Symposium: Walther Penck's contribution to geomorphology', *Ann. Ass. Am. Geogr.* 30, 219–84

WALTON, K. (1966) 'Vertical movements of shorelines in Highland Britain', *Trans. Inst. Br. Geogr.* 39, 1–8

WARWICK, G. T. (1964) 'Dry valleys of the southern Pennines, England', *Erdkunde* 18, 116–23

WATERS, R. S. (1958) 'Morphological mapping', *Geography* 43, 10–17

WOOLDRIDGE, S. W. and D. L. LINTON (1939) 'Structure, surface and drainage in south-east England', *Trans. Inst. Br. Geogr.* 10, 1–124

WRIGHT, W. B. (1914) *The Quaternary ice age*

YOUNG, A. (1972) *Slopes*

RÉSUMÉ. *Géomorphologie au Royaume-Uni depuis la première Guerre Mondiale.* Avant la première Guerre Mondiale, les géomorphologues à la Grande-Bretagne était en grande partie des géologues qui s'intéressèrent à la signification chronologique des hauts plateaux, aux origines des fleuves, et à l'impact de glaciation. Entre les deux guerres, on commença à étudier la géomorphologie aux facultés de géographie dans les universités, et les idées de W. M. Davis et H. Baulig influèrent beaucoup l'intérêt soutenu à la chronologie d'érosion en surface. Depuis la deuxième Guerre Mondiale, une grande augmentation du nombre de géomorphologues, principalement entre les géographes aux universités, fut appariée par une explosion des recherches géomorphologiques. Pendant les premières années de l'après-guerre, des études régionaux de la chronologie d'érosion suivant le modèle combiné par Wooldridge et Linton pour le sud-est d'Angleterre, dominèrent. Plus tard, d'une appréciation tardive de la contribution de W. Penck à la géomorphologie, des études des formes de pente et d'évolution développèrent. Parmi les recherches sur la géomorphologie climatique se trouvèrent des études des phénomènes périglaciaires, et des formes du terrain et des processus dans les régions arides et demi-arides. Il n'est que récemment qu'on s'intéresse aux tropiques humides. Un abord quantitatif à la géomorphologie, avec un intérêt plus grand dans l'étude thème du processus fluvial, répandit des Etats-Unis. Mais le thème glaciaire est toujours bien soutenu et approximativement 36% de toutes recherches géomorphologiques actives ont rapport aux phénomènes glaciaires et périglaciaires. Chenaux d'eau de fonte et terrasses côtières reçoivent d'attention particulière. Il y a une application croissante de la recherche géomorphologique aux problèmes de l'environnement en particulier aux sols, à la stabilite des pentes et à l'évaluation du terrain.

ZUSAMMENFASSUNG. *Geomorphologie in dem Vereinigten Königreich seit dem ersten Weltkriege.* Vor dem ersten Weltkriege waren die Geomorphologen von Britannien meistens Geologen, die sich für die chronologische Bedeutung der Hochebenen, die Ursprünge von Flüssen, und die Wirkung von Vergletscherung interessieren. Zwischen den Kriegen begann man Geomorphologie in geographischen Abteilungen der Universitäten zu studieren, und die Ideen von W. M. Davis und H. Baulig hatten grossen Einfluss auf das andauernde Interesse an der Denudationschronologie. Seit dem zweiten Weltkriege wird eine grosse Zunahme der Zahl der Geomorphologen, nochmals mestens unter den Geographen in den Universitäten angepasst durch eine Explosion geomorphologischer Forschung. Während der frühen Nachkriegsjahre dominierten örtliche Studien von Denudationschronologie im Muster, erdacht von Wooldridge und Linton für Südosten Englands. Später entwickelten sich Abhangsform- und Evolutionsstudien aus einer späten Würdigung des Beitrages zu Geomorphologie von W. Penck. Klimageomorphologie fand Ausdruck in Studien von periglazialen Phänomenen und Landformen und Verfahren in dürren und halbdürren Gebieten. Erst kürzlich zeigt sich viel Interesse für die Feuchttropen. Eine mehr quantitative Methode der Geomorphologie, zusammen mit einem grösseren Interesse für das Studium des Flussverfahrens, verbreitsten sich von den Vereinigten Staaten. Das Glazialthema bleibt jedoch immer noch populär, und ungefähr 36% der ganzen regen geomorphologischen Forschung hat Bezug auf glaziale und periglaziale Phänomene. Schmeltzwasserrinnen und Küstenterrassen ziehen besondere Aufmerksamkeit auf sich. Es gibt eine wachsende Anwendung geomorphologischer Forschung zu den Umweltproblemen, besonders Böden, Abhangsstabilität, und Geländebewertung.

I. WEATHERING

Introduction

DAVID LINTON'S interest in the palaeogeomorphological significance of weathering mantles and tors began many years before the publication of 'The problem of tors', one of his most influential and widely discussed papers (D. L. Linton, 1955). In 1939 while working in the Scottish Highlands he began his field examination of 'the remarkable "tors" that crown several of the high Cairngorm summits and certain other Scottish mountains' (Linton, 1949, p. 31). Having noted that their association with 'old subdued surfaces' and deep weathering pointed to 'long-continued chemical weathering as an essential factor in their production' he interpreted them as elements in a suite of related forms and weathering products which could not have been produced in post-glacial times and could 'hardly have survived the passage across them of even the feeblest streams of moving ice'. Thereafter he used them as indicators of unglaciated enclaves in glaciated regions (Linton, 1949, 1950).

The Linton model of tor production, a two-stage process involving a phase of deep weathering followed by a phase of stripping, was first presented at the XVIIth I.G.U. Congress at Washington in 1952 and elaborated three years later at a meeting of the Royal Geographical Society (Linton, 1952, 1955). Described by Professor S. E. Hollingworth as an outstanding example of careful analysis of field evidence, the 1955 paper not only provoked discussion but also justified fully its author's expressed hope that it would stimulate interest in the tor problem which needed 'close examination and by as many workers as possible'.

The essence of the now well-known hypothesis is contained in the following genetic definitions of the terms 'tor' and 'core-stone' that Linton derived from it: 'A *tor* is a residual mass of bedrock produced below the surface level by a phase of profound rock rotting effected by groundwater and guided by joint systems, followed by a phase of mechanical stripping of the incoherent products of chemical action. Ellipsoidal rock masses produced in the same way but entirely separated from bed-rock are designated *core-stones*. The upper parts of a tor approximate in form to core-stones, and like them may be completely detached, though still perched; the lower portions of a tor approximate to massive joint-bounded blocks. Great variations in form and architectural style thus result from variations in the original joint disposition.' (Linton, 1955, p. 476).

Although it had been developed in regard to the tors of the southwest of England, the universal applicability of the hypothesis was called into question because it appeared seriously inadequate to account for observations derived from extensive field studies of the gritstone tors of the English Pennines and the granite tors of Dartmoor (J. Palmer and J. Radley, 1961; J. Palmer and R. A. Neilson, 1962). But it was not seriously questioned in respect of examples in lower latitudes. Evidence from both the Pennines and Dartmoor led Palmer and his fellow workers to doubt the former existence of deep weathering at tor sites and to propose an alternative hypothesis for the production of Palaeo-arctic tors by frost action on exposed bedrock and solifluction during periglacial phases of the Pleistocene. They agreed with R. A. Pullan (1959) 'that a definition of tors must concentrate on their common origin as residuals of differential weathering and mass movement', and noted that within such a definition sub-classes of tors, i.e., relict 'cold' forms and relict 'warm' forms, might be envisaged (Palmer and Neilson, 1962).

In spite of Linton's closely reasoned rebuttal in 1964 (Linton, 1964) of the arguments advanced in favour of the cryergic origin of the Pennine tors and his demonstration of the survival in the region of relics of former weathering profiles serious doubts remained. It was not until

1971 that they were partially resolved in respect of the Dartmoor tors at least by the consideration of fresh evidence provided by a quantitative examination of the nature of granite decomposition products on the Moor. Textural and mineralogical investigations enabled M. J. Eden and C. P. Green (M. J. Eden and C. P. Green, 1971) to distinguish between the products of pneumatolytic alteration and those of weathering and to identify the weathering products as the 'sandy weathering type' which J. P. Bakker (J. P. Bakker, 1967) attributed to a 'meso-humid subtropical climate.' They recognized the occurrence of a deeply weathered zone but from seismic investigations suggested that it may have been confined to the main river valleys and to the margin of the granite. Thus while accepting that tor formation had been the result of a two-stage process they noted that Linton's general hypothesis required some qualification.

On the other hand serious doubt has been cast on the universal applicability of a single hypothesis of tor formation and of the genetic definitions derived from it by evidence from Victoria Land, Antarctica, 'that several geomorphological processes may act concomitantly on the same slope to produce a variety of tor morphologies, including not only angular and rounded tors but exfoliating corestones, all under a periglacial climate.' (E. Derbyshire, 1972).

REFERENCES

BAKKER, J. P (1957) 'Weathering of granites in different climates, particularly in Europe' in P. MACAR (ed.) *L'Évolution des Versants*, 51-68

DERBYSHIRE, E. (1972) 'Tors, rock weathering and climate in southern Victoria Land, Antarctica', *Inst. Br. Geogr. Spec. Publ.* 4, 93-105

EDEN, M. J. and C. P. GREEN (1971) 'Some aspects of granite weathering and tor formation on Dartmoor, England', *Geogr. Annlr* 53A, 92-9

LINTON, D. L. (1949) 'Unglaciated areas in Scandinavia and Great Britain', *Ir. Geogr.* 2, 25-33

LINTON, D. L. (1950) 'Unglaciated enclaves in glaciated regions', *J. Glaciol.* 1, 451-2

LINTON, D. L. (1952) 'The significance of tors in glaciated lands', *Proc. 17th int. Congr., Int. geogr. Un. (Washington)*, 35-47

LINTON, D. L. (1955) 'The problem of tors', *Geogrl J.* 121, 470-87

LINTON, D. L. (1964) 'The origin of the Pennine tors—An essay in analysis', *Z. Geomorph.* 8, 5*-24*.

PALMER, J. and R. A. NEILSON (1962) 'The origin of granite tors on Dartmoor, Devonshire', *Proc. Yorks. geol. Soc.* 33, 315-40

PALMER, J. and J. RADLEY (1961) 'Gritstone tors of the English Pennines', *Z. Geomorph.* 5, 37-52

PULLAN, R. A. (1959) 'Notes on periglacial phenomena', *Scott. geogr. Mag.* 75, 51-5

Granite landforms: a review of some recurrent problems of interpretation

MICHAEL F. THOMAS

Lecturer in Geography, University of St Andrews

Revised MS received 7 July 1973

ABSTRACT. The central position of granite relief in studies of weathered land surfaces is stressed and major premises that have guided studies of granite landforms are stated. Six topics are selected for further enquiry: the nature and influence of structural control on landform development; the distribution and character of granite regoliths, including discussion of their climatic implications and of the topographic relationships of deep weathering; periodicity and continuity in weathering penetration and regolith removal; the occurrence of basin forms in granite landscapes, including a discussion of special conditions required to produce enclosed basins; the development of spheroidal and domical forms in granites, and the identification of granite landform systems and their interpretation including consideration of the occurrence of distinct multi-concave (basin-form) and multi-convex (dome-form) landscapes. The development of stepped or multi-storey land-scapes is also discussed. Implications for deductive model building in granite terrain are considered in terms of necessary modifications to the formal 'two-stage hypothesis' of tor development. Finally, a model for the development of granite relief in frost-free climates is advanced, assuming a prolonged sub-aerial history during which weathering penetration is regarded as a continuous process, but extensive stripping is attributed to widespread land surface instability resulting from periodic ecological change.

D. L. LINTON's classic enquiry into 'The problem of tors' (1955, 1958) focused attention on the interpretation of granite landforms and revived interest in some fundamental questions which they pose in geomorphology. Although knowledge of granite weathering and its association with tors dates from the nineteenth century (H. T. de la Beche, 1839; T. R. Jones, 1859; J. C. Branner, 1896; R. Chalmers, 1898), and several early studies had advanced hypotheses of tor exhumation from a deep regolith (J. D. Falconer, 1911, 1912; C. A. Cotton, 1917), the significance of such work to geomorphological theory was not widely appreciated. Both Chalmers (1898) and Falconer (1911) considered surviving deep regoliths to be remnants of formerly extensive mantles, and E. J. Wayland (1934) advanced an hypothesis for the lowering of old land surfaces by alternate weathering and stripping of the regolith. But much of this work, as well as studies in German by W. Credner (1931), K. Sapper (1935) and O. Jessen (1936) either remained little known or was seen to be relevant principally to tropical regions, although Jessen (1938) clearly appreciated its importance to studies of Tertiary climates and landforms in Europe. Cotton (1942) was the first to bring many of these tropical studies to a wider English readership, but it was Linton (1955) who emphasized the importance of such work to the understanding of temperate landscapes. His persuasive study opposed the extreme views on normality and uniform-itarianism represented particularly by L. C. King (1948, 1953) who reacted vigorously to many of the implications of Linton's paper (King, 1957, 1958). The concept of pediplanation advocated by King was not found directly applicable to many studies of the humid tropics (J. R. F. Handley, 1952; J. W. Pallister, 1956), although R. W. Clayton (1956) saw clearly the relevance of the German work on weathering to scarp retreat in the savannas. The appearance of J. Büdel's (1957) seminal paper on the contribution of deep weathering to the landforms of both the seasonal tropics and Europe, confirmed the significance of Linton's work in Britain.

A striking feature of many of these studies of weathered land surfaces, and also of the more important of those which shortly followed (B. P. Ruxton and L. Berry, 1957, 1961; J. A. Mabbutt, 1961; C. D. Ollier, 1960) is the use of models of landform evolution developed from studies of crystalline terrains and especially those on granite. In essence Linton's study of tors used a granite model to elucidate the effects of changing morphogenetic systems upon the relative rates of weathering penetration and surface erosion, and clearly emphasized the importance which such a model must have for our understanding of planation surfaces and residual landforms.

Some reasons for the central position of granite relief among these enquiries are possibly: the widespread occurrence of granitic[1] rocks; the striking character of many granite landforms which appear to offer the possibility of clear definition and individual study; the resistance of sound granite to mechanical abrasion, but its susceptibility to physical breakdown and chemical decay under weathering attack; the great contrast in erosional mobility which exists between the unaltered rock and its regolith or saprolite, and the range and diversity of landscapes underlain by granite.

All these features contribute to the variety and fascination of granite scenery. They also offer a model situation for enquiries into the particular effects of weathering and erosion upon landform development, and into the fluctuations in the rates and effectiveness of these processes with changing climate and vegetation or with uplift and tilting of the land surface. However, the construction of general hypotheses on the basis of such enquiries calls into question the validity of the premises from which models of granite relief development have been deduced. It may also be necessary to consider how representative the granite model is in terms of landform development on other rock types.

Some of the premises widely accepted in the study of granite landforms appear to be:

(1) The spatial arrangement and relief development of granite landforms reflects the operation of basic structural controls, including variations in composition, texture and fracture patterns. These lead to large-scale circular or ellipsoidal patterns that reflect the nature and form of granite emplacement and to a rectilinear compartmentation of relief that results from orthogonal fracture patterns.

(2) Granites may possess joint systems resulting from the processes of emplacement and crystallization (R. Balk, 1937), from later diastrophism (C. R. Twidale, 1964) and from relief development itself (C. A. Chapman and R. L. Rioux, 1958).

(3) Granites are commonly found to be stressed, the sources of stress being found among the factors given under (2) above and further jointing of the rock may occur as boundary conditions on rock faces permit the relaxation of stress (E. Gerber and A. E. Scheidegger, 1969). Although simple unloading may permit dilatation in response to vertical stresses, lateral stresses exist which may have more complex origins and effects (Twidale, 1964, 1971).

(4) A major result of stress relaxation is sheeting (exfoliation), a form of curvilinear jointing that displays a general but not necessarily detailed parallelism with the topographic form. Sheeting is considered both to be influenced by and to affect the course of relief development.

(5) Variations in the frequency of joints are, together with the effects of composition and texture, largely responsible for positive and negative elements in the relief and for the appearance of distinctive rock landforms which represent the less jointed rock masses. This principle has been subject to serious qualifications, especially by C. Wahrhaftig (1965).

(6) Although linear erosion may be guided directly by major fractures, the susceptibility of granite to chemical alteration provides the main key to the understanding of granite landforms except perhaps under conditions of extreme cold or aridity. Such granite weathering

involves both the physical disintegration and chemical alteration of the rock, and although local hydrothermal alteration can be demonstrated, these changes more generally result from the effects of ground water.

(7) Granite regoliths are commonly of considerable though very uneven depths and exhibit deep troughs and discrete basins of weathering that may be divided by abrupt rises in the weathering front which sometimes breaks the surface to form tors and domes. It is generally accepted that such patterns develop in response to internal rock weaknesses such as those stated above.

(8) The establishment of rock exposures in the form of tors and domes (bornhardts) results from the stripping of regolith from areas of shallower weathering.

(9) Differential rates of weathering affect exposed and buried rock surfaces, favouring the persistence of rock landforms and their accentuation with time.

(10) In the development of tors and perhaps domes long periods of weathering penetration have been followed by relatively short phases of rapid surface erosion or stripping. This 'two-stage hypothesis' of tor development advocated by Linton (1955) is generally considered to result from climatic changes.

(11) The depth and character of granite regoliths is affected not only by rock properties but also by climate and time, so that properties such as grain size, clay and heavy mineral content may be diagnostic of particular weathering environments (J. P. Bakker, 1967). Because rock breakdown is favoured by warmth and humidity most occurrences of deep weathering and tor landscapes in high latitudes have been attributed to former warm (if not actually tropical) climates.

Several of these propositions are questioned by recent work and it is possible to isolate a number of recurrent problems which affect studies of granite terrain. Some of these may be discussed under the following heads: the nature and influence of structural control; the distribution and character of granite regoliths and their interpretation; periodicity (and continuity) in weathering penetration and regolith removal; the occurrence of basin forms in granite landscapes; the development of spheroidal and domical forms; the identification of granite landform systems and their interpretation.

THE NATURE AND INFLUENCE OF STRUCTURAL CONTROL

In the study of granite terrain in Wyoming a morphostructural explanation for relief contrasts was sought by D. H. Eggler, E. E. Larsen and W. C. Bradley (1969), while F. F. Cunningham (1969) advanced a largely morphogenetic hypothesis. In the former, differences in granite landforms are accounted for largely by variations in the inherent properties of the rocks; in the latter they are seen to be members of a denudational sequence. Such differences of emphasis need not be seen in opposition, but the morphogenetic approach generally depends heavily upon a climatic interpretation of denudational events and both Ollier (1965) and Cunningham in a later paper (1971) argue that the role of climate in the development of granite landforms has been overemphasized. King (1948, 1957) has always held to the view that bornhardt domes occur in response to particular conditions of the rocks and not of climate, adducing as evidence their occurrence in almost every climatic zone. Yet W. A. White (1945) in a perceptive study of the granite domes of the south-east Piedmont of the U.S.A., warned against seeking a single explanation for similar forms which may have converged towards an equilibrium shape from separate origins and via the operation of different processes. This principle of 'equifinality' (L. Von Bertalanffy, 1950) or convergence has since been emphasized by H. Wilhelmy (1958) and Cunningham (1969).

Arguments in favour of detailed structural control over the development of granite landforms are questioned principally by the findings of Wahrhaftig (1965) and J. Hurault (1967), although King (1948) himself denied that the actual boundaries of bornhardts followed joints. Wahrhaftig argued that uplift of the Sierra Nevada in the western U.S.A. led to spurts of stream erosion, resulting in random exposures of granite along river beds. The resistance to erosion of such outcrops led to the development of a complex stepped relief, apparently unrelated to major joint patterns. Hurault (1967) on the other hand acknowledges the importance of dominant fracture patterns, but argues that in tropical regions selective removal of regolith to expose rock faces depends on local and accidental phenomena such as landslides and not upon jointing frequencies. These authors agree with other writers that a reduced rate of weathering will affect exposed rock faces and favour their persistence and growth into prominent landforms. But the randomness of these outcrop patterns has not been rigorously tested. P. Birot (1958) for instance, following studies in Brazil, pointed out that landslides will be favoured by the presence of a well-defined and smooth weathering front beneath the regolith, but plotting the occurrence of such a feature poses great difficulties.

This problem turns to a large extent on the influence of jointing frequencies, for few authors deny the importance of major fractures. The assumption that deeply weathered areas will be more closely jointed underlies Linton's (1955) analysis and has been implicit in many other studies. But joint frequencies beneath a deep regolith are not easily measured, and the observed jointing on tors and domes will be influenced by the forces acting on the hill form itself (Chapman and Rioux, 1958; Gerber and Scheidegger, 1969). Furthermore the appearance of large cores and tors in depressions and basins should also warn against the assumption that these features are consistently more closely jointed than the surrounding terrain.

It is clear that hierarchies or generations of joints exist in a rock mass and that these may influence relief development at different stages. Although weathering may be guided by the major primary fractures, evidence has been adduced for the influence of microfissures (Birot, 1958, 1962), otherwise called micro-cracks (E. B. A. Bisdom, 1967) or potential joints (Chapman, 1958). These may be marked by cracked grains or severed grain boundaries, and although invisible to the eye, can be measured microscopically and can be shown to increase the porosity of the granite and this in turn affects its weatherability.

The complexity of this topic makes it difficult to attempt generalizations. Twidale (1964, 1971) has demonstrated the influence of fracture patterns on both the distribution and ground plan of granite inselbergs in Australia, while J. T. Hack (1966) adduced evidence for the imprint of curved sheeting on the drainage pattern of North Carolina. The apparent lack of such structural controls in the area studied by Wahrhaftig (1965) may be due to the complexity of fracture patterns resulting from repeated diastrophism along the Sierra Nevada. Simpler and more consistent patterns on the shields of Africa and Australia may partially account for the clearer correspondence of landform patterns and geological structure. Vigorous arguments also abound concerning the causes and effects of sheeting (Twidale, 1971). These cannot be explored here, but some aspects of the problem will be considered below.

THE DISTRIBUTION AND CHARACTER OF GRANITE REGOLITHS

Fracture patterns have also been advanced to account for patterns of deep weathering in granites (M. F. Thomas, 1966), but other factors may also be of importance. In particular areas one granite may be altered to considerable depths and another little affected by deep weathering. Eggler, Larsen and Bradley (1969) from studies in Wyoming adduced evidence of slight Precambrian oxidation of biotite in one of two contiguous granites. This they claimed was sufficient to account for deep gruss formation as a result of rapid ground-water penetration of the weakened

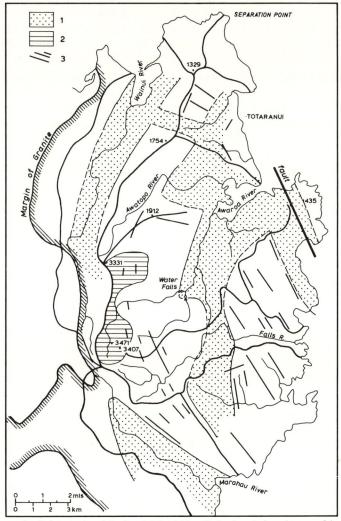

FIGURE 1. The Separation Point granite of the South Island New Zealand. 1, Inferred zones of deep, sandy weathering; 2, summit plateau containing rock floored basins (indicated in Fig. 2); 3, major structural alignments

rock, resulting in contrasting landform systems on the two granites. The occurrence of deep, granular regoliths showing only slight alteration of original minerals is widespread. Its incidence in the Hercynian granites of Europe is well known (J. Demek, 1964; Bakker, 1967; M. J. Eden and C. P. Green, 1971); Ollier (1965) records depths of more than 100 m in the Snowy Mountains of eastern Australia and zones of deep gruss occur in the Separation Point granite of the South Island, New Zealand, discussed below (Fig. 1).

The ages and origins of such deep regoliths pose many problems. The suggestion of Precambrian, biotite oxidation during granite emplacement extends the range of the controversy concerning possible hydrothermal alteration, but, at the same time, Twidale (1971) quotes evidence from R. Barbier (1967) for Precambrian weathering of granite near Tassili in the Sahara.

The diagnosis of hydrothermal alteration still remains problematical in the absence of distinctive mineralization (J. Konta, 1969), and in many instances the case for ground-water weathering rests on the recognition of a distinct weathering front (Ruxton and Berry, 1957; D. Brunsden, 1964). Where this is not seen to occur, doubts must remain, even if the products of alteration could have arisen by ground-water weathering. The recent study of the Dartmoor granite by Eden and Green (1971) contrasts the advanced hydrothermal alteration of the china clay deposits, having a clay plus silt content of 30–72 per cent and a felspar:quartz ratio of 0·20 with the slight alteration in the gruss or growan having a clay plus silt content of 13·5–28·0 per cent and a felspar:quartz ratio of 0·97. This latter deposit they interpret as a product of ground-water weathering in a warm though not tropical climate. Whether all hydrothermal alteration will be marked by advanced decay of the granite is doubtful. Study of regolith from the Separation Point granites reveals a deep sandy gruss with a clay plus silt content of 3·0–10·0 per cent, and a total clay mineral content of around 7 per cent most of which is kaolinite or halloysite. Within this deposit are bands of red material that have a clay plus silt content of 32·0–62·0 per cent and an estimated clay mineral content of more than 70 per cent [2] which is again kaolinitic. The absence of secondary mineralization in this material suggests an origin by ground-water weathering under warm conditions which liberated the iron oxides that impart the strong red colouration (5YR). On the other hand one such exposure was located close to a vein deposit found to be 98 per cent halloysite, and the possibility that this was due to hydrothermal alteration must be considered.

The present climate of the area (mean annual temperature 12°C; annual rainfall around 1500 mm) is probably suitable for gruss formation, and the known recent uplift of this part of New Zealand has contributed to the close jointing and shattering of the rock. On the other hand red weathering is well known from other parts of New Zealand (M. T. Te Punga, 1964) and a slight shift of climate would bring the necessary conditions (an increase in mean annual temperature to perhaps 15°C) to the South Island. It is therefore difficult with existing evidence, which has included infra-red, D.T.A., X-ray diffraction and electron microscope analyses of samples, to evaluate the possible contribution of hydrothermal alteration in any account of the granite weathering in this area. It is clear, however, that any weakening or alteration of the rock from deep-seated causes will be exploited by ground-water weathering, so that the two processes may be difficult to separate from any examination of the superficial zones of granite regoliths.

Climatic implications of regolith characteristics

These and other difficulties make climatic interpretations of particular regolith types hazardous, although J. P. Bakker and T. W. M. Levelt (1964) and Bakker (1967) have used this method widely for the reconstruction of Tertiary climates in central and western Europe. They made a clear distinction between sandy regoliths containing 2–7 per cent of clay less than 2 μm, much of which is kaolinite, and material containing 15–30 per cent clay of mixed kaolinite/illite composition. Bakker (1967) thought the latter represented marginal tropical conditions (20°N.) prevalent during the Miocene in Europe, but did not favour H. Piller's (1951) opinion that the sandy weathering was of contemporary (cool temperate) origin. Bakker (1967) considered it a result of conditions intermediate in character between the climates of the Gulf states and the Mediterranean. It is to be noted that he recognized that this sandy 'dry' weathering type could survive beneath a forest cover during moderate relief development.

Such analyses appear to ignore a number of problems, although Bakker's work is corroborated by much other evidence that cannot be explored here. The weight percentage of clay size particles (less than 2 μm) is subject to several important influences including: the texture and composition of the original rock; the position of the sample in the original weathering profile (E. C. Ruddock, 1967; K. W. Flach *et al.*, 1968); the slope angle on which the deposit was formed (G. West and M. J. Dumbleton, 1970) and the age of the mantle (A. R. Van Wambeke, 1962) as well as the dominant climate and vegetation cover. Furthermore it is clearly established that the amount of clay mineral present will not correspond closely to the weight percentage of particles less than 2 or 4 μm. Ruddock (1967) showed for a site in Ghana that large amounts of kaolinite occurred in material greater than 40 μm at depths of more than 22 m. This amount decreased rapidly towards the surface, but at a depth of 5 m there was 65 per cent kaolinite in the sample, but only 17 per cent was less than 2 μm, most of the clay mineral being found in the silt fraction at this depth. Flach *et al.* (1968) showed how physical breakdown and increased dispersibility affect the clay fraction as the soil surface is approached; while West and Dumbleton (1970) obtained figures for clay percentage in granite soils from West Malaysia that clearly indicate the effect of slope.

Simple comparisons of clay content based on size may therefore be difficult to interpret, especially when obtained from truncated profiles. Moreover, deep gruss formation may be aided by shattering or hydrothermal alteration of the rock, and the disaggregation which follows slight further alteration by ground-water weathering may not be diagnostic of particular climatic conditions. The nature of the clay mineral dominance is of greater significance, but in deep, free draining gruss, kaolin minerals may form over a wide range of external climate. Certainly it is doubtful if the sandy weathering types can be confined closely within climatic limits (Eden and Green, 1971), and as a corollary the phenomenon of deep weathering *per se* should not be taken to indicate former tropical conditions. Other regolith characteristics such as reddening (rubefaction) due to iron oxide liberation, appearance of specific clay minerals, or survival of particular heavy minerals, together with the nature of etching on quartz grains (A. Hillefors, 1971; J. C. Doornkamp and D. Krinsley, 1971) may all have much greater significance for climatic reconstruction, though they may be less directly related to landform development within the granite.

The topographic relationships of deep weathering

Hitherto great emphasis has been placed upon the association of deep weathering with planation surfaces of gentle slope and low relief, but the implication that thick regoliths are absent from other areas is misleading. Sharp juxtapositions of sound and altered rock are encountered widely in both planate and accidented relief. This situation has two components: the wide altitudinal range over which deep regoliths and distinctive rock landforms occur within a given area, and the existence of deep weathering profiles in zones of high internal relief.

In the Snowy Mountains (Ollier, 1965) weathered basins occur hundreds of metres below adjacent rocky summits, while in the Sierra Nevada Wahrhaftig (1965) described a series of steps over a height range of nearly 3000 m. In New Zealand the Separation Point granites are deeply weathered to a sandy gruss along zones followed by the principal valleys and opposite bays excavated into the coast by marine action. The gruss is also found below cols and in steeply dissected country within these zones (Figs 1 and 2). The juxtaposition of tor-crowned summits and weathered cols involving a local relief of 100–200 m is striking. On Dartmoor weathering also follows the main river valleys and Eden and Green (1971) argue that decomposition of the granite has worked outwards from the valley floors in the manner described by R. S. Waters (1957). However, in New Zealand the field evidence favours the incision of the main streams along lines of pre-existing deep regolith, while the lateral penetration of weathering appears to be confined mainly to flat-floored upland basins described below.

Surviving deep regoliths also become confined to lows in the topography in areas which have experienced extensive stripping from interfluves as a result of climatic shifts towards periglacial (Linton, 1955) or semi-arid (Mabbutt, 1961) conditions.

Over moderately dissected terrain, having a relief of less than about 100 m and slopes

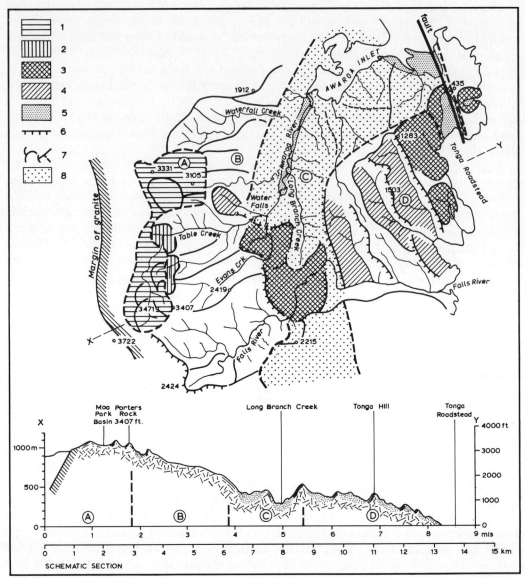

FIGURE 2. Geomorphological zones within the Separation Point granite. Geomorphological zones on map and section: A, summit plateau comprised of partially enclosed rock basins; B, dissected rocky zone; C, zone of deep weathering and valley development, interrupted by rock barriers marked by waterfalls; D, asymmetric ridges descending towards the coast. Morphological features: 1, rock basins of summit plateau; 2, dissected basins retaining rocky rims; 3, partially enclosed basins in weathered granite; 4, dissected weathered terrain on tilted granite ridges; 5, alluvium; 6, major rocky ridges; 7, boundaries of rock basins; 8, inferred zone of deep, sandy weathering

generally below 20°, weathered summits and interfluves are commonly found in forested environments (Thomas, 1966; T. Feininger, 1971). These terrains recall the lower storey landscapes described by Ruxton and Berry (1961) from Hong Kong, and which may be observed in the granite terrain of eastern Australia. In Colombia Feininger (1971) found that the weathering front exhibits 45–78 per cent of the relief of the land surface in areas with a local relief of 100–125 m.

FIGURE 3. Rock cored residual under forest in Johore, West Malaysia. An irregular, shallow regolith, containing cores which are shed on to the surface during denudation, appears to exist in equilibrium with the denudation system

It is clear therefore that the topographic relationships of deep regoliths are highly variable and that contrasting patterns characterize different types of granite landscape. Such differences may be due to one or more of several factors including: prevailing or past climates; degree of relief development and the chronology of changes in such controls.

PERIODICITY AND CONTINUITY IN THE WEATHERING PENETRATION AND REGOLITH REMOVAL

Acceptance of a two-stage hypothesis for tor and dome development implies a periodicity in the formation and removal of a deep regolith. This will not be refuted, but the nature of such oscillations in the morphogenetic system requires scrutiny. Antecedent weathering of long duration has been invoked to account for deep alteration in granites, and other rocks, in both the colder areas of Europe (Linton, 1955; J. Gjessing, 1967; V. Kaitenen, 1969) and in the semi-arid zones of Australia (W. G. Woolnough, 1927; Mabbutt, 1961, 1965; Ollier, 1965). Ollier has favoured Woolnough's (1927) view that a single, prolonged period of weathering penetration has affected most of Australia and pointed to the possibility that former depths of weathering may have been much greater than those which survive today.

In the discussion of such an argument care must be taken to distinguish clearly between the advanced weathering evidenced by duricrusted profiles described from Africa and Australia, and similar depths of disaggregated granite in which weathering marked by chemical changes has hardly begun (Twidale, 1971). The antiquity of many duricrusts can be substantiated from stratigraphic evidence, but there is often no indication that sandy weathering in granites is other than recent in origin. The identification of all deep regoliths and exhumed rock landforms as relict features of ancient landscapes may therefore be grossly misleading except where special circumstances or chronological evidence can be adduced.

Acceptance of continuity in weathering penetration during much of geological time, and its progression today in a wide range of climates, permits a flexibility of argument concerning the development of granite landscapes. In particular it obviates the need to postulate excessive weathering depths in the past, and suggests alternatives to the formal two-stage hypothesis of tor development. Periods of profile formation and of regolith removal should be seen in terms of changing conditions of ground surface stability (B. E. Butler, 1959, 1967). Beneath stable slopes weathering profiles will deepen; on unstable slopes they will be truncated. In addition an equilibrium state of regolith formation and removal may occur, and this appears to persist on many granite scarps in forested environments. Steep wooded slopes on jointed granites in Cameroun, Johore and New South Wales have common features of highly irregular weathering, core formation and a general absence of bare rock faces on slopes of up to 35° (Fig. 3). On such slopes a continuous process of core formation and removal appears to take place, with occasional tors surviving for longer periods. In highly seasonal, tropical climates a similar development appears to prevail, with the formation, destruction and re-formation of tors (Thomas, 1965).

It can be argued that the development of some degree of relief will increase rates of weathering, by accelerating sub-surface water movement. But since there is a general relationship between relief, slope and rate of surface denudation (F. Ahnert, 1970), there must be a condition of relief in any given environment for which the rates of weathering penetration and surface erosion are delicately balanced. Such thresholds are generally unknown, but truncation of weathering profiles is almost certainly effective at lower slope values in areas of open vegetation cover, and especially in the seasonal tropics. Widespread instability of the ground surface however appears to result from disturbance of ecologic conditions, usually during periods of marked climatic change. It seems likely therefore that the amplitude of variation in the rate of surface erosion has been much greater than that for ground-water weathering during the development of many granite landscapes. This situation leads to the periodic exhumation of tors and similar rock landforms, but permits the continued alteration of rock wherever the regolith mantle persists. Pockets of regolith are commonly protected by rock sills or barriers, marking the boundaries of troughs or basins of weathering (Ollier, 1965; Thomas, 1966; Wahrhaftig, 1965), and may survive both relief development and climatic change. This situation favours the continuity of weathering penetration in depressions and possibly leads to the persistence of patterns in the granite landscape over long periods involving the repeated development and decay of individual landforms (Thomas, 1969).

THE OCCURRENCE OF BASIN FORMS IN GRANITE LANDSCAPES

The occurrence of topographic basins in granite terrain presents a number of problems to the geomorphologist. Enclosed basins sometimes containing lakes occur commonly in glaciated terrain, but are also found in extra-glacial areas, where they require special explanation. More common are semi-enclosed basins which although they have a drainage outlet possess poorly drained central floors that may be rocky or more or less deeply weathered. Such basins occur at varying scales (M. B. Thorp, 1967), and appear to be influenced by the properties and structure of the granite.

Discrete depressions within granite weathering profiles (Thomas, 1966; Feininger, 1971) appear to offer a basis for the selective erosion of topographic basins. But the formation and continued deepening of basins of weathering will be opposed by the accumulation of stagnant ground water in which a rising concentration of dissolved ions may lead to equilibria being established in the weathering reactions. However, F. Lelong (1966) has proposed a mechanism of isothermic diffusion of ions, leading to the removal of weathering products from depth and their concentration near the surface. This hypothesis was advanced following detailed studies of ground water in wells sunk into weathered gneiss in Dahomey, and could apply equally to granite.

Surface denudation by running water will normally result only in semi-enclosed topographic basins in which the regolith survives within discrete hollows protected by rock thresholds

(Mabbutt, 1961; Ollier, 1965). Surface depressions which are completely enclosed may form in a variety of ways.

Glacial excavation The form and distribution of many rock basins in crystalline terrain at high latitudes could be accounted for by the selective removal by ice of regolith contained in discrete or interconnecting basins of weathering. Feininger (1971) has recently argued this case strongly, and it would appear to offer a possible explanation for many lake-studded land surfaces of low relief, such as the 'knock and lochan' (Linton, 1959) landscapes of Rannoch Moor, north-west Sutherland and southern Galway.

Deflation This process could operate in areas which have at some time become subject to arid or semi-arid conditions, related either to the mid-latitude deserts or to former periglacial regions. Fines from highly altered regoliths could be exported in dust storms, while coarse material would accumulate in the lee of the depressions. Lunette forms around shallow rock-floored depressions on the Monaro tablelands of New South Wales suggest that these depressions may have been excavated in this way.

Direct removal by weathering processes In basins floored by weathered material, loss of volume during later stages of alteration has been suggested by Twidale (1971), although Ollier (1967) has argued for constant volume alteration in granites during weathering. Loss of ground water into adjacent basins by sub-surface flow, together with the application of Lelong's (1966) diffusion hypothesis might account for such changes. But the development of enclosed depressions by this means would be opposed by surface stream flow. However, rock barriers or sills may inhibit surface erosion within weathered basins, inducing low gradients. As a result low-order streams may be deprived of effective runoff by increased evapotranspiration in the swampy valley floors. Such conditions might lead to the development of small enclosed depressions. In general, however, fluvial activity maintains a link between adjacent depressions.

Derangement of drainage by volcanism Many areas of granite are associated with extensive lava flows: eastern Australia, western U.S.A., and the Jos Plateau of Nigeria are examples. Much of the granite terrain has clearly been exhumed from beneath basalt in these areas, and some enclosed depressions may be inherited from lava-dammed lakes. Such an explanation must invoke one or more of the previously suggested mechanisms, acting in conjunction with an episode of vulcanism.

On the Monaro tableland of eastern New South Wales for instance, it can be shown that extensive stripping of the granite regolith, to form a tor-studded landscape, took place prior to the outpouring of basalt (W. R. Browne, 1964). Partially enclosed basins, previously drained by streams, were frequently dammed by the lavas. In this area and also in Nigeria continued weathering has completely decomposed many of the basalt flows, and also led to further rotting in the granites. Subsequent lowering of the landscape by slope retreat within both basalt and granite regoliths has normally been accompanied by extension of an integrated stream network and the draining of the lava-dammed lakes. But the period of interruption to the surface drainage may have been long enough to allow deepening of basin floors by sub-surface removal of solutes or by deflation.

The volcanic episode is thus invoked to account for the effectiveness of other processes in specific circumstances. The association of granite with basalt flows is common, and the geomorphic implications of this association warrant further enquiry.

FIGURES 4 AND 5. Spheroidal rock cores exposed along the coast of Tasman Bay in the Separation Point granite, New Zealand. The cores are being excavated from more closely jointed granite, occasionally weathered to a sandy gruss. Sheeting is clearly developed on these forms, leading to ambiguity in the distinction between core, tor and dome

Other processes which could be invoked for special cases might include back tilting of drainage systems due to diastrophism, and temporary effects of mass movement.

THE DEVELOPMENT OF SPHEROIDAL AND DOMICAL FORMS IN GRANITE

Domical and spheroidal forms are typical of granite landscapes (Thomas, 1965) and only in regions subject to severe frost or aridity are these characteristics suppressed. A distinction is made by most authors between domes and spheroidal corestones in terms of both scale and origin, and in each case there has been a tendency to seek a single mechanism which will explain the observed form. This approach to the rock landforms of granite is scarcely justified either by theory or observation. The concept of convergence discussed above casts doubt upon hypotheses claiming universality, and White (1945) in his study of the granite domes of the south-east Piedmont of the U.S.A. clearly stated this problem by reference to sheeting as a mechanism of dome formation. The rarity of sheets on the hillslopes he investigated suggested to White that exfoliation could not easily account for the observed form which was none the less very similar to the domes of the Yosemite valley on which sheeting is ubiquitous and very striking.

Furthermore the traditional distinction between dome and tor or corestone, although upheld in a recent study by Ollier (1971), may be difficult to substantiate. Very large spheroidal forms have been described from Nigeria (Thomas, 1965) and Australia (Twidale, 1971), and striking examples are also found in New Zealand (Figs 4 and 5) where sheets are clearly displayed on the rock surface. It is also significant that these 'cores' are contained within a matrix of shattered but little altered granite; they are not typical examples of spheroidal weathering within a deep

FIGURE 6. Development of cores within a sheeted, domical rock mass on the Jos Plateau, Nigeria. It is suggested that a joint hierarchy results in the formation first of the larger dome, but later leads to its subdivision along vertical fractures

regolith. Similar spheroidal fractures have been observed in fresh granite exposed in the tunnels of a diamond mine in Sierra Leone. Conversely cores may be rare or absent from deep regoliths, suggesting that weathering processes alone may not account for the phenomenon.

Recent studies of spheroidal weathering do not solve the basic problem. Bisdom (1967) indicates the role of 'structural micro-cracks' in corestone formation, and Ollier (1971) considers many different mechanisms, finally favouring the physical and chemical effects of ground-water penetration, perhaps involving ionic diffusion. Twidale (1971) quotes several sources suggesting that spheroidal features may represent centres of crystallization, or the centres of compressional stress fields. Cunningham (1971) considers the possibility that some of the larger domes may mark cupola features in the roof of the original granite intrusion.

While it may be necessary to doubt whether spheroidal forms result solely from the operation of weathering processes on cuboid blocks, there can be little doubt that once developed the form persists and may be perfected during further weathering, until the cores are reduced to a size which leads to the disaggregation of the block, possibly as a result of the convergence of micro-crack systems on the core (Bisdom, 1967).

Another feature of multi-convex forms is apparently related to the hierarchical nature of the jointing pattern in granite. Thus corestones may develop within larger domical or spheroidal masses of rock. This can be illustrated from Nigeria (Thomas, 1965, and Fig. 6) and has been seen elsewhere. Most evidence suggests that cores develop as sub-surface features whether in response mainly to mechanical or chemical forces. At the surface, convex forms appear to become diversified by features of surface denudation (Thomas, 1967) or may split along planar joints (Thomas, 1965; Ollier, 1965).

Convexity of profile is occasionally reflected in circularity of plan (Hack, 1966) but more often the larger domes display rectilinear outlines (Twidale, 1964; Thomas, 1967). This does suggest that the larger domes are modified from rectilinear blocks, but like the basin form which occurs on a large scale the convex forms within the granite landscape appear to have their origins at depth, perhaps predetermined by formational and structural features of the granite. The exposure of these convex forms could occur at levels independent in space and time of erosional base level whether they are regarded as random occurrences in the manner of Wahrhaftig (1965) or as structurally controlled masses (Thomas, 1965; Twidale, 1971).

FIGURE 7. Moa Park, a rock floored basin (see Fig. 2) on the summit plateau of the Separation Point granite, New Zealand. Viewed from the tor-studded rim, the basin has a dished form with slopes of less than 1° in the centre. A stream drains from the left towards the rear of the picture and dissection with the formation of convex slopes in the valley can be seen. The stream draining the basin from the right has a very low gradient and marshland communities dominate the basin floor.

THE IDENTIFICATION OF GRANITE LANDFORM SYSTEMS AND THEIR INTERPRETATION

The juxtaposition of topographies developed respectively on rock and regolith has received frequent comment (J. R. Harpum, 1963; Ollier, 1965; Eggler *et al.*, 1969) but variations in the distribution of specific forms such as domes, tors and basins within granite landscapes suggest that a more detailed approach to the identification of granite landform systems is required.

Thus several varieties of multi-concave (basin-form) and multi-convex (dome-form) landscapes exist within which individual rock landforms may occur in characteristic patterns. Differences in morphology are commonly complicated by varying degrees of weathering or exposure of the granite and where granitic rocks outcrop over a wide altitudinal range as in the western Cordillera of the U.S.A. (Wahrhaftig, 1965; Cunningham, 1969, 1971), and in eastern Australia (Ollier, 1965) for instance, stepped landscapes exist within which major hillslopes link upper and lower storeys (Ruxton and Berry, 1961).

Multi-concave (basin-form) landscapes

These occur at different scales (Thorp, 1967), but those which appear to respond to internal structures within a single granite may take several forms:

Enclosed basins   These generally have rocky rims and may possess either rock or regolith covered floors which are generally flat and ill-drained, often containing lakes.

Partially enclosed basins   These retain a central floor of low gradient, either over rock or regolith, but they are drained by a stream which breaches the otherwise enclosing rim (Figs 2 and 7).

Dissected basins   In these the rim remains prominent, but the floor has become dissected by streams.

In all cases corestones and tors may outcrop within the basin as well as on the interfluves, although larger tors are usually located on the rim. The basins with central floors are commonly drained by small first- or second-order streams and cover areas in the order of 1 km². However, dissection may lead to the abstraction of low-order basins which coalesce into larger features that may form granite lowlands of 10 or 100 km². In these cases the basin form as interpreted above is lost and a lowland landscape may develop with multi-convex compartments.

Multi-convex (dome-form) landscapes

Dissected granite terrain comprising multi-convex compartments has been described from many areas including Hong Kong (Ruxton and Berry, 1961) and Guyana, where Hurault (1967) used the terms 'alveolate' and 'cupola-shaped' to describe its form. Such landscapes may take two principal forms:

Multi-convex relief with weathered compartments   In this type of landform system, local relief is usually restricted to 100 m or less. Corestones and tors may outcrop, particularly on lower slopes, but interfluves are deeply weathered. Ruxton and Berry (1961) described this type of landform development in relation to a 'lower storey' landscape, extending from the base of major hillslopes in humid environments, and a similar landscape can be described from some granite embayments in the eastern escarpment of the Great Dividing Range, Australia.

Multi-convex relief with rock-cored compartments   This type of landform appears to be a response to either the massiveness of the granite or a greater magnitude of relief and slope. There are two identifiable sub-types. First, 'dome and cleft' terrain in which the compartments comprise massive bornhardts of widely exposed rock, divided along lines of fracture by deep clefts followed by the drainage. The Cambrian Older Granite suite in West Africa contributes striking examples of this type of terrain, for instance in the Idanre Hills of Nigeria and the Gbenge Hills of Sierra Leone. Secondly, terrain in which a shallow regolith covers most slopes. In a humid climate an increase in relief may lead to the widespread development of such slopes, either along escarpments or as individual compartments. Many residual granite massifs in west Malaysia (Fig. 3) appear to take this form, though the convexity may not extend beyond the upper slopes.

Stepped or multi-storey landscapes

Large-scale and repeated uplift appears to give rise to stepped topography in granites, probably in part as a result of the reduced rates of weathering affecting exposures produced as a consequence of these events (Wahrhaftig, 1965). Examples of granite terrain in which distinct topographies are divided by major hillslopes are numerous. In southern New South Wales basin-form landscapes occur on the Monaro tablelands from which major hillslopes lead down to granite lowlands exhibiting a multi-convex relief with weathered compartments. A comparable sequence may be elaborated for the Separation Point granite of New Zealand (Figs 1 and 2). In this area stripped, rock-floored basins survive on a summit plateau at around 1000 m above sea level. From this plateau the terrain falls in a series of steps eastwards. Dissection of the upland basins has produced a complex relief in response to zones of alteration in the granite on the one hand and the Plio-Pleistocene arching of the mass on the other. Basin-form features have developed within weathered zones and on the 'dip' slopes of tilted granite steps. This has produced a series of relief zones shown in Figures 1 and 2. It is clear that a series of linked landform systems has arisen which reflect the structural framework, weathering patterns, and the tectonic and erosional history of the area.

Continual formation, destruction and re-formation of tors during a prolonged sub-aerial history is evident from many granite areas. On the Monaro of New South Wales, Browne (1964) demonstrated that the stripping of regolith to produce a tor landscape had been well advanced

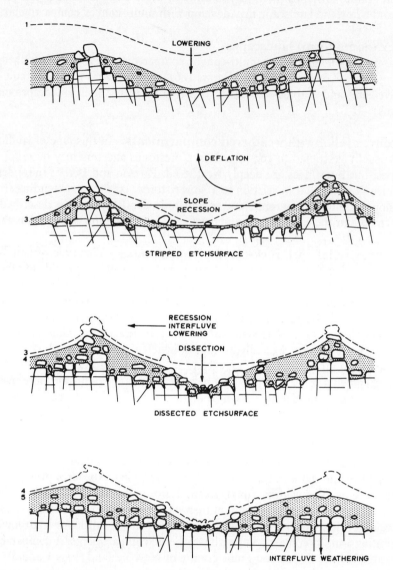

FIGURE 8. Hypothetical sequence in jointed granites during moderate relief development without tilting: 1, unknown former surface; 2, compartmented convex relief with summit tors; 3, excavation of concave basin forms by slope retreat and possible deflation with undermining of some tors; 4, dissection of basin floor and lowering of interfluves; 5, return to convex landscape similar to 1 or 2

prior to the outpouring of Oligocene Basalts. Many of these tor groups are now in a state of decay (Figs 8 and 9) reminiscent of those described from Nigeria (Thomas, 1965). The upland tors of the Separation Point granites may also be of some antiquity, but it is equally clear that spheroidal forms are being excavated at the present time in areas of dissection and along the coast (Figs 4 and 5). Continuity of weathering during the development of granite landscapes thus becomes a concept of central importance, necessary to explain deep weathering in cols and

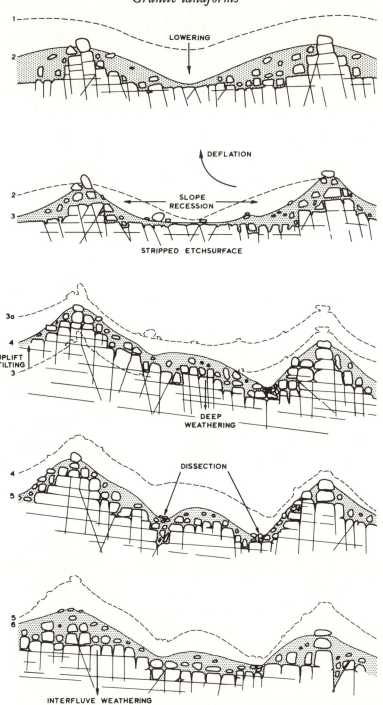

FIGURE 9. Hypothetical sequence in jointed granites during strong relief development accompanied by tilting: 1, unknown former surface; 2, compartmented convex relief with summit tors; 3, 4, surface tilted, leading to dissection of asymmetric ridges; 5, rocky scarps and dissected, weathered back slopes formed; 6, lowering of relief reduces asymmetry, returning landscape to condition similar to 1 or 2. This model appears to fit the observed sequence of forms shown in Figure 2. In this area tilting of the granite appears to be 4–5°; the basins are 1–2 km in width between crests and the internal relief varies from *c.* 100 m in zone A to 200 m in zone D

TABLE I

A model for the development of granite relief

Finely divided rock (joints 0–1 m)	*Rock divided by joints at 1–10 m separation*	*Massive rock* (joints more than 10 m)

Finely divided rock (joints 0–1 m)

Undulating landsurface without tors

Rock divided by joints at 1–10 m separation

Undulating multi-convex landsurface with zones and basins of deep weathering—possibly with occasional tors and exposed corestones—ridge and tor landscape

—*Landsurface instability—from bioclimatic change*(?)
Accelerated erosional stripping from:
1 strong mass movement (high rainfall; periglacial conditions)
2 deflation (aridity—periglacial or tropical) or
3 lateral slope recession (strong seasonality in tropics/sub-tropics)

Basin and tor landscapes (multi-concave landscapes)

Partial basins with integrated drainage (processes 1 and/or 3)

Closed basins with individual summits (process 2 with 3)

—*Landsurface stability—from climatic change*(?)
1 continued slope recession
2 continued (or renewed) weathering penetration

widening of valleys/weathering of ridges

widening of basins/weathering of summits

tors undermined by weathering

—*Landsurface instability—from tectonic change*
1 moderate uplift
2 shallow dissection
3 weathering penetration

deepening of valleys/collapse of summit tors with renewal of outcrops at lower levels

breaching of closed basins/dissection of basin floors/collapse of summit tors with renewal of outcrops at lower levels

in absence of climatic crises or major uplift, ridge and tor landscapes may persist during long peri[...]

Massive rock (joints more than 10 m)

Multi-convex—cupola shaped—land surface with narrow zones of deep weathering

Multi-convex—dome and cleft—landscapes

little morphological change

deepening of clefts/enlargement of domes/emergence of new domes

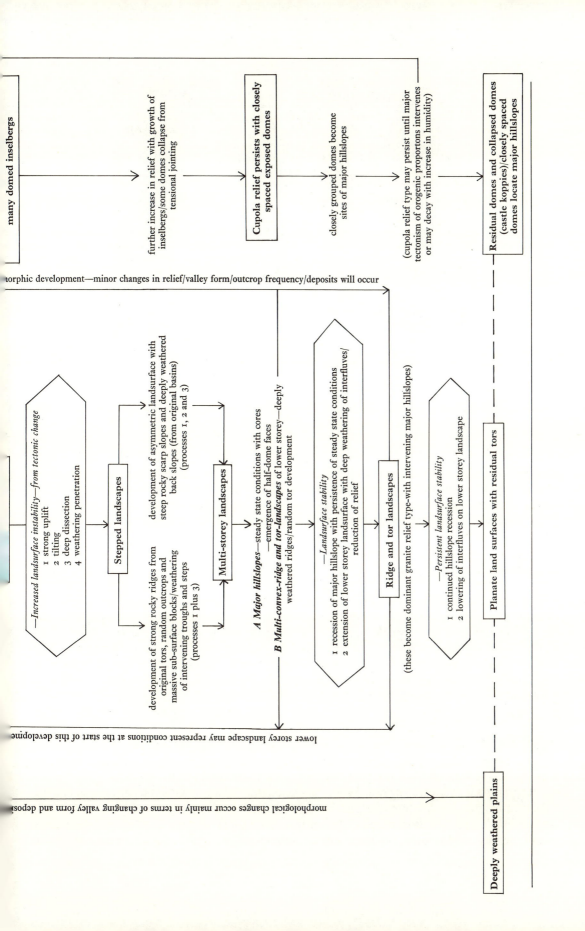

many domed inselbergs

further increase in relief with growth of inselbergs/some domes collapse from tensional jointing

Cupola relief persists with closely spaced exposed domes

closely grouped domes become sites of major hillslopes

(cupola relief type may persist until major tectonism of orogenic proportons intervenes or may decay with increase in humidity)

Residual domes and collapsed domes (castle koppies)/closely spaced domes locate major hillslopes

...morphic development—minor changes in relief/valley form/outcrop frequency/deposits will occur

—Increased landsurface instability—from tectonic change
1 strong uplift
2 tilting
3 deep dissection
4 weathering penetration

Stepped landscapes

development of asymmetric landsurface with steep rocky scarp slopes and deeply weathered back slopes (from original basins) (processes 1, 2 and 3)

Multi-storey landscapes

development of strong rocky ridges from original tors, random outcrops and massive sub-surface blocks/weathering of intervening troughs and steps (processes 1 plus 3)

A Major hillslopes—steady state conditions with cores —emergence of half-dome faces
B Multi-convex-ridge and tor-landscapes of lower storey—deeply weathered ridges/random tor development

—Landsurface stability
1 recession of major hillslope with persistence of steady state conditions
2 extension of lower storey landsurface with deep weathering of interfluves/ reduction of relief

Ridge and tor landscapes

(these become dominant granite relief type—with intervening major hillslopes)

—Persistent landsurface stability
1 continued hillslope recession
2 lowering of interfluves on lower storey landscape

Planate land surfaces with residual tors

lower storey landscape may represent conditions at the start of this developme...

morphological changes occur mainly in terms of changing valley form and deposi...

Deeply weathered plains

FIGURE 10. Decayed tor on the Monaro Tableland of New South Wales. The deep weathering is peripheral, but almost certainly extends beneath the group of corestones

basins hundreds of metres below adjacent summits and ridges, and also the ubiquity of corestones and tors at all levels in many granite landscapes.

SOME IMPLICATIONS FOR DEDUCTIVE MODEL BUILDING

It is clear that the formal 'two-stage' hypothesis for tor development, important though it remains, cannot alone account for the complexities of all tor landscapes. On the one hand tors may arise during systematic slope development in the absence of major ecological change; while on the other hand many tor landscapes appear to have passed through several periods of development and decay.

The analysis of rock landforms in granite should recognize not only features of structure and weathering that are relevant to individual forms, but also the setting of these forms in distinctive landform systems. Such systems are capable of more rigorous description than is offered here and are likely to be functionally and sequentially related throughout any single area of granite terrain. However, in seeking the antecedents of contemporary landscapes, using the methods of comparative morphology, many difficult problems arise. Not the least is the problem of the 'initial conditions' from which the existing landform complexes have developed. The nature of granite emplacement implies complex relationships with country rocks (Cunningham, 1971), and the form and structure of intrusions may contribute to the pattern of terrain in important ways. Early weathering of granites may occur beneath a sedimentary cover (L. K. Jeje, 1972) and the emergence of complex granite landscapes accompany the slow removal of country rock. Possibly few current granite landscapes can be traced directly to inheritance from the intrusive form, but early patterns of weathering and outcrop may persist through several cycles of development and decay of individual landforms (Thomas, 1969), while erosional events may lead to the formation of new patterns apparently unrelated to structure.

In Figures 8 and 9 are suggested some sequences that follow from basic premises about the behaviour of granite under humid or sub-humid conditions in climates generally free from strong frost action. To the premises stated earlier in this paper three others may be offered as conclusions from this analysis. First, weathering penetration should be regarded as a continuous process.

The deepening or truncation of weathering profiles depends less upon variations in the rate of rock decomposition than upon changes in the pace of surface denudation. Secondly, local stripping to expose corestones and tors may occur during slope development under constant conditions, as a function of differential erosion. But extensive stripping to form basin and tor landscapes (stripped etchsurfaces) probably requires a general landsurface instability such as would follow major ecological changes towards arid or cold conditions. Thirdly, many granite landscapes have undergone prolonged sub-aerial development that has involved the formation, destruction and reformation of characteristic forms such as basins, tors and domical compartments. There is thus no single starting point and no inevitable end product in the sequence of landform development in granites.

ACKNOWLEDGEMENTS

The author wishes to thank the Carnegie Institution for finance in support of research in New Zealand; the University of New South Wales for assistance with travel in New South Wales; The Geological Survey of New Zealand for help with the analysis of regolith samples; the Board and Rangers of the Abel Tasman National Park, New Zealand for their help and provision of accommodation; Mr J. A. Davie who drew the maps and diagrams; Mr P. Adamson who provided the photographic prints, and Mrs S. Weaver for her patience in typing the manuscript.

NOTES

1. It is recognized that the use of the term 'granitic rocks' raises problems of definition. Discussion of this question lies outside the scope of this paper, and while many of the points raised here may be most directly relevant to intrusive granites and granodiorites, the arguments advanced can be applied to most other rocks commonly regarded as granitic or granitoid, though not without modification in the case of strongly foliated rocks.

2. This estimate is based on an infra-red analysis carried out by the New Zealand Geological Survey. Some problems of clay mineral identification also arose during this work: X-ray diffraction of the 20–45 μm size range showed dominant kaolinite, but electron microscope photographs of the under 2 μm fraction revealed mainly halloysite in all samples. At present it is not considered possible to base genetic arguments on these observations, but the non-recognition of halloysite by the more conventional methods may affect assumptions made about its presence or absence in earlier studies.

REFERENCES

AHNERT, F. (1970) 'Functional relationships between denudation, relief and uplift in large mid-latitude drainage basins', *Am. J. Sci.* 268, 243–63

BAKKER, J. P. (1967) 'Weathering of granites in different climates' in P. MACAR (ed.) *L'évolution des versants* 51–68

BAKKER, J. P. and T. W. M. LEVELT (1964) 'An enquiry into the problem of a polyclimatic development of peneplains and pediments (etchplains) in Europe during the Senonian and Tertiary Period', *Publs Serv. Carte géol. Luxemb.* 14, 27–75

BALK, R. (1937) *The structural behaviour of the igneous rocks*

BARBIER, R. (1967) 'Nouvelles réflexions sur le problème des pains des sucre a propos d'observations dans le Tassili N'Aljer (Algerie)', *Trav. Lab. Géol. Univ. Grenoble* 43, 15–21

BIROT, P. (1958) 'Les dômes crystallines', *Mém. Docums Cent. Docum. cartogr. géogr.* 6, 1–34

BIROT, P. (1962) *Contribution à l'étude de la désagregation des roches*

BISDOM, E. B. A. (1967) 'The role of micro-crack systems in the spheroidal weathering of an intrusive granite in Galicia (N.W. Spain)', *Geologie Mijnb.* 46, 333–40

BRANNER, J. C. (1896) 'Decomposition of rocks in Brazil', *Bull. geol. Soc. Am.* 7, 255–314

BROWNE, W. R. (1964) 'Grey Billy and the age of tor topography in Monaro, N.S.W.', *Proc. Linn. Soc. N.S.W.* 89 (3), 322–5

BRUNSDEN, D. (1964) 'The origin of decomposed granite on Dartmoor' in I. G. SIMMONS (ed.) *Dartmoor Essays*, 97–116

BÜDEL, J. (1957) 'Die "Doppelten Einebnungsflächen" in den feuchten Tropen', *Z. Geomorph.* 1, 201–28

BUTLER, B. E. (1959) 'Periodic phenomena in landscapes as a basis for soil studies', *Soil Publ. C.S.I.R.O. Aust.* 14

BUTLER, B. E. (1967) 'Soil periodicity in relation to landform development in south eastern Australia' in J. N. JENNINGS and J. A. MABBUTT (eds) *Landform Studies from Australia and New Guinea*, 231–55

CHALMERS, R. (1898) 'The pre-glacial decay of rocks in eastern Canada', *Am. J. Sci.* 5, 273–82

CHAPMAN, C. A. (1958) 'The control of jointing by topography', *J. Geol.* 66, 552–8

CHAPMAN, C. A. and R. L. RIOUX (1958) 'Statistical study of topography, sheeting and jointing in granite, Acadia National Park, Maine', *Am. J. Sci.* 256, 111–27

CLAYTON, R. W. (1956) 'Linear depressions (Bergfüssneiderungen) in savanna landscapes', *Geogrl Stud.* 3, 102–26

COTTON, C. A. (1917) 'Block mountains in New Zealand', *Am. J. Sci.* 44, 249–93

COTTON, C. A. (1942) *Climatic accidents in landscape making*

CREDNER, W. (1931) 'Das Kraftererhältuis morphogenetischer Faktoren und ihr Ausdruck im Formenbild Südostasiens', *Bull. geol. Soc. China* 11, 13–34

CUNNINGHAM, F. F. (1969) 'The Crow tors, Laramie Mountains, Wyoming, U.S.A.', *Z. Geomorph.* 13, 56–74

CUNNINGHAM, F. F. (1971) 'The Silent City of Rocks, a bornhardt landscape in the Cotterel Range, South Idaho, U.S.A.', *Z. Geomorph.* 15, 404–29

DE LA BECHE, H. T. (1839) *Report on the geology of Cornwall, Devon, and West Somerset*

DEMEK, J. (1964) 'Slope development in granite areas of the Bohemian massif, Czechoslovakia', *Z. Geomorph. Suppl.* 5, 82–106

DOORNKAMP, J. C. and D. KRINSLEY (1971) 'Electron microscopy applied to quartz grains from a tropical environment', *Sedimentology* 17, 89–101

EDEN, M. J. and C. P. GREEN (1971) 'Some aspects of granite weathering and tor formation on Dartmoor, England', *Geogr. Annlr* 53A, 92–9

EGGLER, D. H., E. E. LARSON and W. C. BRADLEY (1969) 'Granites, grusses and the Sherman erosion surface, southern Laramie Range, Colorado–Wyoming', *Am. J. Sci.* 267, 510–22

FALCONER, J. D. (1911) *The geology and geography of Northern Nigeria*

FALCONER, J. D. (1912) 'The origin of Kopjes and inselbergs', in *Report of the 82nd meeting of the British Association for the Advancement of Science*, p. 476

FEININGER, T. (1971) 'Chemical weathering and glacial erosion of crystalline rocks and the origin of till', *Prof. Pap. U.S. geol. Surv.* 750-C, C65–C81

FLACH, K. W., J. G. CADY and W. D. NETTLETON (1968) 'Pedogenic alteration of highly weathered parent materials', *Trans. Ninth Int. Congr. Soil Sci.* 4, 343–51

GERBER, E. and A. E. SCHEIDEGGER (1969) 'Stress induced weathering of rock masses', *Eclog. geol. Helv.* 62, 401–15

GJESSING, J. (1967) 'Norway's Paleic surface', *Norsk geogr. Tidsskr.* 21, 69–132

HACK, J. T. (1966) 'Circular patterns and exfoliation in crystalline terrain, Grandfather Mountain area, North Carolina', *Bull. geol. Soc. Am.* 77, 975–87

HANDLEY, J. R. F. (1952) 'The geomorphology of the Nzega area of Tanganyika with special reference to the formation of granite tors', *C. r. 19th Int. geol. Congr. Algiers 1952* fasc. 21, 201–10

HARPUM, J. R. (1963) 'Evolution of granite scenery in Tanganyika', *Rec. geol. Surv. Tanganyika* 10, 39–46

HILLEFORS A. (1971) 'Deep weathered rock material and sand grains', *Lund Stud. Geogr. Ser. A.* 49, 138–64

HURAULT J. (1967) 'L'érosion régressive dans les régions tropicales humides et la genèse des inselbergs granitiques', *Étud. Photo-interpret.* 3

JEJE, L. K. (1972) 'Landform development at the boundary of sedimentary and crystalline rocks in south-western Nigeria', *J. trop. Geogr.* 34, 25–33

JESSEN, O. (1936) *Reisen und Forschungen in Angola* (Berlin)

JESSEN, O. (1938) 'Tertiarklima und Mittelgebirgsmorphologie', *Z. Ges. Erdk. Berl.* 41–2, 36–49

JONES, T. R. (1859) 'Notes on some granite tors', *The Geologist* 2, 301–12

KAITANEN, V. (1969) 'A geographical study of the morphogenesis of northern Lapland', *Fennia* 99, 1–85

KING, L. C. (1948) 'A theory of bornhardts', *Geogrl J.* 112, 83–7

KING, L. C. (1953) 'Canons of landscape evolution', *Bull. geol. Soc. Am.* 64, 721–52

KING, L. C. (1957) 'The uniformitarian nature of hillslopes', *Trans. Edinb. geol. Soc.* 17, 81–102

KING, L. C. (1958) 'Correspondence: the problems of tors', *Geogrl J.* 124, 289–91

KONTA, J. (1969) 'Comparison of the proofs of hydrothermal and supergene kaolinization in two areas of Europe', *Proc. Int. Clay Conf. Tokyo 1969* 1, 281–90

LELONG, F. (1966) 'Régime des nappes phréatiques continues dans les formations d'alteration tropicale. Conséquences pour la pédogenèse', *Sciences Terre* 11, 203–44

LINTON, D. L. (1955) 'The problem of tors', *Geogrl J.* 121, 470–87

LINTON, D. L. (1958) 'Correspondence: the problem of tors', *Geogrl J.* 124, 289–91

LINTON, D. L. (1959) 'Morphological contrasts between eastern and western Scotland' in R. MILLER and J. W. WATSON (eds) *Geographical Essays in Memory of Alan G. Ogilvie*, 116–45

MABBUTT, J. A. (1961) 'A stripped landsurface in Western Australia', *Trans. Inst. Br. Geogr.* 29, 101–14

MABBUTT, J. A. (1965) 'The weathered landsurface of central Australia', *Z. Geomorph.* 9, 82–114

OLLIER, C. D. (1960) 'The inselbergs of Uganda', *Z. Geomorph.* 4, 43–52

OLLIER, C. D. (1965) 'Some features of granite weathering in Australia', *Z. Geomorph.* 9, 285–304

OLLIER, C. D. (1967) 'Spheroidal weathering, exfoliation and constant volume alteration', *Z. Geomorph.* 11, 285–304

OLLIER, C. D. (1971) 'Causes of spheroidal weathering', *Earth-Sci. Rev.* 7, 127–41

PALLISTER, J. W. (1956) 'Slope development in Buganda', *Geogrl J.* 122, 80–7

PILLER, H. (1951) 'Über Verwitterungsbildungen des blockengranits Nördlich St Andreasberg', *Heidelb. Beitr. Miner. Petrogr.* 2, 498–522

RUDDOCK, E. C. (1967) 'Residual soils of the Kumasi district in Ghana', *Géotechnique* 17, 359–77

RUXTON, B. P. and L. BERRY (1957) 'Weathering of granite and associated erosional features in Hong Kong', *Bull. geol. Soc. Am.* 68, 1263–92

RUXTON, B. P. and L. BERRY (1961) 'Weathering profiles and geomorphic position on granite in two tropical regions', *Revue Géomorph. dyn.* 12, 16–31

SAPPER, K. (1935) 'Geomorphologie de feuchten Tröpen', *Geogr. Schr.* 7, 1–150

TE PUNGA, M. T. (1964) 'Relict red weathered regolith at Wellington, New Zealand', *N.Z. Jl Geol. Geophys.* 7, 314–39

THOMAS, M. F. (1965) 'Some aspects of the geomorphology of domes and tors in Nigeria', *Z. Geomorph.* 9, 63–81

THOMAS, M. F. (1966) 'Some geomorphological implications of deep weathering patterns in crystalline rocks in Nigeria', *Trans. Inst. Br. Geogr.* 40, 173–93

THOMAS, M. F. (1967) 'A bornhardt dome in the plains near Oyo, Western Nigeria', *Z. Geomorph.* 11, 239–61

THOMAS, M. F. (1969) 'Some outstanding problems in the interpretation of the geomorphology of tropical shields' in *Geomorphology in a tropical environment, Occ. Publ. Br. Geomorphol. Res. Grp* 5, 42–9

THORP, M. B. (1967) 'Closed basins in younger granite massifs, northern Nigeria', *Z. Geomorph.* 11, 459–80

TWIDALE, C. R. (1964) 'A contribution to the general theory of domed inselbergs', *Trans. Inst. Br. Geogr.* 34, 91–113

TWIDALE, C. R. (1971) *Structural landforms*

VAN WAMBEKE, A. R. (1962) 'Criteria for classifying tropical soils by age', *J. Soil Sci.* 13, 124–32

VON BERTALANFFY, L. (1950) 'An outline of general systems theory', *Br. J. Phil. Sci.* 1, 134–65

WAHRHAFTIG, C. (1965) 'Stepped topography of the southern Sierra Nevada, California', *Bull. geol. Soc. Am.* 76, 1165–90

WATERS, R. S. (1957) 'Differential weathering and erosion on oldlands', *Geogrl J.* 123, 501–9

WAYLAND, E. J. (1934) 'Peneplains and some other erosional platforms', *Geol. Surv. Annual Rept. Bull.* Notes 1, 74, 366

WEST, G. and M. J. DUMBLETON (1970) 'The mineralogy of tropical weathering illustrated by some West Malaysian soils', *Q. J. Eng. Geol.* 3, 25–40

WHITE, W. A. (1945) 'The origin of granite domes in the south-east piedmont', *J. Geol.* 53, 276–82

WILHELMY, H. (1958) *Klimamorphologie der Massengesteine*

WOOLNOUGH, W. G. (1918) 'Physiographic significance of laterite in Western Australia', *Geol. Mag.* 6, 385–93

WOOLNOUGH, W. G. (1927) 'The duricrust of Australia', *J. Proc. R. Soc. N.S.W.* 61, 25–53

RÉSUMÉ. *Les terrains du granit: une revue de quelques problemes récurrents de l'interprétation.* La place centrale du relief granitique aux études des terrains profondément alterés est soulignée et des prémises majeures qu'ont guidé les études des reliefs granitiques sont posés ici. Les six sujets qui suivent sont choisis à plus ample informées: L'essence de et l'influence de la contrainte structurale sur le terrain; la répartition et le caractère des altérites (regoliths) granitiques, y compris une discussion de leurs insinuations climatiques et des rapports de désagrégation profond du relief; l'activité périodique et la continuité quant à la pénétration de l'altération et à l'enlèvement de regolith; l'occurrence des reliefs granitiques taillés en forme de bassin, embrassant une discussion des conditions particulières nécessaires de causer des bassins renfermés; le développement des formes à dômes et spheroïdes aux granits; l'identification des systèmes des terrains granitiques et l'interprétation d'eux. C'est dans cette partie qu'on examine l'occurrence des reliefs caractérisés par tous les deux formes 'multi-concaves' (en formes des bassins) et 'multi-convexes' (en formes des dômes). Aussi l'on agit le développement des terrains échelonnés ou à 'multi-étages'. On considère les insinuations pour ce qui concerne la déduction du modèle aux reliefs granitiques en fonction des modifications requises pour la hypothèse formele des 'deux périodes du developpement' des 'tors' (boules). Enfin, on met en avant une modèle pour le développement du relief granitique aux climats débarrassés de la gelée, en supposant qu'on admet une histoire prolongée et sous-aérienne pendet que la pénétration de l'alteration soit un cours continu, mais on attribue le décapage étendu à l'instabilité étendue du terrain par suite du changement du climat biologique.

FIG. 1. Le Separation Point granit de l'île du Sud, Nouvelle Zélande
1. Les zones inférées du arenisation profonds
2. Le plateau du sommet contenant de formes du bassin avec les fonds rocheux (indiqué de Figure 2.)
3. Des alignements structurales majeurs
FIG. 2. Les Zones Géomorphologiques dans le Separation Point granit
A. Le plateau du sommet compris des bassins rocheux renfermés en partie
B. La zone rocheuse et disséquée
C. La zone de l'alteration profonde et du développement de la vallée, contenant les barrières rocheuses et des chutes d'eau
D. Des crètes asymmetriques qui descendent vers la côte
   Des traits morphologiques sont indiqués:
   1. Les bassins rocheux de plateau du sommet
   2. Les bassins disséqués retenant les bords rocheux
   3. Les bassins (renfermés en partie) au granit altéré
   4. Le relief altéré et disséqué sur les crètes granitiques inclinées
   5. Alluvions
   6. Les crètes rocheuses et majeures
   7. Les bords des bassins rocheux
   8. La zone inférée du arenisation profond
FIG. 3. Le relief résiduel du noyau rocheux sous le forêt, en Johore, Malaise
FIGS 4 et 5. Les noyaux rocheux comme des boules exposées le long de la côte de La Baie Tasman, Nouvelle Zélande
FIG. 6. Le développement des noyaux dans une dôme rocheuse sur le Plateau de Jos, Nigeria
FIG. 7. Le bassin avec le fond rocheux (le Moa Park, voie Figure 2) sur le plateau du sommet du Separation Point granit Nouvelle Zélande
FIG. 8. La succession hypothétique pour le développement du relief dans les granits fissurés pendent la periode disséquée modéré et sans inclinée

1. La surface antérieure et inconnue
2. Le relief convexe et compartimenté avec des boules du sommet
3. Le fouillement des bassins concaves par les pentes reculées avec le possibilité de la deflation et aussi la désagrégation des quelques inselberges (tors)
4. La dissection du fond du bassin avec l'abaissement des interfluves
5. Le relief retourne aux formes convexes comme 1 ou 2

FIG. 9. La succession hypothétique pour le développement du relief dans les granits fissurés pendent le periode disséquée energique accompagnée de basculement

1.   La surface antérieure et inconnue
2.   Le relief convex et compartimenté avec des boules du sommet
3 et 4.   La dissection et formation des crêtes asymetriques consécutif à la surface inclinée
5.   La formation des escarpements rocheux et des pentes disséquées à derrière
6.   L'abaissement du relief reduit l'asymetrie et le terrain retourna à la morphologie comme 1 ou 2

Il semble que cette modèle soit d'accord avec la succession observée dans la figure 2. Dans cette région, le basculement du granit semble qu'il suit 4–5°; des bassins on 1–2 km de large entre les crêtes, et le hauteur du relief s'écarte de 100 m dans la Zone A a 200 m dans la Zone D

FIG. 10. L'inselberge (tor) decomposé sur le Plateau de Monaro, New South Wales, Australie

ZUSAMMENFASSUNG. *Granitlandformen: ein Bericht einiger wiederkehrender Probleme der Interpretation.* Die zentrale Position des Granitreliefs in Studien verwitterter Landoberflächen ist hervorgehoben und Hauptprämisse, welche Studien der Granitlandformen geleitet haben, sind dargebracht. Sechs Themen sind für weitere Erkundigungen ausgewählt: die Natur und der Einfluss struktureller Kontrolle auf die Entwicklung von Landformen; die Verteilung und der Charakter von Granit Regolithen, eine Erörterung ihrer klimatischen Folgerungen enthaltend, wie auch der toprgraphischen Beziehungen tiefer Verwitterungen; periodische und anhaltende Witterungsdurchdringung und Regolithentfernung; das Vorkommen von Wannenformen in Granitlandschaften, mit einer Erörterung besonderer notwendiger Bedingungen eingeschlossene Wannen zu produzieren; die Entwicklung sphäroidischer und kuppelförmigenr Formen in Graniten, und die Identifikation von Granitlandformsystemen und iherer Interpretation, enthaltend darin sind Betrachtungen der Vorkommen bestimmter multikonkaver (Wannenform) Landschaften und multikonvexer (Kuppelform) Landschaften. Die Entwicklung von Stufen- oder Vieletagenlandschaften ist auch diskutiert. Implikationen für deduktiven Musterbau im Granitgelände sind betrachtet in Form von notwendigen Modifikationen zur formellen 'Zwei Stufen Hypothese' der Felsturmentwicklung. Letztlich, ein Muster für die Entwicklung von Granitrelief in frostfreien Klimas ist vorgebracht, angenommen eine langanhaltende sub-luftige Geschichte, während der die Witterungseindringung als andauernder Prozess angesehen ist, aber umfassendes Abreiben ist der weitverbreiteten Instabilität der Landoberfläche zugeschrieben, welche sich von periodischen ekologischen Veränderungen ergibt.

ABB. 1. 'Separation Point' Granit von der Südinsel Neuseelands. 1, Folgende Zonen von tiefer, sandiger Verwitterung; 2, Gipfelhochebene, eine Felsbodenwanne enthaltend (angezeigt in Abb. 2); 3, haupt-strukturelle Absteckungen

ABB. 2. Geomorphologische Zonen innerhalb des 'Separation Point' Granits. Geomorphologische Zonen auf Karte und Sektion: A, Gipfelhochebene, teilweise eingeschlossene Felswannen enthaltend; B, zerschnittene felsige Zone; C, Zone von tiefer Verwitterung und Talentwicklung, unterbrochen von Felssperren, markiert durch Wasserfälle; D, asymmetrische Kämme zur Küste hin herabfallend. Morphologische Besonderheiten: 1, Felswannen der Gipfelhochebene; 2, zerschnittene Wannen felsige Ränder beibehaltend; 3, teilweise eingeschlossene Wannen in verwittertem Granit; 4, zerschnittenes verwittertes Gelände auf schiefen Granitrücken; 5, Alluvium; 6, Hauptfelsrücken; 7, Grenzen von Felswannen; 8, folgende Zone von tiefer, sandiger Verwitterung

ABB. 3. Felskerniger Rückstand unter Wald in Johor, West Malysien. Ein unregelmässiges, flaches Regolith, welches Kerne enthält, die während Denudation auf die Oberfläche verbreitet werden, es erscheint im Gleichgewicht mit dem Denudationssystem zu existieren

ABB. 4 und 5. Sphäroidische Felskerne, entlang der Küste der Tasmanischen Bucht blossgestellt, im 'Separation Point' Granit, Neuseeland. Die Kerne werden von dichter klüftiger Granit ausgehoben, zeitweise verwittert zu einem sandigen Grus. Plattenformen sind auf diesen Formen klar entwickelt, was zu Zweideutigkeit in der Unterscheidung zwischen Kern, Kelstrum und Kuppel führt

ABB. 6. Entwicklung von Kernen, innerhalb plattenförmiger, kuppelartiger Felsmasse auf dem Jos Plateau, Nigeria. Es ist vorgeschlagen, dass eine Klüftenhierachie in der Formation erst einer grösseren Kuppel erfolgt, aber später zu ihrer Unterteilung entlang vertikaler Bruchlinien führt

ABB. 7. Moa Park, eine felsgrundige Wanne (s.o. Abb. 2) suf der Gipfelhochebene des 'Separation Point' Granits, Neuseeland. Gesehen vom felsturmpfostigen Rand, die Wanne hat einer Tellerform sit Hängen von weniger also 1° im Zentrum. Ein Wasserlauf läuft von links zum Hintergrund des Blides ab und allmähliche Zergliederung mit der Formation von Konvexhängen in dem Tal kann gesehen werden. Der Wasserlauf, welcher die Wanne von rechts entwässert, hat eine sehr flach Neigung und Gemeinschaften des Marschlandes dominieren den Wannenboden

ABB. 8. Hypothetische Folge in klüftigem Granit während mässiger Reliefentwicklung ohne Neigung: 1, unbekannter früherer Oberfläche; 2, abgeteiltes konvex Relief mit Gipfelfelstürmen; 3, Ausgrabung konkaver Wannenformen durch Schrägungsrückzug und möglicher Entleerung mit Untergrabung einiger Felstürme; 4, Zergliederung des Wannengrundes und Heruntersetzen von Riedeln; 5, Rückkehr zu konvex Landschaften ähnlich zu 1 oder 2

ABB. 9. Hypothetische Folge in klüftigen Graniten während starker Reliefentwicklung, verbunden mit Neigung; 1, unbekannte frühere Oberfläche; 2, abgeteiltes konvex Relief mit Gipfelfelstürmen; 3, 4, Oberflächenneigung, zu Zergliederung und Formierung von asymmetrischen Ketten führend; 5, felsige Stirne und zergliederte, verwitterte Stufenflächen formiert; 6, Heruntersetzen von Relief reduziert Asymmetrie, Rückkehr der Landschaft zum Zustand ähnlich zu 1 oder 2. Dieses Muster scheint der beobachteten Folge von Formen in Ann. 2 gezeigt zu folgen. In diesem Gebiet scheint die Neigung von Granit zwischen 4°–5° zu liegen; die Wannen sind 1–2 km in Breite und zwischen Kämmen und das innere Relief variert von c. 100 m in Zone A bis zu 200 m in Zone D

ABB. 10. Verfallene Felstürme auf dem Monaro Tafelland von Neusüd Wales, Australien. Die tiefe Verwitterung ist ouf der Peripherie, aber mit beinaher Gewissheit dehnt sie sich unter der Gruppe von Kernsteinen aus

The geomorphological importance of jointing in the Dartmoor granite

A. J. W. GERRARD

Lecturer in Geography, University of Birmingham

Revised MS received 5 July 1973

ABSTRACT. The Dartmoor granite is traversed by three sets of prominent joints, two vertical and one horizontal. The horizontal joints or pseudo-bedding planes appear to conform to the general contours of the land surface and it is suggested that they have evolved in sympathy with the land surface as denudation of overlying material gradually reduced the primary confining pressures of the granite. The result is a series of 'unloading' or dilatation domes delimited in the first instance by the evolving drainage net. Analysis of the vertical jointing shows a slight favouring of north–south and east–west directions which are probably equivalent to primary joints resulting from original stress phenomena. It is this jointing that has been picked out by weathering agencies and stream courses. Tors are found crowning many of the unloading domes and their distribution suggests that their formation relies on further incision of the streams into the domes. This would release more compressive stresses remaining in the granite and allow the opening up of the vertical jointing which would then permit the penetration of weathering agencies. At the same time, it is possible that contour joints were also formed by small-scale sliding along pseudo-bedding planes. The summit tors owe their form to the combined action of sub-surface chemical weathering and cryo-nival processes. It is thus concluded that the influence of jointing in the physical landscape of Dartmoor has been considerable.

THE significance of joints as influencing factors in the development of granite landforms has been stressed by most workers and in this respect the granites of Dartmoor are no exception. Granite in an unjointed state possesses a low porosity and permeability and it is only the well-developed joint systems that allow the easy passage of water. In addition, sound unjointed granite is a comparatively strong rock and it is mostly the jointing that renders it susceptible to erosion.

Granite is normally traversed by three sets of joints, two vertical or steeply inclined and one horizontal or gently dipping. These, intersecting approximately at right angles and with varying frequency, give the granite tors both their general similarity and their internal variation. Thus tors range in form from massive to lamellar depending on whether they are dominated by vertical or horizontal joints.[1] However, many statements concerning jointing belie the true complexity of the situation. Thus 'the joint in the rock . . . provides much yet describes little'.[2]

The aim of this paper is to examine some of this complexity with respect to the Dartmoor granite and to assess the role jointing has played in the evolution of the physical landscape. To achieve this the author's field investigations are equated with a review of previous workers' findings.

JOINT SYSTEMS

Joints have been defined, following C. A. Chapman,[3] as fractures in the rock along which there has been no visible movement parallel to the surface of the break. Joint systems which at first sight appear reasonably simple are found to be not so on closer inspection. Joints vary not only in frequency and orientation but also in their degree of openness and whether they consist of single or many partings. Also, within any one exposure some are undoubtedly of primary origin whereas others are secondary in nature. Even within the primary or formative jointing it may be possible to differentiate between those joints with a regional trend and those locally developed. Variation can therefore be expected even over short distances.[4]

The most ubiquitous joints wherever granite is exposed are the curved sheet joints often called pseudo-bedding planes which impart a dome-like structure to the landscape. This sheeting, exposed on some of the steeper slopes, is seen to be approximately horizontal again in the floors of the major streams. Thus although Dartmoor presents two contrasting landscape types, the comparatively gentle convex slopes of the higher central areas and the deeply incised adjacent valley systems, a connection between the two may be provided by the pseudo-bedding planes. It would seem reasonable then to agree with C. R. Twidale who has suggested that this contrasted relationship between upper convexity and incised valley slopes should be of some significance in the as yet largely unravelled erosional history of Dartmoor.[5] Thus the relationships exhibited by these pseudo-bedding planes are of the utmost importance in understanding the Dartmoor scene and will therefore be analysed first.

Pseudo-bedding planes

No better description of this jointing exists than that provided by A. Brammall. 'The steeper faces of many tor-masses have the aspect of bedded flags, fluted horizontally and more or less regularly from top to bottom. . . . Sheeted, lenticular or cake-like bedding grades into an irregular pillow structure frequently in one and the same tor mass'.[6]

The simple solution is to suggest that these joints arose in response to stresses in the granite at the time of emplacement and would thus correspond to the L joints in the classification of H. Cloos.[7] However although some, especially those exhibiting subsequent mineralization, are undoubtedly of this type, many present perplexing problems, the most obvious of which is the apparent coincidence between the attitude of the joints and the form of the ground surface. This relationship although not perfect is so close as to persuade many workers that the two are related cause to effect.

A difference of opinion exists however as to which is the causative factor. Thus H. T. De La Beche[8] has suggested that the attitudes of the joints reflect the form of the original roof of the intrusion while C. A. McMahon[9] concluded that the joints were due entirely to subaerial weathering. Worth attributed the undulations of the pseudo-bedding planes to the moulding of the igneous mass to the undersides of the folds in the original sedimentary cover[10] while Brammall[11] and later the officers of the Geological Survey[12] argue in favour of the joints being original structural features. Alternatively R. S. Waters[13] has argued forcibly for the development of these joints in response to the evolving topography. Finally a composite view has been presented by C. S. Exley[14] working on the St Austell granite where in his view the joints represent planes of weakness already initiated by consolidation stresses in the granite but which are not apparent as such until the removal of the overlying rock.

The idea that the pseudo-bedding planes are original structural features would seem to demand that the topography was so adjusted to structure that, as Waters has put it, 'every . . . stream must be regarded as occupying a structural valley'.[15] This would seem extremely unlikely. It has been shown by previous workers that the Dartmoor region has suffered more denudation than the Cornish masses,[16] that only the deeper parts of the mineral lodes remain[17] and that the Dartmoor granite has contributed large quantities of detritus to the sediments of south-east England.[18] The removal of this inferred thickness of granite would make it unlikely that the main rivers still occupied original downfolds in the granite.

G. K. Gilbert[19] in his examination of sheeting and granite domes in the High Sierras thought that some process of pressure release or unloading as erosion progressed was most consistent with the facts. The parts of the granite successively exposed at the surface by denudation would be in a condition of potential expansion and strain would be relieved by the separation of layers of rock approximately parallel to the land surface. This general idea of the dilatation of igneous masses after reduction of the primary confining pressure has also been suggested by R. Farmin[20] and accepted by R. H. Jahns.[21] Recent work has also shown that these divisional planes can develop remarkably quickly[22] and it is therefore not difficult to envisage the pseudo-bedding planes on Dartmoor originating in this way.

Progressive lowering of the surface would cause the rock to attempt to expand in all directions and although expansion horizontally would be hindered by the rock itself there would be a certain amount of vertical expansion leading to rupture. Thus pseudo-bedding planes would be formed but with the rock still essentially in a state of compression.[23] Domes or whale-back ridges would develop and be delimited by the evolving drainage net with each dome or compartment evolving independently of its neighbour.[24] The domes being in a state of compression would become even more resistant to erosion and would then be imprinted on the landscape for a long time. Some of this jointing is, however, undoubtedly equivalent to the primary (L) jointing of Cloos[25] especially in those instances, as at Meldon Hill, Gidley Tor, Little Mis Tor and elsewhere, where there appears to be an inverse relationship between topography and jointing.[26] But the evidence would seem to suggest that the majority of the pseudo-bedding planes are secondary features in the granite and can best be explained by a process of dilatation or unloading on the removal of a substantial weight of overlying rock.[27]

The differentiation of the landscape into unloading domes must be intimately connected with the configuration of the drainage net and the formation of basins or depressions. Since it has been recognized that jointing is a structural element which can and does exert a control over the details of topography[28] a number of workers have suggested that many of the Dartmoor streams are joint controlled. If this is so then the possibility exists that the vertical joints are the major controllers of the occurrence and configuration of the domes.

Vertical or steeply dipping joints

Observations of vertical jointing have been made on exposures of the 'giant' or 'tor' granite[29] which forms the majority of the exposures. Statements concerning the orientation and frequency of vertical jointing in this granite are widespread throughout the literature. Thus in 1839 De La Beche[30] argued that 80 per cent of the joint directions differed by only 14 degrees from a direction of N 25°W and that a further 15 per cent differed from this same direction by only 14 to 20 degrees. Therefore 95 per cent of the vertical jointing was ascribed to directions between N 45°W. and N 5°W. A little later A. Sedgwick and R. I. Murchison[31] argued for a mean direction north to south while G. W. Ormerod in 1869[32] agreeing in part with De La Beche claimed a certain preponderance for joints bearing north-north-west and east-west.

Worth,[33] basing his conclusions on 795 joint directions measured at 178 stations, has questioned the validity of all these assertions. There was certainly little evidence in his observations that north-north-west was a favoured direction. He found that the favoured arc of De La Beche contained only 15·1 per cent of his observations whereas had the distribution been even it should have contained 15·55 per cent. Analysing his own results in arcs of 15 degrees, Worth found that east-west was most favoured with 10·7 per cent. Much later, F. G. H. Blyth[34] working in north-east Dartmoor found 11 per cent of the vertical joints aligned along a 150 to 330 degree line, again emphasizing the approximate north-north-west to south-south-east line. In this case however, there is some evidence to suggest that the jointing in this area is anomalous being so closely connected with the north-west to south-east Lustleigh–Sticklepath fault zones. More recent still has been the work of the Geological Survey in the Okehampton area.[35] In a very exhaustive study they found a maximum of 13 per cent of the joints aligned north–south with a further 8 per cent aligned east–west. North–north–west to south–south–east was the second commonest direction with 10 per cent of the joints.

Thus there appear to be conflicting statements. The analysis presented here is based on 990 observations at 118 localities. These observations refer to the dominant joint direction with a direction being classed dominant if at least five well-defined joints are aligned in accord with it. Although the direction of every vertical joint has not been measured there is no reason to suspect that the results do not present a good approximation of the true situation.

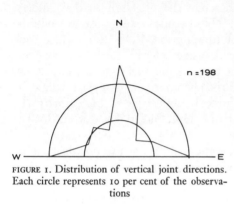

FIGURE 1. Distribution of vertical joint directions. Each circle represents 10 per cent of the observations

As shown in Figure 1, there is a slight preponderance of north–south jointing (25 per cent) followed by an east–west direction (19 per cent). This distribution, although statistically different from a regular distribution, does also emphasize the fact that vertical joints can be found aligned along almost every direction.[36] It is because of this that simple statements concerning joint directions can be misleading. Thus although there is some favouring of the north–south direction it is also clear that many different types of jointing might be represented in these figures. Much of this jointing was undoubtedly formed at an early stage and some of the apparent complexity may be simplified by a consideration of the types of primary joints common to most igneous masses.

Types of vertical jointing

In any igneous rock much of the vertical jointing is of primary origin related to the mode of rock emplacement and the associated stresses and flow phenomena.

The most usual analysis of primary vertical jointing is based on the classification and nomenclature proposed by Cloos.[37] Joints lying perpendicular to the flow lines, Q or cross joints, are tensional open joints liable to be infilled with more fluid magma residues. S joints strike parallel to the flow lines and are best developed where the flow lines approach the horizontal which is often the case near the roof of large intrusions. Thus the major sets of vertical joints may be explained in terms of the formation of the granite. In addition, two rather more uncommon sets could be of primary origin. Diagonal joints, formed at an angle of 45 degrees to the pressure directions, can occur as also can folds or quasi-anticlinals due to tensional forces directed away from the strike of the Q joints.

This classification provides a valuable working framework within which to consider the Dartmoor situation. Recent work[38] has established the form of the granite batholith by geophysical methods and has shown that it differs little from that postulated by earlier workers.[39] It seems that the granite magma was confined between two parallel thrusts; the southerly being the Lizard–Dodman–Start thrust and the northerly a thrust extending through the Tintagel area. The Dartmoor granite was thus subject to a thrust on the south side trending east–west which led Exley[40] to suggest that north–south and east–west joints were formed in response to this thrust. In a similar way the jointing in the other granitic masses of south-west England are assumed to be associated with the direction of this thrust (Fig. 2). The suggestion is therefore that the majority of the north–south and east–west jointing is of primary origin, a fact that is of importance in the evolution of the physical landscape.

Diagonal jointing seems to be essentially absent and pseudo-anticlinals rare although Brammall[41] believes some are present in the Widecombe–Hameldown area and Blyth[42] has suggested that some of the jointing in north-east Dartmoor may represent Riedel shears[43] associated with

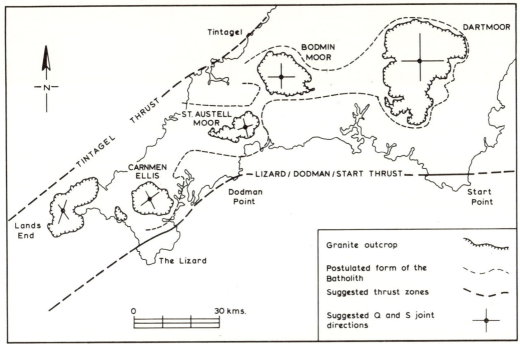

FIGURE 2. Relationship between the major thrust directions and the orientation of vertical jointing in the granites of south-west England. (After C S. Exley, 'Magmatic differentiation and alteration in the St Austell granite', *Q. Jl geol. Soc. Lond.* 114 (1958), 197–230 and K. F. G. Hosking, 'Permo-Carboniferous and later primary mineralization of Cornwall and south-west Devon' in *Present views of some aspects of the geology of Cornwall and Devon* (1964), 201–36)

the north-north-west to south-south-east faulting. Irrespective of their mode of formation, it is important that the primary joints be differentiated from later jointing because of the former's suggested influence over the development of the drainage net, the delimitation of upstanding areas from depressions by guiding the action of weathering agencies and the formation of tors.

Control of drainage by jointing

The considerable local accordance between valley and joint directions, noted by J. Palmer and R. A. Neilson,[44] has been stressed by Brammall where he argues that 'joint systems in the granite and local jointing are often strikingly concordant with the trend of valleys and of the ridges between them'.[45] A close inspection of the drainage shows that many streams possess portions of their courses which are remarkably constant in direction[46] and that a large number also possess sharp changes in direction of approximately ninety degrees.[47] Some of these abrupt changes of direction may be referred to river capture but in the majority of cases this is not a feasible explanation. To examine the possibility of joint-controlled stream courses the drainage systems have been analysed in detail.

The drainage was taken as that portrayed in blue on the 1:25 000 Ordnance Survey maps. The streams were differentiated according to the A. N. Strahler[48] modification of R. E. Horton's[49] ordering system. The length and direction of every stream course possessing a constant direction for at least 800 m was measured. First-order streams often possess relatively constant directions for their entire lengths but streams of higher orders vary more in direction. It was hoped to treat streams of different orders separately but the sparsity of streams of third or higher orders precluded this. Therefore a two-fold division of first-order streams and higher has been adopted.

The results of this analysis are shown in Figure 3. The similarity between the distribution of jointing and stream courses is striking. It is, however, a big step from this to conclude unequivocally that the one is a cause of the other. But this, in conjunction with other evidence, seems to suggest that some of the jointing has been responsible for guiding the evolution of the drainage systems. Certainly the fact that the granites of south-west England have suffered considerable weakening along the joint patterns by late stage hydro-thermal activity would make this feasible.

EVOLUTION OF THE PHYSICAL LANDSCAPE

The thesis presented so far is that the drainage net, in part largely governed by major jointing and lines of weakness, has determined the spatial arrangement of the domes and ridges which were created in response to the release of primary confining pressures. This has resulted in a differentiation of the landscape into positive and negative areas[50] in which it is possible to note a hierarchy of relief elements from small to large depending on the type and intensity of jointing and on the response of the drainage to these structural features.[51]

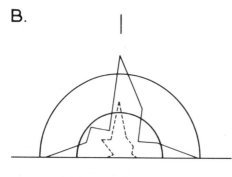

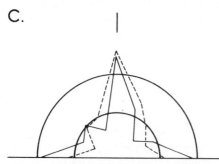

The arrangement of the tors also suggests that they are intimately connected with the form and evolution of the unloading domes. Tors are found in essentially three relationships to these domes and their morphology suggests that slightly different mechanisms have been instrumental in their formation. The 'classic' Dartmoor tors are those found crowning the ridges and summits, hereafter referred to as summit tors, e.g. Great Mis Tor, Greater Staple Tor and Hay Tor. Secondly, there is a composite group encompassing valley-side and spur tors. These are often found at the break of slope between the upper convexity and the maximum valley-side slope segment and like the summit tors are upstanding on all sides, e.g. Vixen Tor on the valley slopes of the River Walkham and Black Tors in the Meldon valley. Finally there are tors, less impressive in terms of dimensions, which seem to emerge from the flanks of low convex hills. These possess a free face often on only one side and are rarely more than 7 m in height, e.g. Heckwood, Hucken and Feather Tors. They also possess, to varying degrees, exposures of sheeted granite at their bases.

Summit tors

The Dartmoor landscape shows evidence of denudational agencies dating back to at least the beginning of the Tertiary period. Comparatively level surfaces and accordant summit heights led to an early recognition of erosion surfaces especially when the possibility of large-scale erosion cycles was suggested

FIGURE 3. Relation of jointing to stream directions. Black lines represent joints, dotted total stream length. A. First-order streams B. Greater than first order streams C. All streams. Each circle represents 10 per cent of the observations

for the Tertiary period. Present opinion favours the existence of three subaerial surfaces above approximately 300 metres.[52]

The summit or upper surface is represented by remnants ranging in height from 500 to 650 m, with subsequent cycles of erosion leading to the formation of new erosion surfaces between 500 and 300 m. At these latter stages there are indications that the evolving drainage systems were gradually becoming adjusted to structures within the granite. The formation of these new erosion surfaces would have led to the development of a new series of unloading domes which is in agreement with Waters who has suggested that the undulations of the pseudo-bedding planes reflect a mid-Tertiary peneplain.[53] Most workers have tended to regard these surfaces as normal sub-aerial peneplains but it is not inconceivable that processes such as modified pedi-planation or savanna planation have been involved.[54]

There is certainly some evidence to suggest that chemical weathering was more intense during this period.[55] C. M. Bristow[56] has shown the presence of a weathering mantle beneath late Oligocene sediments in the Petrockstow basin, 15 km north-west of Okehampton, and has suggested that the sediments themselves have originated in a weathering mantle developed under humid sub-tropical or warm temperate conditions. In addition, the type of weathering found in the Okehampton area seems also to be consistent with conditions of this type[57] as might much of the weathered granitic material found on Dartmoor.[58]

Comparisons with areas presently experiencing these conditions suggest that a variable thickness of chemically weathered material would have been formed.[59] This variability in thickness would be expected first because the granite varies quite considerably in its composition and texture and therefore its weathering potential and secondly because zones of granite were already ripe for weathering before being exposed. Although much has been removed by subsequent erosion a variable amount of weathered material *in situ* can still be found on the valley sides. Thus thicknesses of up to 4 m exist on the valley slopes of the Cowsic, Meavy, Dart and Cherrybrook but this must be contrasted with the upstanding tors and the exposed sheets of granite such as the Slipper Stones in the West Okement valley and similar features on Hamel Down and near Bellever Bridge.

The theories of D. L. Linton[60] and Palmer and Neilson,[61] taken singly or in combination, explain many of the features of the summit tors. As mentioned above, there is evidence to suggest that sub-surface chemical weathering has been active in the past. Also there is undoubted evidence that cryo-nival processes have been instrumental in fashioning the tors to a greater or lesser degree.[62] Even so, several perplexing problems remain.

It has been argued by both authorities that the tors and summit areas remain upstanding because of the paucity of vertical jointing. This would be expected if the granite domes were in a state of compression as this would tend to close some of the vertical jointing. Yet close inspection of these tors

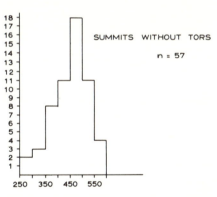

SUMMITS WITHOUT TORS

n = 57

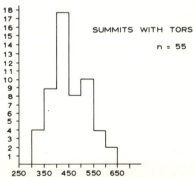

SUMMITS WITH TORS

n = 55

FIGURE 4. Summit height frequency distribution. (Horizontal scales are summit heights in metres; vertical scales are the number of summits recorded in each class)

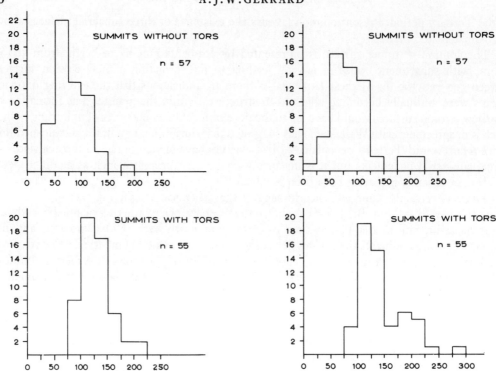

FIGURE 5. Frequency histograms of relative relief within 800 m for summits with and without tors. (Horizontal scales are relative relief values in metres; vertical scales are number of observations recorded in each class)

FIGURE 6. Frequency histograms of maximum relative relief for summits with and without tors. (Horizontal scales are relative relief values in metres; vertical scales are number of observations recorded in each class)

reveals in many instances a very dense network of joints. Allied to this is the problem concerning the large gaps of the order of a hundred metres that exist among tor groups. It has been customary to ascribe the missing portions of these 'associated'[63] or 'avenue'[64] tors to more closely spaced jointing, yet it is difficult to imagine jointing more closely spaced than that exposed on nearby tors.

The distribution of the summit tors is also perplexing. Why should tors have developed on some ridges and unloading domes and not on others? Part of the explanation may be related to variations in the granite and in other cases masses of clitter attest perhaps to the former presence of tors long since destroyed. But these facts seem insufficient to explain the majority of instances where tors are lacking. The possibility obviously exists of sub-surface tors waiting to be revealed, but what little evidence is available seems to point against this. Also, if cryo-nival processes alone are responsible for the production of tors, it is difficult to see why the processes should be so selective in their action.

As pointed out by Palmer and Neilson[65] there is no significant difference in altitude between those summits possessing and those lacking tors (Fig. 4). The two distributions merely reflect the three major erosion levels present although the majority of the tors are found on the two higher levels. An analysis of the relative relief around tor-crowned summits and around other summit areas does however, provide striking differences (Figs 5 and 6). Unless these differences are a mere coincidence, they would seem to be geomorphologically significant. Tors are found mostly on those summits and ridges with a relative relief greater than 100 m within a distance of 800 m. This distinction is still valid when the maximum possible relative relief is considered (Fig. 6)

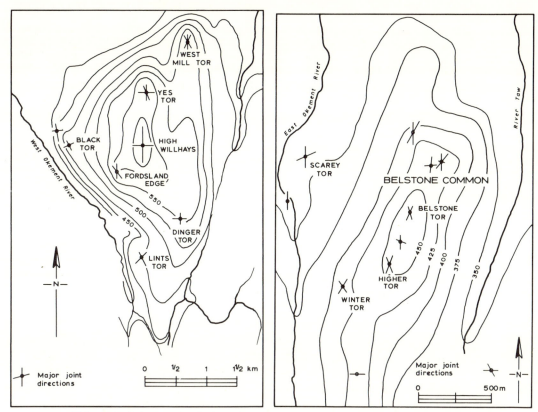

FIGURE 7. Major vertical joint directions in the vicinity of Yes Tor, northwest Dartmoor. Contour heights in metres
FIGURE 8. Major vertical joint directions in the vicinity of north Dartmoor. Contour heights in metres

but the distinction over 800 m would seem to suggest that average slope angle is an important factor as well as relative relief. Taken in conjunction, it seems that relative relief and average slope should figure in any explanation of the distribution of tors.

It is the joints that have been exploited to give the tors their characteristic shapes and for maximum effect these joints have to be essentially open. If the idea that the unloading domes were in a state of compression when formed is accepted, then some other mechanism must be responsible for further opening up the vertical joints and thus allowing the penetration of weathering agencies. Assuming the marked differences in relative relief to be significant, then the mechanism involved could be incision into the domes by 'rejuvenated' streams.[66] This incision into the domes and ridges would allow compressive stresses still present in the granite to be released and allow further jointing to develop. It seems, on mechanical principles, that the jointing so initiated would be most intense in the summit areas which would help to explain the frequency of jointing. Joints could also be expected to form on the flanks of the domes roughly in accord with their configurations, i.e. parallel and at right angles to the contours.

The incision of streams into the upland areas with the consequent increase in slope angles may have been also sufficient to impart some instability to the jointed granite. The granite may have been susceptible to sliding along the pseudo-bedding planes, a fact noted elsewhere by R. Balk[67] and many civil engineers.[68] Thus joints initially formed because of renewed release of pressure may have been further widened by small-scale sliding on the pseudo-bedding planes and it is not inconceivable that additional joints could originate in the way described by Chapman.[69]

Close analysis suggests that jointing of a similar nature is present. Many of the joint directions exposed on the flanks of Yes Tor and High Willhays (Fig. 7) are aligned either approximately parallel to or at right angles to the contours. Thus the jointing gradually changes direction between Black Tors through Fordsland Ledge to Dinger and Lints Tors roughly in accord with the change in aspect of the maximum slope. Similarly the same general relationship is observed on and around the flanks of Belstone Common (Fig. 8) and at other localities such as Leeden Tor, North Hessary Tor, Great Mis Tor, Hay and Saddle Tors. Because of the widely dispersed nature of the outcrops these ideas can only be tentative but the relationship between this topography and configuration of the slopes is striking. If some of these joints were not formed in this manner then this is another instance of the physical landscape being extremely well adapted to structure and raises the problems mentioned earlier in connection with the pseudo-bedding planes.

The formation of many of the tors would thus seem to have been a continuous process from the mid-Tertiary period onwards. The unloading domes would have formed during successive periods of planation when something approaching a type of etchplain may have existed. A reconstruction of the present distribution of weathered material and exposures of sheeted granite is very similar to incised etchplains described in other parts of the world.[70] Tors would probably have developed as further incision opened up the vertical joints to the action of chemical weathering agents and would accord with the ideas of Linton.[71] Subsequently during the cold periods of the Pleistocene the summit tors would have been modified by cryo-nival processes. It appears possible that these cryo-nival processes may have affected tors below the former basal surface of weathering. The height of most summit tors may therefore be in excess of the maximum depth of the chemically-altered regolith.

The influence of chemical weathering seems to have been less in the formation of both the valley-side tors and the small cliffs of granite flanking low domes. The latter, with prominent pseudo-bedding planes, few vertical joints and exposed sheets of granite at their base, exhibit many of the characteristics of frost-riven cliffs as described by J. Demek in Bohemia.[72]

Thus, directly and indirectly, the types of jointing within the Dartmoor granite have determined both the appearance and evolution of many of the physical features. It is also seen that the apparently diverging views of tor formation are encompassed within the general theme of joint controlled landscapes.

CONCLUSION

A re-examination of earlier work coupled with detailed analysis of the jointing within the Dartmoor granite has provided a possible synthesis with which to view the evolution of the physical landscape. In this synthesis, the horizontal and vertically developed joint systems are of the utmost importance, first in guiding the developing drainage net and thus delimiting the evolving unloading domes and secondly in governing the form and distribution of the tors. Some of this jointing is of primary origin but the possibility obviously exists that much, both horizontal and vertical, has developed in response to the changing configuration of the landscape.

Many of the ideas must, however, remain tentative. More information is required concerning the intensity of jointing and its direction between the presently available exposures. Also, more detailed knowledge is required of both joint forming mechanisms and the response of jointed rock to certain erosional forces. The use of complex mathematical models may provide some information of the latter.

ACKNOWLEDGEMENTS

The author gratefully acknowledges the early help and encouragement of Dr D. Brunsden and the helpful suggestions of the late Professor D. L. Linton. Thanks are also due to the photographic staff and drawing office of the Department of Geography, University of Birmingham.

NOTES

1. R. H. WORTH, 'The physical geography of Dartmoor', *Rep. Trans. Devon. Ass. Advmt Sci.* 62 (1930), 49–115
2. C. M. NEVIN, *Principles of structural geology*, New York (1942), 131
3. C. A. CHAPMAN, 'Control of jointing by topography', *J. Geol.* 66 (1958), 552–8
4. R. D. TERZAGHI, 'Some sources of error in joint surveys', *Géotechnique* 15 (1965), 287–304
5. C. R. TWIDALE, *Structural landforms* (1971)
6. A. BRAMMALL, 'The Dartmoor granite', *Proc. Geol. Ass.* 37 (1926), 261
7. H. CLOOS, *Einführung in die Geologie* Berlin (1936)
8. H. T. DE LA BECHE, *Report on the geology of Cornwall, Devon and West Somerset* (1839)
9. C. A. McMAHON, 'Notes on Dartmoor', *Q. Jl geol. Soc. Lond.* 49 (1893), 385–97
10. R. H. WORTH, op. cit.
11. A. BRAMMALL, op. cit.
12. E. A. EDMONDS ET AL., 'Geology of the country around Okehampton', *Mem. geol. Surv. U.K.* (1968)
13. R. S. WATERS, 'Pseudo-bedding in the Dartmoor granite', *Trans. R. geol. Soc. Corn.* 18 (1954), 456–62
14. C. S. EXLEY, 'Magmatic differentiation and alteration in the St Austell granite', *Q. Jl geol. Soc. Lond.* 114 (1958), 197–230
15. R. S. WATERS, op. cit., 458
16. A. BRAMMALL and H. F. HARWOOD, 'The Dartmoor granite', *Mineralog. Mag.* 20 (1923), 39
17. A. BRAMMALL, 'Notes on fissure phenomena and lode-trend in the Dartmoor granite', *Trans. R. geol. Soc. Corn.* 16 (1927), 15
18. A. W. GROVES, 'The unroofing of the Dartmoor granite and the distribution of its detritus in the sediments of southern England', *Q. Jl geol. Soc. Lond.* 87 (1931), 62
19. G. K. GILBERT, 'Domes and dome structures of the High Sierra', *Bull. geol. Soc. Am.* 15 (1904), 29–36
20. R. FARMIN, 'Hypogene exfoliation in rock masses', *J. Geol.* 45 (1937), 625–35
21. R. H. JAHNS, 'Sheet structure in granites: its origin and use as a measure of glacial erosion in New England', *J. Geol.* 51 (1943), 71–98
22. For an account see W. S. WHITE, 'Rock bursts in the granite quarries at Barre, Vermont', *Circ. U.S. geol. Surv.* 13 (1946)
23. For a detailed discussion see T. N. DALE, 'The commercial granites of New England', *Bull. U.S. geol. Surv.* 738 (1923) and C. R. TWIDALE op. cit.
24. An arrangement noted elsewhere by B. P. RUXTON and L. BERRY, 'Basal rock surface on weathered granitic rocks', *Proc. Geol. Ass.* 70 (1959), 285–90
25. H. CLOOS, op. cit.
26. Noted by C. R. TWIDALE, op. cit.
27. It has been estimated by A. W. GROVES, op. cit., that 1600 m of rock has been removed since the Permian
28. The literature concerning this subject prior to 1914 has been discussed by I. D. SCOTT, 'The spacing of fracture systems and its influence on the relief of the land', *Beitr. Geophys.* 13 (1914), 163–81 and 241–60. The literature after this has been summarized by S. JUDSON and G. W. ANDREWS, 'Pattern and form of some valleys in the Driftless Area, Wisconsin', *J. Geol.* 63 (1955), 328–36
29. For a description of the types of granite see W. R. DEARMAN, 'Dartmoor: its geological setting' in I. G. SIMMONS (ed.) *Dartmoor Essays* (1964), 1–29
30. H. T. DE LA BECHE, op. cit.
31. A. SEDGWICK and R. I. MURCHISON, 'On the physical structure of Devonshire and on the subdivisions and geologic relations of its older stratified deposits', *Trans. Geol. Soc.* 2nd ser. 5 (1840), 633–704
32. G. W. ORMEROD, 'On some results arising from the bedding, joints and spheroidal structure of the granite on the east side of Dartmoor', *Q. Jl geol. Soc. Lond.* 25 (1869), 273–80
33. R. H. WORTH, op. cit.
34. F. G. H. BLYTH, 'The structure of the north-eastern tract of the Dartmoor granite', *Q. Jl geol. Soc. Lond.* 118 (1962), 435–53
35. E. A. EDMONDS ET AL., op. cit.
36. R. H. WORTH, op. cit., has described the situation at Bellever Tor where joints can be found aligned along just about every conceivable direction.
37. H. CLOOS, op. cit.
38. M. H. P. BOTT, A. A. DAY and D. MASSON-SMITH, 'The geological interpretation of gravity and magnetic surveys in Devon and Cornwall', *Phil. Trans. R. Soc.* 251A (1958), 161–91
39. K. F. G. HOSKING, 'Fissure systems and mineralization in Cornwall', *Trans. R. geol. Soc. Corn.* 18 (1949), 9–49
40. C. E. EXLEY, op. cit.

41. A. BRAMMALL, op. cit. (1926)

42. F. G. H. BLYTH, op. cit.

43. Originally described by W. RIEDEL, 'Zur Mechanik geologischer Brucherscheinungen', *Zentbl. Miner. Geol. Paläont.* B 354 and later verified by J. S. LEE, C. H. CHEN and M. T. LEE, 'Experiments with clay on shear fractures', *Bull. geol. Soc. China* 33 (1948), 25

44. J. PALMER and R. A. NEILSON, 'The origin of granite tors on Dartmoor, Devonshire', *Proc. Yorks. geol. Soc.* 33 (1962), 315–40

45. A. BRAMMALL, op. cit. (1926), 260

46. F. G. H. BLYTH, op. cit., has emphasized that many streams appear to flow north–south

47. Although it is normally assumed that constant stream directions suggest structural control it is possible for curved portions to be structurally guided as shown by J. T. HACK, 'Circular patterns and exfoliation in crystalline terrane, Grandfather Mountain Area, North Carolina', *Bull. geol. Soc. Am.* 77 (1966), 975–86

48. A. N. STRAHLER, 'Hypsometric (area-altitude) analysis of erosional topography', *Bull. geol. Soc. Am.* 63 (1952) 1117–42

49. R. E. HORTON, 'Erosional development of streams and their drainage basins; hydrophysical approach to quantitative morphology', *Bull. geol. Soc. Am.* 56 (1945), 275–370

50. Terms first used by R. S. WATERS, 'Differential weathering in oldlands', *Geogrl J.* 123 (1957), 503–13

51. For a summary of the literature on repeating patterns and relief hierarchy imposed by jointing see W. H. HOBBS, 'Repeating patterns in the relief and in the structure of the land', *Bull. geol. Soc. Am.* 22 (1911), 123–76

52. For a survey see D. BRUNSDEN, *Dartmoor* (British Landscapes through Maps) Sheffield, Geographical Association (1968)

53. R. S. WATERS, op. cit. (1954)

54. Suggested by A. R. ORME, 'The geomorphology of southern Dartmoor' in I. G. SIMMONS (ed.) *Dartmoor Essays* (1964), 31–72

55. D. L. LINTON, 'The problems of tors', *Geogrl J.* 121 (1955), 470–87 and 'Tertiary landscape evolution' in J. WREFORD WATSON and J. B. SISSONS (eds) *British Isles: a systematic geography* (1964), 110–30

56. C. M. BRISTOW, 'The derivation of the Tertiary sediments in the Petrockstow Basin, North Devon', *Proc. Ussher Soc.* 2 (1968), 29–35

57. P. G. FOOKES, W. R. DEARMAN and J. A. FRANKLIN, 'Some engineering aspects of rock weathering with field examples from Dartmoor and elsewhere', *Q.J. Eng. Geol.* 4 (3) (1971), 139–85

58. For a good summary see R. BRUNSDEN, 'The origin of decomposed granite on Dartmoor' in I. G. SIMMONS (ed.) *Dartmoor Essays* (1964), 97–116

59. M. F. THOMAS, 'An approach to some problems of landform analysis in tropical environments' in J. B. WHITTOW and J. P. D. WOOD (eds) *Essays in geography for Austin Miller* (1965), 118–44

60. D. L. LINTON, op. cit. (1955)

61. J. PALMER and R. A. NEILSON, op. cit.

62. For a summary of the evidence for cryo-nival processes on Dartmoor see R. S. WATERS, 'The Pleistocene legacy to the geomorphology of Dartmoor' in I. G. SIMMONS (ed.) *Dartmoor Essays* (1964), 39–57

63. G. W. ORMEROD, op. cit.

64. R. H. WORTH, op. cit.

65. J. PALMER and R. A. NEILSON, op. cit.

66. Rejuvenation is here taken in its widest sense. The role of rejuvenated streams was first stressed by D. L. LINTON in a paper read to the 17th International Geographical Congress, Washington (1952)

67. R. BALK, 'Disintegration of glacial cliffs', *J. Geomorph.* 2 (1939), 303–34

68. I. TERZAGHI, 'Dam foundations on sheeted granite', *Géotechnique* 12 (3), (1962), 199–208

69. C. A. CHAPMAN, op. cit.

70. M. F. THOMAS, op. cit.

71. D. L. LINTON, op. cit. (1955)

72. J. DEMEK, 'Slope development in granite areas of the Bohemian Massif (Czechoslovakia)', *Z. Geomorph.* Supp. 5 (1964), 82–106 and 'Castle koppies and tors in the Bohemian Highland (Czechoslovakia)', *Biul. peryglac.* 14 (1964), 195–216

RÉSUMÉ. *L'importance géomorphologique des joints au granit de Dartmoor.* Le granit de Dartmoor est traversé par trois groupes de joints prononcés, deux verticaux et un horizontal. Les joints horizontaux apparaissent se conformer au profil du terrain et on suggère qu'ils sont évolus pareillement à la surface de la terre à mesure que l'enlèvement de la matière superposée diminue peu à peu les pressions primaires du granit. Le résultat est un série de dômes de 'déchargement' ou de dilatation, délimitées en premier lieu par le réseau de drainage développant. L'analyse des joints verticaux indique une légère préférence pour les directions nord-sud et est-ouest, qui sont probablement équivalents aux joints primaires provenants de la tension originale. Ces joints sont altérés par l'action des agents atmosphériques et cours d'eau. Beaucoup des dômes de dilatation sont couronnés de tors, et leur distribution suggère que leur formation exige plus d'incision par les ruisseaux dans les dômes. Cela libère plus des efforts de compression restants au granit et permis l'ouverture des joints verticaux à la pénétration par les agents d'altération. Au même temps, c'est possible que les profils des joints sont formés par petits dégagements le long des larges plaques horizontaux. Les tors doivent leur forme aux processus chimiques des actions météoriques sou-terrains et aux actions périglaciaires. Alors on conclut que l'influence des joints sur le terrain physique de Dartmoor a été considérable.

FIG. 1. Distribution de direction des joints verticaux. Chaque cercle représente 10% des observations

FIG. 2. Connexité entre la direction des poussées majeures et l'orientation des joints verticaux aux granites du sud-ouest de l'Angleterre. (Après C. S. Exley, 'Magmatic differentiation and alteration in the St Austell granite', *Q. Jl geol. Soc. Lond.* 114 (1958), 197–230 et K. F. G. Hosking, 'Permo-Carboniferous and later primary mineralization of Cornwall and south-west Devon', in *Present views of some aspects of the geology of Cornwall and Devon* (1964), 201–46)

FIG. 3. Connexité entre les joints et les directions des ruisseaux. Les traits noirs indiquent les joints, les traits discontinues la longueur totale des ruisseaux. A. Ruisseaux de premier ordre. B. Ruisseaux plus grand que premier ordre. C. Tous ruisseaux. Chaque cercle représente 10% des observations

FIG. 4. Distribution de la fréquence de hauteur de sommet. (Échelles horizontales représentent hauteur du sommet en mètres; échelles verticales représentent le nombre de sommets dans chaque catégorie)

FIG. 5. Histogrammes de la fréquence de relief relatif entre 800 m (Échelles horizontales représentent valeurs de relief relatif en mètres; échelles verticales représentent le nombre des observations dans chaque classe)

FIG. 6. Histogrammes de la fréquence de maximum relief relatif (échelles horizontales représentent valeurs de relief relative en mètres; échelles verticales représentent le nombre des observations dans chaque classe)

FIG. 7. Direction de majeurs joints verticaux près de Yes Tor, Dartmoor Nord-Ouest. Courbes de niveau en mètres

FIG. 8. Direction de majeurs joints verticaux à Dartmoor nord. Courbes de niveau en mètres

---

ZUSAMMENFASSUNG. *Die geomorphologische Bedeutung der Klüftung im Granit von Dartmoor.* Der Granit von Dartmoor ist durch drei Reihe markanten Kluftsystemen durchkreuzt; zwei senkrechten und ein wagerechter. Die waterechten Klüfte scheinen die allgemeine Schichtlinie des Bodens zu folgen; es wird beantragt dass sie sich in Übereinstimmung mit dem Boden entfaltet haben während der Abträgung des überliegenden Materials die beschränkenden Drücke des Granits nach und nach vermindert hat. Der Erfolg ist eine Reihe 'Entlastungs-' oder Dilatations-dome, die in der ersten Instanz durch das sich entwickelnde Entwässerungssystem begrenzt werden. Analyse der senkrechten Klüftung zeigt eine geringe Begünstigung der Nord-Sud und Ost-West Richtungen, welchen vielleicht einstimmen mit einer primären Klüftung die aus den ursprunglichen Druckphenomene entstanden ist. Diese Kluftsysteme sind durch verwitternde Wurkungen und Flüsse hervorgehoben worden. Viele Entlastungsdome sind durch Felsburgen bekront und ihre Verbreitung deutet an, dass ihre Schöpfung eine weitere Fluss-einschnitt in die Dome fordert. Das wird mehr noch im Granit bleibenden, zusammenpressenden Drücke entlassen und die Öffnung der senkrechten Klüfte erlauben, wass also das Eindringen der Verwitterung permitiert. Zu gleicher Zeit ist es auch möglich, dass kleine Bewegungen die senkrechten Absonderungen entlang die Schichtklüfte verursacht haben. Die Felsburgen verdanken ihre Gestalt der Zusammenwirkung der unterirdischen chemischen Verwitterung und periglazialen Bewegungen. Es lässt sich deshalb nachweisen, dass der Einfluss der Klüftung für die physikalische Landschaft von Dartmoor sehr bedeutend gewesen ist.

ABB. 1. Verbreitung der senkrechten Kluftrichtungen. Jeder Zirkel repräsentiert 10% der Aufzeichnungen

ABB. 2. Verhältnis zwischen Richtungen des Hauptdrucks und Orientierung der senkrechten Klüftung in den Granite von sud-west England. (nach C. S. Exley, 'Magmatic differentiation and alteration in the St Austell granite', *Q. Jl Geol. Soc. Lond.* 114 (1958), 197–230, und K. F. G. Hosking, 'Permo-Carboniferous and later primary mineralization of Cornwall and south-west Devon', in *Present views of some aspects of the geology of Cornwall and Devon* (1964), 201–46)

ABB. 3. Verhältnis zwischen Klüftung und Flussrichtung. Schwarze Linien repräsentieren Klüfte, strichpunktierte Linien sind gesamte Flusslänge. A. Flüsse von erster Klasse. B. Flüsse grösser als erste Klasse. C. Alle Flüsse. Jeder Zirkel repräsentiert 10% der Aufzeichnungen

ABB. 4. Graphische Darstellung der Gipfelhöhe. (Horizontale Einteilungen sind Gipfelhöhe in Meter; vertikale Einteilungen sind Nummer der Gipfel, die in jeder Klasse aufgezeichnet sind)

ABB. 5. Graphische Darstellung des relativen Reliefs innerhalb 800 m. (Horizontale Einteilungen sind relatieve Reliefe in Meter; vertikale Einteilungen sind Nummer der Aufzeichnungen in jeder Klasse)

ABB. 6. Graphische Darstellung des maximum relativen Reliefs. (Horizontale Einteilungen sind relative Reliefe in Meter; vertikale Einteilungen sind Nummer der Aufzeichnungen in jeder Klasse)

ABB. 7. Richtungen der senkrechten Hauptklüfte in der Umgebung von Yes Tor, Nord-west Dartmoor. Schichtlinien in Meter

ABB. 8. Richtungen der senkrechten Hauptklüfte in der Umgebung von Nord Dartmoor. Schichtlinien in Meter

Granite tors in the Sudeten Mountains

ALFRED JAHN

Professor of Physical Geography, University of Wrocław

MS received 9 October 1972

ABSTRACT. Tors are numerous and widely distributed on the Karkonosze Mountains and in the Jelenia Góra basin. Their shapes and locations are related to the density and pattern of jointing in the two varieties of granite, equiangular and porphyritic, which beyond the tors carry a predominantly coarse-grained regolith ($> 50\ \mu$m) akin to growan and interpreted as a subtropical weathering product. It pre-dates the oldest glaciation. Tors and regolith are explicable in terms of the Linton concept of two-stage tor development. The selective, subsurface weathering is attributed to the latter part of the Neogene period but the tors were exhumed during Pleistocene periglacial phases subsequent to the dissection of Tertiary planation surfaces. Thus their density varies directly with the density of the valley network. They are elements of the meso-relief brought into prominence by denudation processes which operated progressively upward to rejuvenate the slopes. Consequently the youngest tors rise from the highest parts of the Karkonosze slopes.

FEW publications in the world's literature on geomorphology have generated so much interest and discussion as the 1955 paper by D. L. Linton on 'The problem of tors'. Its impact was due not so much to the interesting concept put forward by Linton, because the essence of his hypothesis of a two-stage process of tor evolution had been suggested by other authors, as to the fact that hitherto no one had elaborated the concept and supported it by reasoned argument so explicitly as did this distinguished geomorphologist whose premature death is deeply regretted.

The Sudeten Mountains include an extensive massif called Karkonosze (Riesengebirge) which occupies the border zone between Poland and Czechoslovakia. This massif has acquired a reputation for its numerous tors which exhibit a great variety of forms and attain heights of as much as 25 m. Similar tors occur in the adjoining Jelenia Góra basin. Both mountains and basin belong to the same geological unit, the *Karkonosze Massif*, and both are built of granites that will be referred to below as the Karkonosze granite.

For a long time both the granite and the tors have been the object of detailed investigations; particular mention should be made of the very extensive studies by G. Gürich in 1914 and the classical treatise published in 1925 by H. Cloos. Since the last war I have for many years been investigating the Karkonosze tors and in this research I had the assistance, among other scientists, of the late Professor P. Bakker of Amsterdam. In 1962 I published my study on 'The origin of granite tors' in which in principle I endorsed Linton's theory in respect of the development of the Karkonosze examples. When, upon my invitation, the author of this theory visited the Karkonosze in 1970 we had occasion to discuss very thoroughly the tors seen in the Sudeten Mountains and to correlate them with the classical model of tors occurring on Dartmoor.

CHARACTERISTICS OF THE KARKONOSZE TORS

Distribution

The tors occur all over the massif, from its base at 300 m absolute elevation up to its ridge at about 1400 m above sea level. A characteristic feature is their disposition on convex landforms of ridges and slopes; often they rise from convex breaks of slope. Only in the Jelenia Góra basin do they appear at relatively low elevations. Their position indicates that they are residual forms.

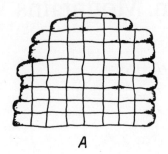 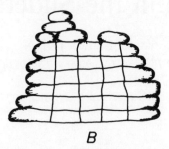 

A   B   C

FIGURE 1. The types of Karkonosze tors
A—table hill, B—tower, C—loose blocks

Shape

The tors appear in a wide variety of forms; yet one might distinguish three principal types of shape (Fig. 1): (a) table-hills, (b) towers, (c) rounded loose blocks. The table-hills occur in the western Karkonosze, the towers in the eastern part of these mountains and spheroidal blocks in the Jelenia Góra basin.

This regular association between tor shape and tor distribution is an indication that the tors developed in conformity to the structure of the granite, that is, to the density and pattern of its joints (quader joints and globate joints).

Weathering products

M. Borkowska (1966) has determined the composition of the Karkonosze granite as follows:

	%
quartz	30–40
potash feldspar	20–30
plagioclase	20–30
biotite	some 5

Chlorite, muscovite, hornblende and iron oxides are also present. On the average, the ratio of quartz to feldspars is $1 \cdot 0 : 1 \cdot 5$ and, in extreme cases, $1 \cdot 0$ to $1 \cdot 0$ or $1 \cdot 0$ to $2 \cdot 0$. Two granite varieties may be distinguished: equigranular granite (in the main Karkonosze ridge) and a porphyritic granite (in the Karkonosze slopes and the Jelenia Góra basin).

Owing to physical weathering a loosening of the crystals is taking place but all further textural and structural features of the rock, such as jointing, are fully maintained. This regolith product is here called 'granite grit'. Its thickness varies from a few to a dozen or so meters.

Symptoms of chemical weathering are most often observed on plagioclase crystals; they often turn opaque and disintegrate into a white powder. On the other hand, weathered potash feldspar (orthoclase) is rarely seen save along certain fissures.

The mechanical composition of the granite grit is also variable and very dependent on crystal size. From 70 to 90 per cent of the material consists of crystals larger than 50 μm and from 5 to 20 per cent of particles between 2 and 50 μm. A loamy fraction of less than 2 μm size is rarely present; where it exists it amounts at most to 2 or 3 per cent. Clay minerals (illite, montmorillonite, kaolinite) account for less than 1 per cent of the weathered material.

The above data indicate that the weathering products associated with the tors belong to what in England is called 'growan'. In 1960 J. P. Bakker (1960) was ready to agree that the weathering products of the Karkonosze Massif were not a product of a very warm climate; later he (Bakker, 1967) defined this type of weathering as 'sandy deep weathering'. It may be noted that in some parts of the Sudeten Mountains, outside of the Karkonosze Massif, kaolin covers occur which are the products of weathering in a tropical or subtropical climate. Moreover J. Sekyra (1964) asserts that remnants of kaolin covers are met with on the Czechoslovak side of the Karkonosze; however, to my knowledge all the covers present on these mountains are represented by weathering products of the 'growan' type. The climate in which these covers developed must certainly have been warmer than it is today, but it may not have been a fully tropical one.

The age of the regolith covers is indicated, among other features, by the fact that in the Jelenia Góra basin the granite grit is overlain by deposits of the oldest glaciation. Hence they must have developed towards the decline of the Tertiary and the rise of the Pleistocene. Still, by virtue of their petrographic character, it also seems possible that part of these covers might go back to interglacial periods.

THE ORIGIN OF THE KARKONOSZE TORS

Any discussion of the origin of the tors seen in the Karkonosze Massif and the Jelenia Góra basin must include consideration of their relation to both land relief and geological structure.

Relation to land relief

The occurrence of the tors on ridges and on breaks of slope implies that they represent residual forms. The ridge tors may have originated by slope recession, which created the sharp ridge forms, and by dissection of the ridge. This mode of tor origin by the development of pediments is an older theory put forward in 1936 by K. Bryan. But the tors might also have originated from the development of altiplanation terraces as was assumed for the Alaskan mountains by H. M. Eakin (1916) and for the Ural chain by S. G. Boč and J. J. Krasnov (1951). Ridge dissection as a mechanism of tor formation has been postulated by J. Palmer and J. Radley (1961).

All the above suppositions based on the location of the tors would suggest that as residual forms they have been developing during a single phase of evolution.

Relation to geological structure

Much evidence resulting from this relation points to a possible different interpretation of the origin of the tors. Their formation depends closely upon geological structure, their shape is associated with the way the rocks are jointed (H. Cloos, 1925) and they correspond to the hardness of the particular parts of the granite (B. Dumanowski, 1961). Hence the shape of the tors tallys with the local structure of the granite. Further it might be claimed that underneath the present-day cover the tor shape is ready-made within the layer of weathered granite which shows thicknesses of up to 30 m.

In the Jelenia Góra basin the granite disintegrates into spheroidal forms (the structure of lava cooling) and visible are typical 'core stones' as termed by Linton (1955)—blocks of hard granite surrounded by granite grit (Fig. 2). This is the interior appearance shown by sub-surface tors. In this same basin area the exposed tors look the same; they are large rounded rock blocks lying one on top of another. In the Karkonosze Massif where granite weathering proceeds not into spheroidal shapes but along vertical and horizontal joints, sub-surface tors may be encountered in the shape of pillars covered by a granite regolith (Figs 3 and 4). Similar sub-surface tor forms have been observed here by T. Czudek *et al.* (1964).

A further argument in favour of a two-stage development of the tors is their situation; most often they appear on level surfaces of hard granite. Their walls rise above this surface without any transition forms (Fig. 5). Between the tors and their surrounding surfaces there lie neither

FIGURE 2. Spheroidal structures of granite. Core-stones surrounded by weathered loose granite mass. Jelenia Góra basin

FIGURE 3. Typical example of granite tower-shaped tors in the Karkonosze Mts. The tors are 25 m high

rock rubble nor larger blocks; no traces of disintegration are visible. Applying Linton's (1955) terminology I am inclined to consider level surfaces of this type surrounding the tors as 'basal platforms'—meaning the existence of a sharp boundary between fully disintegrated rock and non-weathered rock. I want to add here that at Acapulco in Mexico I observed this sort of boundary between a granite regolith and non-weathered granite. This clearly indicates that the tors did not develop as residual surface forms by retreat of the rock walls because no traces of

FIGURE 4. Blocks of hard granite inside weathered rock. The slopes of the Karkonosze Mts near Karpacz

FIGURE 5. Flat rocky surface surrounding the tors on the Karkonosze ridge

wall disintegration can be observed. This being so there remains but one explanation: the tor was a ready-made form hidden within the weathered granite cover and it assumed its present appearance due to exhumation.

All the above arguments tend to endorse the theory of a two-stage development of the tors observed in the Karkonosze Massif and the Jelenia Góra basin. The research results presented above speak in favour of a general tor theory which is increasingly gaining support by a variety of authors. It might be recalled here that it was in 1953 that J. Hövermann put forward his two-phase theory of the development of the tors he studied in the Harz Mountains. This author ascribed the symptoms of deep weathering of the granite to the tropical climate of the Tertiary.

After D. L. Linton published 'The problem of tors', an intense controversy set in; there was no lack of opponents to his theory. A recent study published by M. J. Eden and C. P. Green (1971) contributes to the debate, but in principle these authors accept Linton's theory regarding the origin of the Dartmoor tors.

Further considerations

The Karkonosze tors whose appearance confirms the concept of deep weathering of the granite and of a two-stage development, show certain specific features which must be considered against the background of Linton's theory.

The age of deep weathering I find it impossible to assume interglacial periods to be the principal phases in which this weathering took place. More decidedly than in the past (Jahn, 1962) I am inclined to believe that the principal period of selective weathering in which the tors were set apart was the decline of the Tertiary. I must also mention that differences of opinion exist regarding the age of the tors known in Europe, i.e. those of the Harz Mountains which J. Hövermann (1953) holds to be of Tertiary age and those seen in the classical Dartmoor area (note the list of opinions reported in the 1971 study of Eden and Green, pp. 96–7). The controversy centres on climatic conditions—whether the climate was hot, humid tropical or rather temperate subtropical. The regolith of the Karkonosze granite points to the latter type of climate. However, in view of the disintegration of the plagioclases observed here and the presence of clay minerals, I presume that the climate in which the Karkonosze granite grit did develop was sufficiently humid and hot to create conditions favourable to feldspar weathering. This would point to the decline of the Tertiary which terminated the protracted Palaeogene and Miocene phase in which the widespread kaolin deposits of the Sudeten Mountains developed.

The date of exhumation The agencies exhuming the tors have been water and solifluc-tion, both of which operated for the most part in the periglacial periods. But less important is the type of the stripping medium than the lack of a vegetation cover. The tors were laid bare during a climate in which no forest growth covered the Karkonosze slopes, and it was conditions like these which prevailed during the periglacial climate.

The tors are not of identical age. My investigation of the Karkonosze Pleistocene covers indicated that during the Quaternary slope evolution must have proceeded in an upward direc-tion. In consequence, slopes situated at altitudes greater than 1000 m are of younger age than slopes in the Jelenia Góra basin at 400 m near the mountain base where the ancient Tertiary slope has survived (A. Jahn, 1965). Owing to slope rejuvenation in an upward direction, the younger tors, exhumed at a later date, are situated higher up on the ridges, while the older tors lie farther down. After the last glaciation which left glacier cirques in the mountains the youngest post-glacial tors developed near these cirques.

Being sub-aerial forms the tors suffered relatively minor resculpturing. What did take place was a rounding of the granite blocks, producing 'wool sacks' and, particularly, the development

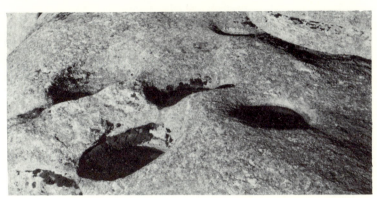

FIGURE 6. Weathering pit on tors in the Karkonosze Mts

of characteristic weathering kettles. (The Polish name suggested by J. P. Bakker as a comprehensive term is 'kociolki', the German term 'Opferkessel'.) Most of these small kettles developed in the periglacial period or during the post-glacial (Fig. 6).

The morphological setting of tor development   In the Harz mountains most tors are associated with valleys.* There it appears that the deep weathering which led to the formation of core-stones and tors took place mostly along valleys. Waters (1957) advanced a similar theory regarding the Dartmoor tors, while Eden and Green wrote: 'The . . . explanation of the distribution of tors suggests that the process of deep weathering was localized, in most cases, in or adjacent to river valleys, where the tors now commonly occur' (1971, p. 98). In the Karkonosze Massif the tors occur on ridges, near valleys and along edges of widespread morphological levels. Their origin is due to the stripping of sub-surface core-stones; but this applies only to sites where the morphological conditions of mountain development, like the growth of a river system and of the valley network, happened to be auspicious for the process. In other words, for ridge tors to be produced ridges had to develop first; and ridge development presupposes the evolution of valleys. Hence we observe the characteristic phenomenon, that in the Karkonosze Massif the number of tors increases with the growing density of the valley network. In consequence it is my belief—in contrast to Linton's theory—that the tors are second-stage forms and that they should be regarded as meso-forms subordinated to macro-forms, that is, to ridges and valleys. The tors are not residual hills—macro-forms—but rather elements of the meso-relief within the range of residual landforms.

In the Karkonosze Massif we are faced with ancient planation surfaces but we do not encounter any tors which can be attributed to the Tertiary stages of mountain planation. The tors are associated with landforms created by the dissection of these surfaces and the formation of valleys. This is why they occur on ridges, on slope breaks and on edges of old morphological levels to which denudation processes starting from the valleys have extended upwards. This denudation removed the regolith deposits and laid bare the tors. These processes proceeded in accordance with cyclic changes in the Quaternary climate. The last glaciation, as well as the last periglacial period, brought about the exhumation of the youngest tors rising from the highest parts of the Karkonosze slopes.

* This I observed while on an excursion to the Harz Mountains in April 1972 in the company of Professor and Mrs Hövermann.

REFERENCES

BAKKER, J. P. (1960) 'Some observations in connection with recent Dutch investigations about granite weathering and slope development in different climates and climate changes', *Z. Geomorph.* Supplementband 1, 69–92

BAKKER, J. P. (1967) 'Weathering of granites in different climates particularly in Europe', *Les congrés et colloques de l'Université de Liège*, 40 (*L'évolution des versants*), 51–68

BOČ, S. G. and J. J. KRASNOV (1951) 'Process golcovoho vyravnivanija i obrazovanija nagornych terras', *Příroda* 5, 25–35

BORKOWSKA, M. (1966) 'Petrografia granitu Karkonoszy' (Résumé: Petrographie du granite des Karkonosze), *Geol. Sudetica* 2, 7–119

BRYAN, K. (1936) 'Processes at Granite Gap, New Mexico', *Z. Geomorph.* 9, 125–35

CLOOS, H. (1925) *Einführung in die tektonische Behandlung magmatischer Erscheinungen (Granittektonik), Teil I. Das Riesengebirge in Schlesien* (Berlin)

CZUDEK, T., J. DEMEK, P. MARVAN, V. PANOS and J. RAUŠER (1964) 'Verwitterungs-und Abtragungsforman des Granits in der Böhmischen Masse, *Petermanns Geogr. Mitt.* 108, 182–92

DUMANOWSKI, B. (1961) 'Cover deposits of the Karkonosze Mountains' in A. JAHN (ed.) *Studies in geology of the Sudetic mountains*, Zeszyty Nauk. Uniwer. Wrocławskiego, B8, 31–56

EAKIN, H. M. (1916) 'The Yukon–Koyukuk Region, Alaska', *Bull. U.S. geol. Surv.* 631, 1–85

EDEN, M. J. and C. P. GREEN (1971) 'Some aspects of granite weathering and tor formation on Dartmoor, England', *Geogr. Annlr* 53A, 92–9

GÜRICH G. (1914) 'Die geologischen Naturdenkmäler des Riesengebirges', *Beitr. NatDenkmPflege* B. IV, H.3, 141–324

HÖVERMANN, J. (1953) 'Die periglaziale Erscheinungen im Harz', *Göttinger geogr. Abh.* 14, 1–40

JAHN, A. (1962) 'Geneza skalek granitowych' (Summary: Origin of granite tors), *Czas. geogr.* 33, 19–44

JAHN, A. (1965) 'Formy i procesy stokowe w Karkonoszach' (Summary: Slope forms and processes in the Karkonoše Mountains), *Opera Corcont.* 2, 7–16

LINTON, D. L. (1955) 'The problem of tors', *Geogrl J.* 121, 470–87

PALMER, J. and J. RADLEY (1961) 'Gritstone tors of the English Pennines', *Z. Geomorph.* 5, 37–51

SEKYRA, J. (1964) 'Kvarterně geologické a geomorfologické problemy Karkonošského Krystalinika' (Summary: Quaternary-Geological and Geomorphological Problems of the Crystalline complexes of the Karkonoše Mountains), *Opera Corcont.* 1, 7–24

WATERS, R. S. (1957) 'Differential weathering and erosion on oldlands', *Geogrl J.* 123, 503–9

RÉSUMÉ. *Les éminences coniques de granite dans les montagnes Sudetens.* Les éminences coniques sont nombreuses et largement répandues sur les montagnes Karkonosze et dans le bassin de Jelenia Góra. Les formes et les emplacements ont rapport à la densité et le type de fissuration des deux genres de granit, equi-anguleux et porphyrique, qui au delà des éminences coniques portent un régolite, pour la plupart à grain grossier (> 51 µm) et ressemblant à 'growan', et on l'intérprète comme produit d'action météorique presque tropicale. Il vint avant la glaciation la plus ancienne. On peut expliquer les éminences coniques et le régolite en fonction du concept de Linton d'un développement des éminences coniques à deux étapes. On attribue les actions météoriques sélectives sous la surface à la fin de la période Néogène mais les éminences coniques furent exhumées pendant les phases périglaciaires du Pléistocène consécutif à la dissection des surfaces d'aplanissement Tertiaire. Ainsi, la densité varie directement selon la densité du réseau de vallée. Elles sont des éléments du méso-relief ressortis par les processus d'érosion de surface qui opérèrent progressivement vers le haut à rajeunir les pentes. Par conséquent, les éminences coniques le plus jeunes se lèvent des points les plus hauts des pentes de Karkonosze.

FIG. 1. Les types des éminences coniques de Karkonosze. A—colline en table, B—tour, C—blocs détachés

FIG. 2. Les structures sphéroïdales de granite. Noyau—des pierres entourées par une masse ébouleuse de granit altérée par les intempéries dans le bassin de Jelenia Góra

FIG. 3. Exemple typique des éminences coniques de granit formées comme tours dans les montagnes Karkonosze. Les éminences coniques sont hautes de 25m

FIG. 4. Blocs de granit due en dedans de roche altérée par les intempéries. Les pentes des montagnes Karkonosze, près de Karpacz

FIG. 5. Surface platte et rocheuse entourant les éminences coniques sur l'arête de Karkonosze

FIG. 6. Fosse d'action météorique sur des éminences coniques dans les montagnes Karkonosze

ZUSAMMENFASSUNG. *Granitfelstürme im Sudetengebirge.* Man findet zahlreiche und weitverbreitete Felstürme auf dem Karkonoszegebirge und im 'Jelenia Góra' Becken. Ihre Formen und Lagen haben Bezug auf die Dichte und das Klüftigkeitsmuster der zwei Granitsorten, gleichwinkelig und porphyritisch, die jenseits der Felstürme ein überweigend grobfaseriges Regolith tragen (>5 µm), sehr ähnlich zu 'growan' (verwittertem Granit) und der als ein subtropisches Verwitterungserzeugnis angesehen wird. Er datiert vor der ältesten Vergletscherung. Man kann die Felstürme und das Regolith im Sinne des Begriffs von Linton von einer zweistufigen Felsturmentwicklung erklären. Man schreibt die ausgewählte Unterflächenverwitterung dem späteren Teile der Neogänzeit zu, aber die Felstürme wurden während der periglazialen Phasen des Pleistozäns infolge der Zergliederung der tertiären Einebnungsoberflächen ans Tageslicht gebracht. Ihre Dichte variiert also direkt mit der Dichte des Talnetzes. Die sind Elemente des Mesoreliefs, die von Denudationsverfahrene hervorgehoben wurden, die allmählich die Abhänge verjüngend aufwärts liefen. Deshalb ragen die jüngsten Felstürme von den höchsten Teilen der Karkonoszeabhängen empor.

ABB. 1. Die Karkonosze Felsturmtypen. A—Tafelhügel, B—Turm, C—lose Blöcke

ABB. 2. Sphäroidische Granitstrukturen. Kern—Steine von verwitterter lockerer Granitmasse im 'Jelenia Góra' Becken umgeben

ABB. 3. Typisches Beispiel der turmförmigen Granitfelstürme im Karkonoszegebirge. Die Felstürme sind 25 m hoch

ABB. 4. Harte Granitblöcke innerhalb verwittertes Felsens. Abhänge des Karkonoszegebirges in der Nähe von Karpacz

ABB. 5. Flache felsige Oberfläche, die die Felstürme auf die Karkonosze Bergkette umgibt

ABB. 6. Verwitterungsgrube auf den Felstürmen im Karkonoszegebirge

II. RATES OF DENUDATION

Introduction

IN 'The Everlasting Hills', his Presidential Address to Section E of the British Association (Linton, 1957), David Linton drew attention to 'the mutability of what, to everyday sight, is everlasting' and showed by reasoned argument 'that the carving and even the destruction of mountain ranges is, by the geological time-scale, a relatively rapid process'. From a consideration of three major unconformities of the British stratigraphic record he estimated the erosive work accomplished during the intervals they represent. From the sub-Eocene unconformity in south-east England he concluded that 500 m of soft sediments could be removed and base-levelling completed in less than 10 million years; from the sub-Permian unconformity and Permian lithologies in Devon he suggested that the Hercynian mountains had been reduced to a pene-plain (or pediplain) in 30 million years, and from the post-Caledonian unconformity in the Scottish Highlands that 15 million years had been sufficient to unroof and expose deeply the Foyers granite pluton. More generally he inferred 'that on a continental margin an erosion cycle can be carried to peneplanation in twenty to forty million years according to the magnitude of the initial relief'. Thus did he affirm that 'the seventy million years of post-Cretaceous time in Great Britain and Ireland have been ample for two erosion cycles to proceed to peneplanation across hard and soft rocks alike' and suggest that the present land form was unlikely to exhibit relics of any pre-Tertiary erosion surface unless like the sub-Eocene plane in the North Downs or the sub-Triassic hills of Charnwood Forest they had been exhumed during Tertiary times (Linton, 1957, pp. 62, 67).

Linton's estimates of denudation rates, based as they were on a perceptive interpretation of the stratigraphic record and for areas of heavy Pleistocene precipitation in north-western Britain the forms of glacial erosion, preceded by a few years those derived from field measurements. In many respects they represent the limits to which it was possible to go by reasoned argument without putting 'more weight on any type of evidence than it can properly sustain' (Linton, 1957, p. 66). Nevertheless they were soon to be validated.

When he returned in 1966 to the question of rates of erosion it was in the context of a consideration of the nature and origin of the striking contrasts between terrestrial and lunar morphology (Linton, 1966). In this thought-provoking paper (of which in a letter to one of the editors he wrote 'I am inclined to think it one of the most important things I have written') he considered the relations between denudation and isostatic movements and their implications for base-levelling and peneplanation. He noted that the denudation rates determined by S. A. Schumm (Schumm, 1963) were comparable with his own 1957 estimates and accepted Schumm's conclusion that peneplanation was not only possible but was very likely to have occurred between successive orogenies in a given region. 'But the resulting surface is not quite the Davisian pene-plain in which a stable, quasi-permanent relationship between the land surface and the under-lying rock structures was assumed. Such a stable relationship cannot, in the long run, be achieved. Isostatic uplift in the source regions leads to renewed erosion stripping the continental mass down to ever lower structural levels. The folded mountains that first caused the rivers to run may in the fullness of time disappear entirely and be replaced at the surface by their metamorphic under-structures. Alternatively we may regard the peneplain as a surface that may begin to approximate fairly closely to base-level in an altitudinal sense but tectonically as a surface through which progressively lower zones of rock structures rise intermittently to the surface and disappear by erosion.

The 'complete' destruction of a land mass by erosion is thus not a meaningful concept, since the crust material of which the land mass is made will continually 'rise to base-level'. But the removal by erosion of a volume equal to *or exceeding* that of the original relief of a continent is quite possible and, it seems to me, inevitable. Base-levelling in the Davisian sense has been questioned since he required for it long periods of still-stand. Base-levelling as I see it is rather made inevitable by the fact that still-stand is in a literal sense impossible. Isostatic re-adjustments will, in any phase, be greatest in the areas or regions where unloading is greatest and so will continue to be significant as long as there is significant relief to destroy' (Linton, 1966, p. 261).

Nevertheless a critical point is reached at which the mechanism of step by step isostatic re-adjustment of the crustal imbalance created by large-scale relief destruction and sediment transfer (some 125 million km^3 in from 50 to 150 million years) 'gives way rather abruptly to new processes leading to the availability of energy in greatly increased amount' (p. 263). Linton quoting H. S. Yoder (Yoder, 1955) suggests the point is reached in geosynclinal sedimentation when the water built into such clay minerals as montmorillonite and kaolinite by hydrolysis in the regolith and soil is released at depth with the reconstitution of the original silicate and alumino-silicate minerals. 'If the water is able to escape the reaction may accelerate and the situation may become critical. Fluxed by high pressure steam, and other hyperfusibles, the newly formed felspars and ferro-magnesian minerals would constitute a melt that could migrate under compression as magma. The phenomena of orogenesis would have been initiated' (Linton, 1966, p. 264).

No brief summary can do justice to the elegance of the arguments that Linton used to demonstrate the relations between denudation and diastrophism and to sustain his hypothesis that both are manifestations of the distributions of solar energy through atmosphere, hydrosphere and lithosphere. It will fall to others to demonstrate the compatibility or otherwise of the concept with those of the new global tectonics; for its implication that rates of denudation in a given area ultimately determine the length of its inter-orogenic phases is too revolutionary to be ignored.

REFERENCES

LINTON, D. L. (1957) 'The everlasting hills', *Advmt Sci.* 14, 58–67

LINTON, D. L. (1966) 'Geomorphology by the light of the Moon', *Tidschr. K. ned. aardrijksk.Genoot.* 83, 249–65

SCHUMM, S. A. (1963) 'The disparity between present rates of denudation and orogeny', *Prof. Pap. U.S. geol. Surv.* 454 H

YODER, H. S. (1955) 'Role of Water in metamorphosis' in A. POLDERVAART (ed.) *Crust of the Earth (A Symposium)*, 505–24

The rate of slope retreat

ANTHONY YOUNG

Reader in Environmental Sciences, University of East Anglia

Revised MS received 11 June 1973

ABSTRACT. The results of 109 studies of observed rates of soil creep, solifluction, surface wash, solution, landsliding, cliff retreat, slope retreat and total denudation are summarized. Creep moves more soil downslope than wash in humid temperate climates, wash is the dominant process in semi-arid and savanna climates, whilst both creep and wash are moderately rapid under rainforest. Solution is an important process in all humid climates. The relative importance of periglacial denudation in Quaternary slope retreat in the temperate zone has not been established; nor, for all climatic zones, has the relative work accomplished by catastrophic and continuous processes. In future work on surface processes there is a need for a system of stratified areal sampling in the siting of recording stations.

D. L. LINTON's Presidential Address to Section E of the British Association in 1956, 'The everlasting hills', carried a different significance for the two audiences to whom it was directed. For the lay public, its message was that the apparent immutability of natural landscapes, as expressed in the Biblical quotation from which its title is taken, is illusory, and results from the shortness of the human lifetime by comparison with geological time. Such had been a tenet of geomorphology since the time of Hutton and Playfair. But for the professional audience there was a deeper, closely-reasoned, argument. This was that the time-scale on which landform evolution had commonly been considered was too long. To speak of landforms as having originated in early Tertiary, or even Triassic, times was to ignore both the amount of ground surface modification accomplished during the Tertiary era and the great extent, and often radical nature, of Pleistocene change. Linton himself went as far back as the early Tertiary in seeking to explain the initiation of the present British drainage pattern on an eastward-sloping cover of Cretaceous rocks. But he also frequently demonstrated Pleistocene origins, as when attributing the radial drainage of the English Lake District to glacial action, as opposed to the older explanation of superimposition (Linton, 1957b). He was not alone in these arguments. At about the same time W. D. Thornbury (1954) had given as one of nine fundamental concepts of geomorphology, 'Little of the earth's topography is older than Tertiary and most of it no older than Pleistocene'.

At that time there were few measurements of surface processes, and arguments were based on evidence drawn from the form of the land and from associated deposits. In the period 1956–72 over 100 studies of the absolute rates of surface processes and slope retreat have been published, providing a new quantitative basis for landform studies. It is the aim of this paper to summarize these observations, and to discuss some problems to which they give rise.

HISTORICAL DEVELOPMENT OF THE MEASUREMENT OF SURFACE PROCESSES

The rate of denudation was an active question from the 1860s onwards, but many nineteenth-century estimates rested on incorrect assumptions. In 1880, A. Geikie estimated the rate of rock weathering by changes on dated tombstones in Edinburgh churchyards. The first reliable estimate of the overall rate of ground loss came in 1909, from a summary of river sediment loads in the United States by R. B. Dole and H. Stabler. In the 1930s, records of surface wash were made for a different purpose, and only subsequently entered geomorphological cognizance: these were the soil erosion experiments of the U.S. Soil Conservation Service. They measured erosion on plots 22·2 m long, and commonly included a control plot under natural vegetation, intended to demonstrate how erosion was increased by unwise cropping practices, but also providing records of absolute rates of wash.

The earliest measurement of surface processes on slopes was Charles Darwin's celebrated observation (1881) of soil movement on a $9\frac{1}{2}°$ slope caused by worm casting, which yields a value (expressed in modern units) of 0·52 cm³/cm/yr. Other isolated studies were a record of the movement of stones on Alpine screes, using tin traps, by S. Morawetz (1932); and an estimate by S. R. Capps (1941), based on geological evidence, that post-glacial creep of till amounted to 135 m in 30 000 years. Noteworthy among the few attempts to determine slope retreat over a long period is L. C. King's (1956) estimate of the retreat of the Drakensburg scarp, South Africa, by 240 km in *c*. 150 million years.

The pioneer study in the modern period of slope process measurements was by J. Michaud (1950), who began observations of the movement of painted stones in the French Alps in 1947. Surface wash was recorded on badlands by S. A. Schumm (1956; records commenced 1952) and under rainforest by G. Rougerie (1956). Solifluction was first measured by P. J. Williams (1957), frost creep by J. Schmid (1955), soil creep by A. Young (1960) and the rate of cliff retreat, based on scree accumulation, by A. Rapp (1957). Recordings of the rates of landslide movement were made by A. T. Grove (1953) and H. Mortensen and J. Hövermann (1956). The first attempts to record quantitatively all the slope processes within a stream catchment were those of H. Jäckli (1957) in Switzerland, and Rapp's (1960a) study of Kärkevagge, Sweden, the comprehensiveness of which has since been equalled only by M. I. Iveronova (1969) in the Kirghiz S.S.R., and O. Slaymaker (1972; as unpublished thesis, 1967) in Wales. It is only more recently that the instrumentally difficult feat of recording subsurface wash has been accomplished, by R. Z. Whipkey (1965) in the U.S.A. and E. J. Roose (1968) in the Ivory Coast. Noteworthy among current work is the Vigil Network System, inaugurated by the U.S. Geological Survey, which includes hillslope erosion among the types of data collected (W. W. Emmett and R. F. Hadley, 1968).

RATES OF SURFACE PROCESSES ON SLOPES

Available records of surface processes are summarized in Tables I–V. Movements of rocks on screes are not included. Various assumptions, not detailed for reasons of space, have been made in converting the original data to a uniform presentation. The following units are employed:

Linear downslope movement of the ground surface, or uppermost layer of the regolith: mm/yr.
Volumetric downslope movement of regolith: cm³/cm/yr (Young, 1960).
Slope retreat, or ground loss (perpendicular to the ground surface): Bubnoff units (B).
1 B = 1 mm/1000 yr, equivalent to 1 m per million years, 1 μm per year, and 1 m³/km²/yr (A. G. Fischer, 1969; J. M. Erickson, 1969).

Soil creep (Table I)

There is a concurrence of results that in humid temperate climates, on slopes of 15–30°, surface movement is of the order of 1–3 mm/yr and volumetric movement of 0·1–10·0 cm³/cm/yr; under the matted grasslands of Britain, values from three independent studies fall in the range 0·5–2·7 cm³/cm/yr. Rates become higher, but still of the order of 10 cm³/cm/yr, both towards the polar margin of the temperate zone, where frost creep causes ground heave of up to several centimetres, and in semi-arid and humid tropical climates. Movement is largely confined to the upper 20 cm of regolith. (The exceptional value obtained by E. Kojan (1967) clearly involves continuous creep, movement extending to depths of several metres.)

TABLE I

Observed rates of soil creep

Climate	Rock	Slope angle	Movement of surface or upper 5 cm mm/yr	Volumetric movement cm³/cm/yr	Location	Source
Cold temperate maritime	Till	30–40°	6[a]	15·0	Southern Alaska	Barr and Swanston, 1970
Temperate maritime	Palaeozoic sedimentaries	25°	1–2	0·5	Northern England	Young, 1960, 1963a
Temperate maritime	Palaeozoic shales	17°	2	2·1	Scotland	Kirkby, 1964, 1967
Temperate maritime	—	—	11	—	New Zealand	Owens, 1969
Temperate maritime[b]	Lias clay	4°	0·04[b]	3·2	Central England	Chandler and Pook, 1971
Temperate maritime	Palaeozoic sedimentaries	—	—	0·3	Wales	Slaymaker, 1972
Temperate maritime	Gneiss	17–28°	0·5	2·7	Washington D.C., U.S.A.	Leopold and Emmett, 1972
Temperate continental	Till	—	4·5[c]	0·2	Idaho, U.S.A.	Capps, 1941
Temperate continental	—	20°	1	6·0	Ohio, U.S.A.	Everett, 1963
Temperate continental	Permian sedimentaries	22°	1–3	5·7–8·4	Tatar, U.S.S.R.	Dedkov and Duglav, 1967
Temperate continental	Palaeozoic sedimentaries	25°	5	—	Wisconsin, U.S.A.	Black, 1969
Temperate continental	Granite, limestone	21–23°	3·6–7·6	—	New Mexico, Wyoming, U.S.A.	Leopold and Emmett, 1972
Warm temperate	—	17°	—	1·3	Maryland, U.S.A.	Carson and Kirkby, 1972
Warm temperate	Granite and sandstone	15°	—	1·9–3·2	N.S.W., Australia	Williams, 1973
Mediterranean	—	19°	10–50	650	California, U.S.A.	Kojan, 1967
Semi-arid	Mesozoic shale	20°	6–12	—	Colorado, U.S.A.	Schumm, 1964
Semi-arid	Unconsolidated	Steep	5	4·9	New Mexico, U.S.A.	Leopold *et al.*, 1966
Semi-arid	Alluvium, shale	25–30°	3·0–8·6	—	New Mexico, Wyoming, U.S.A.	Leopold and Emmett, 1972
Savanna	Granite and sandstone	16°	—	4·4–7·3	Northern Territory, Australia	Williams, 1973
Rainforest	Shale and sandstone	10°	5	12·4	Malaya	Eyles and Ho, 1970
Rainforest	Sandstone, volcanic rocks	18°	4	7·5–9·1	Puerto Rico	Lewis, L. A., unpublished

Notes: [a] Solifluction? [b] Geological evidence, mean rate over 10 000 years. [c] Geological evidence, mean rate over 30 000 years

TABLE II

Observed rates of solifluction

Climate	Movement of surface or upper 5 cm		Volumetric movement cm³/cm/yr	Location	Source
	Mean mm/yr	Range mm/yr			
Polar maritime: frost-debris zone	5	0–10	5	Lappland, Sweden	Rudberg, 1964, 1970
Polar maritime	40	0–80	50	Lappland, Sweden	Rapp, 1960a
Polar maritime	—	6–52	—	Spitzbergen	Büdel, 1961
Polar maritime: tundra zone	20	0–60	25	Lappland, Sweden	Rudberg, 1964, 1970
Polar maritime	—	10–200	—	Sweden	Rapp, 1966
Polar maritime	—	10–100	—	Alaska	Everett, 1966
Polar maritime	9[a]	—	—	Greenland	Washburn, 1967, 1970
Polar maritime	37[b]	—	—	Greenland	Washburn, 1967, 1970
Polar maritime	—	10–100	—	Spitzbergen	Büdel, 1968
Polar maritime	—	2–16	—	Yukon Terr., Canada	Price, 1972
Polar continental	—	0–100	50	Quebec, Canada	Williams, 1966
Montane	10	2–60	—	Alps	Pissart, 1964
Montane	25	0–337	—	Tasmania	Caine, 1968
Montane	300	—	—	Kirghiz S.S.R.	Iveronova, 1969
Montane	2[c]	—	—	Colorado, U.S.A.	Benedict, 1970

Notes: [a] Dry sites. [b] Wet sites. [c] Turf-banked lobe, radiocarbon dating evidence, mean rate over 2400 years

On theoretical grounds it is to be expected that the rate of creep is proportional to the sine of the slope angle. An interesting anomaly is that none of the records of soil creep have positively demonstrated this relation. It has been found only for the uncommon process of surficial rock creep (Schumm, 1967).

Solifluction (Table II)

Solifluction comprises two types of movement, frost creep by the heave mechanism, and gelifluction with the nature of flow. As frost creep is also a cause of soil creep there is no sharp dividing line, and intermediate situations occur (e.g. D. J. Barr and D. N. Swanston, 1970). Rates of solifluction are 10–100 times faster than soil creep, commonly with surface movements of 100 mm/yr and volumetric transfers of 50 cm³/cm/yr. Movement is faster in the tundra zone than the frost-debris zone, on wet sites than dry, and on solifluction lobes as compared with the surrounding slope. It is widely found that movement extends to 50 cm depth. Measurements tend to be made where the process is particularly active, and not enough is known about its spatial distribution to be confident that the high recorded rates are typical for slopes as a whole.

Surface wash (Table III)

Surface wash comprises both rainsplash and surface flow, the former being an agent of transport as well as detachment. Methodological problems arise in relating the rates of downslope sediment transport and ground lowering. The latter may be directly recorded by reference poles, but this is only practicable where retreat is rapid. The more common method is to collect transported sediment in traps. If the traps are below enclosed areas (as in soil erosion plots), direct conversion is possible but there is a disturbance to natural conditions. If the traps are located on unenclosed slopes, differences of interpretation are possible over what is the source area of the sediment. One approach is to divide the difference in sediment between successive traps by the area between them; this may show zones of apparent accumulation (J. R. G. Townshend, 1970). The question

TABLE III

Observed rates of surface wash

Climate	Vegetation	Slope angle	Volumetric movement $cm^3/cm/yr$	Ground lowering B	Location	Source
Polar	Bare	—	—	1–10	Spitzbergen	Jahn, 1961
Temperate maritime	Grass	25°	0·08	—	Northern England	Young, 1960
Temperate maritime	Grass	—	0·09	—	Scotland	Kirkby, 1967
Temperate maritime	Tussock grass	—	—	10	New Zealand	Soons and Rayner, 1968
Temperate maritime	Tussock grass	22–32°	0·13	2	New Zealand	Soons, 1971
Temperate maritime	Tussock grass	36°	5·6	19	New Zealand	Soons, 1971
Temperate maritime	Bare	5°	—	10 000[a]	Wales	Bridges and Harding, 1971
Temperate maritime	Bare	43°	1·5	—	Wales	Slaymaker, 1972
Temperate continental	Bare	—	—	>230 000[b]	New Jersey, U.S.A.	Schumm, 1956
Temperate continental	Grass	—	—	5–0·05	Poland	Starkel, 1962
Temperate continental	Grass	—	—	5	Poland	Gerlach, 1963
Temperate continental	Mainly grass	—	—	10–60	U.S.A.	Smith and Stamey, 1965
Temperate continental	Forest	—	—	0·03	Poland	Gerlach, 1967
Temperate continental	Bare	36–39°	—	100	Poland	Dumanowski, 1970
Temperate continental	Forest	—	—	0·1	Poland	Gerlach, 1970
Temperate continental	Forest, grass	21–23°	—	0–508	New Mexico, Wyoming, U.S.A.	Leopold and Emmett, 1972
Temperate continental	Bare	Steep	—	900[b]	Alberta, Canada	Campbell, 1970
Temperate continental	Bare	45–50°	—	100 000[b]	Crimea, U.S.S.R.	Blagovolin and Tsvetkov, 1972
Warm temperate	—	—	—	50–100	N.S.W., Australia	Williams, 1968, 1973
Humid sub-tropical	—	—	—	19	Mississippi, U.S.A.	Ursic, 1963
Mediterranean	—	29–35°	—	90	Southern France	Gabert, 1964
Mediterranean	—	42°	—	29	California, U.S.A.	Krammes, 1963
Mediterranean	—	14°	—	253	California, U.S.A.	Krammes, 1963
Mediterranean	Garrigue	25–35°	—	0·4	Southern France	Birot, 1970
Mediterranean	Garrigue	—	—	75	Southern France	Birot, 1970
Semi-arid	—	—	—	2 000	Colorado, U.S.A.	Schumm, 1964
Semi-arid	Bare	Steep	—	6 400–8 200	New Mexico, U.S.A.	Leopold et al., 1966
Semi-arid	—	—	4·2	1·2	Arizona, U.S.A.	Carson and Kirkby, 1972
Semi-arid	Pinon, sage	25–35°	—	7 600–11 700	New Mexico, Wyoming, U.S.A.	Leopold and Emmett, 1972
Arid	Bare	—	0·3	—	Arizona, U. S.A.	Carson and Kirkby, 1972
Savanna	Savanna	—	—	1·6	Senegal	Roose, 1967
Savanna	Savanna	—	—	39	N. Territory, Australia	Williams, 1968, 1973
Savanna	Cerrado, grass	—	0·6–9·6	—	Mato Grosso, Brazil	Townshend, 1970
Rainforest	Forest	—	—	500–1 500	Ivory Coast	Rougerie, 1956
Rainforest	Forest	15–22°	1	—	Malaya	Young, unpublished
Rainforest	Forest	26–30°	4	—	Malaya	Young, unpublished
Rainforest	Montane forest	Steep	—	260	Tanzania	Rapp et al., 1972

Notes: [a] Vegetation destroyed by air pollution. [b] Badlands. For explanation of Bubnoff units (B) see text p. 66

TABLE IV

Some estimates of the rate of solution

Method	Climate	Rock	Ground lowering Limestone B	Other rocks B	Location	Source
Carbonate in rivers, world summary	Various	Limestones	25-100	—	World	Nicod, 1970
Silica in rivers, world summary	Various	Various	—	1-6	World	Douglas, 1969
Dissolved river load, U.S.A. summary	At 1000 mm/yr rainfall	Various	—	30	U.S.A.	Langbein and Dawdy, 1964;
	At 500 mm/yr rainfall	Various	—	2-15	U.S.A.	Leopold, Wolman and Miller, 1964; cf. Carson and Kirkby, 1972, p. 238
Dissolved river load	Polar maritime	Metamorphic	—	5-10	Scandinavia	Rapp, 1960a
Dissolved river load	Temperate continental	Sandstones	—	26	Polish Carpathians	Gerlach, 1967
Dissolved river load	Temperate continental	Till	—	30	New Hampshire, U.S.A.	Johnson et al., 1968
Dissolved river load	Temperate continental	Limestone; various	32	18	Poland	Pulina, 1972
Dissolved river load	Temperate continental	Sedimentaries	—	2	Maryland, U.S.A.	Cleaves et al., 1970
Dissolved river load	Mediterranean	Dolomite; granite	17-21	1-2	California, U.S.A.	Marchand, 1971
Dissolved matter in slope wash	Temperate montane	Limestone	28	—	Austrian Alps	Bauer, 1964
Dissolved matter in soil solution	Temperate maritime	Chalk	17	—	Southern England	Perrin, 1965
Theoretical calculation	At 10°C, 1000 mm/yr rainfall	Limestone; igneous	50	5	—	Carson and Kirkby, 1972

TABLE V

Observed rates of slope retreat from landslides

Climate	Ground lowering B	Location	Source
Polar	36	Lappland, Sweden	Rapp, 1960a
Montane	109	Kirghiz S.S.R.	Iveronova, 1969
Temperate maritime	11	Verdal, Norway	Jørstad, 1968
Semi-arid	204	California, U.S.A.	Bailey and Rice, 1969
Rainforest	227	New Guinea	Simonett, 1967

is affected by whether the rate of wash is subject to control by detachment or control by transport (Young, 1972, 66).

Absolute rates of surface wash vary greatly with the vegetation cover. With a grass cover in temperate maritime climates, sediment movement of the order of 0·1 cm³/cm/yr and ground lowering of 1–10 B are typical, whilst values as low as 0·1 B have been reported for forested slopes in eastern Europe. An incomplete vegetation cover is associated with values of the order of 100 B. This rises to over 1000 B, i.e. 1 mm/yr, wherever vegetation is sparse, whether this condition is man-induced or arises naturally as in semi-arid climates, on badlands and on very steep slopes.

Solution (Table IV)

Most data on solution loss has been derived from dissolved matter in river water and thus includes channel erosion, and on limestones solution in subterranean channels, as well as material derived from slopes. The limited evidence available concurs with *a priori* reasoning in appearing to suggest that rates are higher on limestones than on other rocks, possibly by an order of magnitude. There is divergence of opinion about the relative rates of limestone solution in different climates (M. M. Sweeting, 1972). F. Bauer (1964) found the solution rate on bare limestone rock to be less than half that under a soil and vegetation cover, a result consistent with K. Terzaghi's (1913 and 1958) hypothesis for the growth of bare rock residuals.

It is likely that in many environments as much material is removed from slopes in solution as by all other processes (Jäckli, 1957; Rapp, 1960a; T. Gerlach, 1967). Moreover, solution leads directly to ground lowering; process-response models indicate that for similar assumed rates of solution loss and processes of downslope transport, slope retreat is considerably greater in the case of solution, the difference increasing as the dimensions of the slope become greater (Young, 1963b). Hence measurement of solution loss on slopes is the greatest gap in current knowledge of surface processes.

Landslides (Table V)

To obtain rates of slope retreat from observation of the size and frequency of landslides involves difficult spatial and temporal sampling problems, for they occur mainly during exceptional weather conditions with a low recurrence interval. In converting to average rates of slope retreat there is an assumption that in the long period the slides are acting areally, rather than tending towards linear dissection. The high rates of slope retreat are in part accounted for by the fact that such observations are made only on slopes where landslides are active, and should be set against the numerous slopes for which there is no evidence of rapid mass-movements.

TABLE VI
Rates of cliff retreat

Climate	Rock	Retreat B	Location	Source
Polar	Metamorphic	40–150	Lappland, Sweden	Rapp, 1957, 1960a
Polar	Sandstone, limestone	20–500	Spitzbergen	Rapp, 1960b
Polar	Igneous	20	Yukon Terr., Canada	Gray, 1972
Polar	Metasedimentary	70	Yukon Terr., Canada	Gray, 1972
Montane	Metamorphic	700–1000	Austrian Alps	Poser, 1954
Temperate maritime	Sandstone	500	Germany	Rühl, 1914
Temperate maritime	Igneous	200–900	Scandinavia	Rapp, 1964
Temperate continental	Conglomerate, sandstone	650 000	Crimea, U.S.S.R.	Blagovolin and Tsvetkov, 1972
Semi-arid	Shale, sandstone	2 000–13 000	Colorado, U.S.A.	Schumm and Chorley, 1966
Arid	Sandstone	0·04	North Africa	Mortensen and Hövermann, 1956
Arid	Sandstone	200	Colorado, U.S.A.	Mortensen and Hövermann, 1956
Arid	Sandstone	600	Colorado, U.S.A.	Schumm and Chorley, 1966
Rainforest	Granite, gneiss	2 000–20 000	Brazil	Freise, 1932

Examples of marine cliff retreat

Climate	Rock	Retreat B	Location	Source
Temperate maritime	Limestone, marl	4 000–6 000	Gotland, Sweden	Rudberg, 1967
Temperate maritime	Chalk	670 000	Sussex, England	May, 1971
Temperate maritime	Till	100 000–200 000	Lake Vättern, Sweden	Norman, 1964
Temperate maritime	Till	1 000 000	Eastern England	Williams, 1956; Steers, 1953
Temperate maritime	Till	300 000	Norfolk, England	Cambers, G., unpublished

TABLE VII
Rates of slope retreat

Method	Climate	Rock	Retreat B	Location	Source
Radiocarbon dating	Temperate montane	Flysch, weak beds	268	Carpathians, Poland	Klimazewski, 1971
Volcanic flow dating	Temperate maritime	Igneous	<4	Massif Central, France	Bout et al., 1960
Volcanic flow dating	Temperate maritime	Calcareous marl	40	Massif Central, France	Bout et al., 1960
Direct measurement	Temperate continental	Unconsolidated: badlands	8 750–15 000	South Dakota, U.S.A.	Schumm, 1962
Radioactive dating	Warm temperate	Basalt	950	N.S.W., Australia	Caine and Jennings, 1968
Geological	Humid sub-tropical	Various	1 270	Drakensberg, South Africa	King, 1956
Radiocarbon dating	Mediterranean	Landslide debris	314–711	California, U.S.A.	Stout, 1969
Tree rings	Mediterranean/ montane	Dolomite: gentle slopes	152	California, U.S.A.	Lamarche, 1968
Tree rings	Mediterranean/ montane	Dolomite: steep slopes	1 220	California, U.S.A.	Lamarche, 1968
Radioactive dating	Rainforest	Andesite lava	80–500	Papua	Ruxton and McDougall, 1967

RATES OF SLOPE RETREAT

Cliff retreat (Table VI)

Comparison with marine cliffs suggests that with respect to rates of retreat there are three groups of rocks, each separated from the next by one or more orders of magnitude. These are first, igneous and metamorphic rocks together with hard Palaeozoic sedimentaries, in which cliffs may show little detectable change in a lifetime; secondly, Mesozoic and lithified Tertiary

TABLE VIII

Some estimates of average world and continental rates of mean denudation

Area	Ground lowering B	Source
World	24	Judson, 1968
World	30	Corbel, 1959
World	84	Ritter, 1967
World	69	Holeman, 1968
World, excluding Asia	19	Holeman, 1968
Continents, excluding Asia	9–32	Holeman, 1968
Asia	202	Holeman, 1968
Europe	5–25	Fournier, 1960
Europe	28	Strakhov, 1967
North America	43	Strakhov, 1967
World, plains and moderate relief	12–58	Corbel, 1959
World, plains and moderate relief	20–81	Young, 1969
World, plains and moderate relief	16–195	Ahnert, 1970
World, mountains and steep relief	92–800	Corbel, 1959
World, mountains and steep relief	92–800	Corbel, 1959
World, mountains and steep relief	400–750	Fournier, 1960
World, mountains and steep relief	92–750	Young, 1969
World, mountains and steep relief	300–430	Ahnert, 1970

sedimentaries, with intermediate rates of retreat; and thirdly, non-lithified sediments and drift deposits, particularly boulder-clay, with cliffs showing visible change annually. Most non-marine cliffs belong to the first group, since without basal erosion those in softer rocks are rapidly eliminated by slope replacement. Rates of the order of 100–1000 B, or 0·1–1·0 mm/yr, are common. The fastest retreat on any slope, exceeding even those of badlands, is a loss equal to a layer of average thickness 2·6 m over 4 years recorded for a Jurassic sandstone slope in the Crimean Mountains, on which 'slope processes are extremely active' (N. S. Blagovolin and D. G. Tsvetkov, 1972). By contrast, the freshness of inscriptions on many Egyptian monuments (albeit on rocks preselected for their resistance) suggests rates as low as 1 B.

Slope retreat (Table VII)
Direct recording of slope retreat is possible only on badlands. Otherwise, opportunities for absolute dating of the former position of a slope are rare. Most values in Table VII are of the order of 100–1000 B, but some of the slopes are partly cliffed and there is undoubtedly a sampling bias towards rapidly-retreating slopes.

Overall rates of denudation (Table VIII)
There have been a number of estimates of overall denudation rates, including both river erosion and slope processes, some of which are shown in Table VIII; some of these are themselves summaries of 50 or more sources. The main types of evidence are dissolved and suspended river load, reservoir sedimentation, and rates of geological sedimentation. There is a general consensus that 50 B is the correct order of magnitude for non-mountainous areas and for the world as a whole, and 500 B for mountainous relief. Regarded as, respectively, 50 and 500 m per million years, these values provide a quantitative standard of reference for comparison with denudation chronologies. It has been suggested, however, that agriculture and other activities of man have

increased world denudation rates 2–3 times above the geological norm (S. Judson, 1968; R. H. Meade, 1969), or as much as 10 times in the tropics (A. Rapp *et al.*, 1972).

DISCUSSION

There are four main variables affecting rates of surface processes and slope retreat: climate, rock type, vegetation and slope angle. Even the broadest of classifications—10 climates each with its climax vegetation, 10 rock types and 5 angle classes—yields some 500 environments, well exceeding the number of existing process studies. Hence only a few generalizations on the relative importance of different processes in given environments can yet be made, other than from *a priori* reasoning.

On slopes above a critical angle, *c.* 35° for most rocks but considerably less for clays, rapid mass-movements accomplish greater denudation than slow, continuously-acting processes; these are termed unstable slopes. It is not known, however, whether the converse applies to slopes below such an angle.

In humid temperate climates, soil creep, although slow, moves more material than the very low values recorded for surface wash. Where the vegetation cover is incomplete but some rain occurs, particularly in Mediterranean, savanna and semi-arid regions, surface wash is the dominant process, even although absolute rates of creep are higher than in temperate latitudes. In the rainforest environment solution is probably a major process; creep and wash both occur at substantial rates, but it is not known which accomplishes greater downslope transport. In polar and montane regions solifluction is certainly important locally, but reservations concerning its distribution have been noted above, and in the studies of Rapp (1960a) and Iveronova (1969) it does not emerge as the dominant process.

An unresolved problem peculiar to the temperate zone is the extent to which present slopes owe their form to processes operating under periglacial conditions. This includes the questions: (i) of the total slope retreat during the Quaternary, how much has been achieved under cold and how much under warm climates? (ii) to what extent were 'inter-glacial slopes' altered during the last glaciation? and (iii) how much change in form has occurred since periglacial conditions were replaced by warmer climates? These questions depend in turn first, upon the relative rates of slope retreat under polar and temperate climates, and secondly, on the durations of successive climatic regimes. Deposits of demonstrably periglacial origin occur on many slopes in the temperate zone, including in the Southern Hemisphere (e.g. A. B. Costin and H. A. Polach, 1971). This fact, together with the established greater speed of solifluction as compared with soil creep, has led to suggestions that slopes of the temperate zone are dominated by relict periglacial forms. The apparent freshness of many relict forms lends support to this view (e.g. Rapp, 1967; but cf. R. F. Black, 1969). However, the time elapsed since the last glacial retreat is short, relative to the average duration of inter-glacial periods, and hence no great alteration of periglacial forms is to be expected. This problem can be approached by establishing an absolute chronology of climatic change, and thus of surface processes, for a given area under study, combining this with present-day process rates in comparable climates, and comparing the results of such calculations with observed surface form and deposits. For the general case, it has not yet been determined whether slope retreat in the temperate zone during the Quaternary has taken place mainly during periods of periglacial conditions.

A general problem is that of magnitude and frequency with respect to processes on slopes. Many studies have described catastrophic events in which a large volume of denudation has been accomplished during a single exceptional storm. Such occurrences most commonly involve debris avalanching, although other processes, particularly surface wash, may also be involved. For the case of fluvial processes, it has been established that more than half the sediment transport is achieved by moderate flows with a recurrence interval of less than one year (L. B. Leopold, M. G. Wolman and J. P. Miller, 1964, 67ff). The mechanisms by which sediment is moved on slopes are so different to those of rivers that arguments by analogy are not strong. On some slopes, it appears that more work is accomplished during events with a recurrence-interval of the order of 10–50 years than during the intervening periods; but this applies only to slopes where catastrophic denudation has been observed, and there are many others for which no such records exist. Most rapid mass-movements tend to dissect slopes, yet many slopes show no signs of such dissection. Long-term monitoring of selected slopes, necessarily by institutions rather than individuals, is the surest method of studying the problem, albeit, definitive answers will require half a century of observation. An indirect approach is possible by comparing denudation accomplished under known climatic conditions, from normal to exceptional, with the duration and frequency of such conditions as indicated by climatic records.

CONCLUSION

Direct evidence of the rate of slope retreat is only available in exceptional circumstances. Knowledge must come mainly from observation of the present rates of surface processes. In view of the wide diversity of slope environments, there is opportunity for many more studies of individual processes. There is in particular a need for observations of the relative rates of all processes in the rainforest environment; the relative extent of slopes undergoing rapid as compared with slow denudation in polar climates; and of the absolute rate of solution on siliceous rocks in all climates.

If future studies of processes are to make the maximum contribution to knowledge, one particular feature of experimental design is needed: this is that the location of recording sites should be based upon a system of stratified areal sampling. The region under study is divided into units, on the basis of surface form, micro-relief, regolith properties and vegetation cover; the areas of these units are measured, and recording sites located in each. By comparing the area and distribution of the units with the process rates recorded, a regional balance of regolith movement can be established. The fact that process rates on apparently similar sites are found to have a large variance calls for statistically-controlled replication of sites, but in no way reduces the need for areal stratification. Such studies would serve not only to establish average rates of slope retreat, but also to indicate the circumstances under which it becomes exceptionally rapid. Greater knowledge of spatial distribution and temporal variability of surface processes are the principal needs in future work directed towards establishing rates of slope retreat.

REFERENCES

AHNERT, F. (1970) 'Functional relationships between denudation, relief and uplift in large mid-latitude drainage basins', *Am. J. Sci.* 268, 243–63
BAILEY, R. G. and R. M. RICE (1969) 'Soil slippage: an indicator of slope instability on chaparral watersheds of southern California' *Prof. Geogr.* 21, 172–7
BARR, D. J. and D. N. SWANSON (1970) 'Measurement of creep in a shallow, slide-prone till soil', *Am. J. Sci.* 269, 467–80
BAUER, F. (1964) 'Kalkabtragungsmessungen in den österreichischen Kalkhochalpen', *Erdkunde* 8, 95–102
BENEDICT, J. B. (1970) 'Downslope soil movement in a Colorado alpine region: rates, processes and climatic significance', *Arctic Alpine Res.* 2, 165–226
BIROT, P. (1970) 'Etude quantitatif des processus érosifs agissant sur les versants', *Z. Geomorph. Suppl.* 9, 10–43
BLACK, R. F. (1969) 'Slopes in southwestern Wisconsin, U.S.A., periglacial or temperate?', *Biul. peryglac.* 18, 69–82
BLAGOVOLIN, N. S. and D. G. TSVETKOV (1972) 'The use of repeated ground photo-grammetric survey for studying the dynamics of slopes', in *International Geography 1972*, 9–10
BOUT, P., M. DERRUAU and A. FEL (1960) 'Utilisation des cônes et des coulées volcaniques du Massif Central français pour évaluer le recul des versants crystallins', *Z. Geomorph. Suppl.* 2, 133–9
BRIDGES, E. M. and D. M. HARDING (1971) 'Micro-erosion processes and factors affecting slope development in the lower Swansea valley', *Spec. Publ. Inst. Br. Geogr.* 3, 65–80
BÜDEL, J. (1961) 'Die Abstragungsvorgänge auf Spitzbergen im Umkreis der Barentinsel', *Verh. dt. GeogrTags* 33, 337–75
BÜDEL, J. (1968) 'Hang- und Talbildung in Südest-Spitzbergen', *Eiszeitalter Gegen.* 19, 240–3
CAINE, N. (1968) 'The log-normal distribution and rates of soil movement: an example', *Revue Géomorph. dyn.* 18, 1–7

CAINE, N. and J. N. JENNINGS (1968) 'Some blockstreams of the Toolong Range Kosciusko State Park, New South Wales',
J. Proc. roy. Soc. New S. Wales 101, 93–103

CAMBERS, G. (unpublished) Univ. East Anglia, Norwich, England

CAMPBELL, I. A. (1970) 'Erosion rates in the Steveville Badlands, Alberta', *Can. Geogr.* 14, 202–16

CAPPS, S. R. (1941) 'Observations of the rate of creep in Idaho', *Am. J. Sci.* 239, 25–32

CARSON, M. A. and M. J. KIRKBY (1972) *Hillslope form and process*

CHANDLER, R. J. and M. J. POOK (1971) 'Creep movement on low gradient clay slopes since the Late Glacial', *Nature, Lond.*
229, 399–400

CLEAVES, E. T., A. E. GODFREY and O. P. BRICKER (1970) 'Geochemical balance of a small watershed and its geomorphic
implications', *Bull. geol. Soc. Am.* 81, 3015–32

CORBEL, J. (1959) 'Vitesse de l'érosion', *Z. Geomorph.* 3, 1–28

COSTIN, A. B. and H. A. POLACH (1971) 'Slope deposits in the Snowy Mountains, south-eastern Australia', *Quaternary Res.*
1, 228–35

DARWIN, C. (1881) *The formation of vegetable mould through the action of worms*

DEDKOV, A. V. and V. A. DUGLAV (1972) 'Slow movements of soil masses on grassed slopes', *Nat. Lending Library Transl.*
Progr. RTS 7548 (Russian original *Izv. Akad. Nauk SSSR ser. Geogr.* (1967) 90–3)

DOLE, R. B. and H. STABLER (1909) 'Denudation', *Water Supply Pap. U.S. geol. Surv.* 234, 78–93

DOUGLAS, I. (1969) 'The efficiency of humid tropical denudation systems', *Trans. Inst. Br. Geogr.* 46, 1–16

DUMANOWSKI, B. (1970) 'Examples of geomorphological investigations in Poland based on quantitative studies', *Geogr.*
Polonica 18, 199–206

EMMETT, W. W. and R. F. HADLEY (1968) 'The vigil Network—Part C, Preservation and access of data', *Circ. U.S. geol.*
Surv. 460-C

ERICKSON, J. M. (1969) 'Geological rate units', *Compass* 47, 5–9

EVERETT, K. R. (1963) 'Slope movement, Neotoma Valley, southern Ohio', *Rep. Inst. polar Stud. Ohio State Univ.* 6

EVERETT, K. R. (1966) 'Slope movement and related phenomena' in N. J. WILIMOVSKY and J. N. WOLFE (eds) *Environment*
of the Cape Thompson region 172–220

EYLES, R. J. and R. HO (1970) 'Soil creep on a humid tropical slope', *J. trop. Geogr.* 31, 40–2

FISCHER, A. G. (1969) 'Geological time-distance rates: the Bubnoff unit', *Bull. geol. Soc. Am.* 80, 549–52

FOURNIER, F. (1960) *Climat et érosion*

FREISE, F. W. (1932) 'Beobachtungen über Erosion an Urwaldgebirgsflüssen des Brasilianischen Staats Rio de Janeiro', *Z.*
Geomorph. 7, 1–9

GABERT, P. (1964) 'Premiers resultats des mesures d'érosion sur des parcelles expérimentales dans la région d'Aix-en-
Provence (Bouches du Rhône - France)', *Z. Geomorph. Suppl.* 5, 213–14

GEIKIE, A. (1880) 'Rockweathering, as illustrated in Edinburgh churchyards', *Proc. R. Soc. Edinb.* 10, 518–32

GERLACH, T. (1963) 'Les terrasses de culture comme indice des modifications des versants cultivés', in *Neue Beiträge zur*
internationalen Hangforschung (*Nachrichten der Akademie der wissenschaften in Göttingen, II Math.-Phys. Klasse*)
239–49

GERLACH, T. (1967) 'Évolutions actuelles des versants dans les Carpathes, d'après l'example d'observations fixes', in P.
MACAR (ed.) *L'évolution des versants*, 129–38

GERLACH, T. (1970) 'Etat actual et méthodes de recherches sur les processus morphogénétiques actuels sur le fond des étages
climatiques et végétaux dans les Carpathes Polonaises', *Stud. Geomorph. Carpatho-Balcanica* 4, 47–63

GRAY, J. T. (1972) 'Postglacial rock wall recession in the Ogilvie and Wernecke Mountains, central Yukon Territory', in
International Geography 1972, 24–6

GROVE, A. T. (1953) 'Account of a mudflow on Bredon Hill, Worcs, April 1951', *Proc. Geol. Ass.* 64, 10–13

HOLEMAN, J. N. (1968) 'The sediment yield of the major rivers of the world', *Wat. Resour. Res.* 4, 737–47

IVERONOVA, M. I. (1972) 'An attempt at the quantitative analysis of contemporary denudation processes', *Nat. Lending*
Library Transl. Progr. RTS 7436 (Russian original *Izv. Akad. Nauk SSSR Ser. Geogr.* (1969) 13–24)

JÄCKLI, H. (1957) 'Gegenwartsgeologie des bundnerischen Rheingebietes—ein Beitrag zur exogenen Dynamik Alpiner
Gebirgslandschaften', *Beitr. Geol. Schweiz Geotechnische Serie* 36

JAHN, A. (1961) 'Quantitative analysis of some periglacial processes in Spitzbergen', *Panstwowe Wydawnictno Naukowe,*
Warsaw, Geophysics Geogr. and Geol. II B 5.

JOHNSON, N. M., G. E. LIKENS, F. H. BORMANN and R. S. PIERCE (1968) 'Rate of chemical weathering of silicate minerals
in New Hampshire', *Geochim. cosmochim. Acta* 32, 531–45

JØRSTAD, F. A. (1968) 'Leirskred i Norge', *Norsk geogr. Tidskr.* 22, 214–9

JUDSON, S. (1968) 'Erosion of the land, or what's happening to our continents', *Am. Scient.* 56, 356–74

KING, L. C. (1956) 'Drakensburg scarp of South Africa: a clarification', *Bull. geol. Soc. Am.* 67, 121–2

KIRKBY, M. J. (1964) in 'Slope profiles: a symposium', *Geogrl J.* 130, 86

KIRKBY, M. J. (1967) 'Measurement and theory of soil creep', *J. Geol.* 75, 359–78

KLIMAZEWSKI, M. (1971) 'The effect of solifluction processes on the formation of mountain slopes in the Beskidy (Flysch
Carpathians)', *Folia Quatern.* 38

KOJAN, E. (1967) 'Mechanics and rates of natural soil creep', in *Proceedings fifth Annual Engineering geology and soils engineering symposium, Pocatello, Idaho* 233–53

KRAMMES, J. S. (1963) 'Seasonal debris movement from steep mountainside slopes in southern California', *Misc. Publs U.S.*
Dep. Agric. 970, 85–8

LAMARCHE, V. C. (1968) 'Rates of slope degradation as determined from botanical evidence, White Mountains, California', *Prof. Pap. U.S. geol. Surv.* 352-I, 341–77

LANGBEIN, W. B. and D. R. DAWDY (1964) 'Occurrence of dissolved solids in surface water in the United States', *Prof. Pap. U.S. geol. Surv.* 501-D, 115–17

LEOPOLD, L. B. and W. W. EMMETT (1972) 'Some rates of geomorphological processes', *Geogr. Polonica* 23, 27–35

LEOPOLD, L. B., W. W. EMMETT and R. M. MYRICK (1966) 'Channel and hillslope processes in a semiarid area', *Prof. Pap. U.S. geol. Surv.* 352-G, 193–253

LEOPOLD, L. B., M. G. WOLMAN and J. P. MILLER (1964) *Fluvial processes in geomorphology*

LEWIS, L. A. (unpublished) Clark Univ., Worcester, Mass., U.S.A.

LINTON, D. L. (1957a) 'The everlasting hills', *Advmt Sci.* 14, 58–67

LINTON, D. L. (1957b) 'Radiating valleys in glaciated lands', *Tijdschr. K. ned. aardrijksk. Genoot.* 74, 297–312

MARCHAND, D. E. (1971) 'Rates and modes of denudation, White Mountains, eastern California', *Am. J. Sci.* 270, 109–35

MAY, V. J. (1971) 'The retreat of Chalk cliffs', *Geogrl J.* 137, 203–6

MEADE, R. H. (1969) 'Errors in using modern stream-load data to estimate natural rates of erosion', *Bull. geol. Soc. Am.* 80, 1265–74

MICHAUD, J. (1950) 'Emploi de marques dans l'étude des mouvements du sol', *Revue Géomorph. dyn.* 1, 180–94

MORAWETZ, S. (1932) 'Eine Art von Abtragungsvorgang', *Petermanns Geogr. Mitt.* 78, 231–3

MORTENSEN, H. and J. HÖVERMANN (1956) 'Der Bergrutsch an der Mackenröder Spitze bei Göttingen', in *Premier rapport de la Commission pour l'étude des versants, préparé pour le Congrès Internationale de Géographie de Rio-de-Janeiro*, 2–8

NICOD, J. (1970) 'Sur la vitesse d'évolution au cours du Quaternaire de quelques formes karstiques superficielles', *Annls Géogr.* 79, 311–24

NORMAN, J. O. (1964) 'Lake Vättern. Investigations on shore and bottom morphology', *Geogr. Annlr* 46, 1–238

OWENS, I. F. (1969) 'Causes and rates of soil creep in the Chilton Valley, Cass, New Zealand', *Arctic Alpine Res.* 1, 213–20

PERRIN, R. M. S. (1965) 'The use of drainage water analyses in soil studies', in *Nottingham Univ. Easter Sch. Agr. Sci. Proc., 11th Easter Sch.* 73–96

PISSART, A. (1964) 'Vitesses des mouvements du sol en Chambeyron (Basses Alpes)', *Biul. peryglac.* 14, 303–9

POSER, H. (1954) 'Die Periglacial-Erscheinungen in der Umgebung der Gletscher des Zemmgrundes (Zillertaler Alpen)', *Göttinger geogr. Abh.* 15, 125–80

PRICE, L. W. (1972) 'Solifluction rates in the Ruby Range, Yukon Territory: a preliminary report', in *International Geography* (1972), 56–8

PULINA, M. (1972) 'A comment on present-day chemical denudation in Poland', *Geogr. Polonica* 23, 45–62

RAPP, A. (1957) 'Studien über Schutthalden in Lappland und auf Spitzbergen', *Z. Geomorph.* 1, 179–200

RAPP, A. (1960a) 'Recent development of mountain slopes in Kärkevagge and surroundings, northern Scandinavia', *Geogr. Annlr* 42, 65–200

RAPP, A. (1960b) 'Talus slopes and mountain walls at Tempelfjorden, Spitzbergen', *Skri. Norsk Polarinst.* 119

RAPP, A. (1964) 'Recordings of mass wasting in the Scandinavian mountains', *Z. Geomorph. Suppl.* 5, 204–5

RAPP, A. (1966) 'Solifluction and avalanches in the Scandinavian mountains' in *Proc. Permafrost Int. Conf. 1963* Wash. D.C., 150–4

RAPP, A. (1967) 'Pleistocene activity and Holocene stability of hillslopes, with examples from Scandinavia and Pennsylvania' in P. MACAR (ed.) *L'évolution des versants*, 229–44

RAPP, A., V. AXELSON, L. BERRY and D. H. MURRAY-RUST (1972) 'Soil erosion and sediment transport in the Morogoro river catchment, Tanzania', *Geogr. Annlr* 54A, 125–55

RITTER, D. F. (1967) 'Rates of denudation', *J. geol. Educ.* 15, 154–9

ROOSE, E. (1967) 'Dix années de mesure de l'érosion et du ruissellement au Sénégal', *Agron. trop. Nogent* 22, 123–52

ROOSE, E. (1968) 'Un dispositif de mesure du lessivage oblique dans les sols en place', *Cah. Orstom. Pédol.* 6, 235–49

ROUGERIE, G. (1956) Etudes des modes d'érosion et du façonnement des versants en Côte d'Ivoire equatoriale', in *Premier rapport de la Commission pour l'étude des versants, Union Géographique Internationale* (Amsterdam), 136–44

RUDBERG, S. (1964) 'Slow mass movement processes and slope development in the Norra Storkjäll area, southern Swedish Lappland', *Z. Geomorph. Suppl.* 5, 192–203

RUDBERG, S. (1967) 'The cliff coast of Gotland and the rate of cliff retreat', *Geogr. Annlr* 49A, 283–99

RUDBERG, S. (1970) 'Recent quantitative work on slope processes in Scandinavia', *Z. Geomorph. Suppl.* 9, 44–56

RÜHL, A. (1914) 'Der Einfluss von Verwitterung und Erosion auf die Bodengestaltung (1910–12)', *Geogr. Jb.* 28, 315–46

RUXTON, B. P. and I. McDOUGALL (1967) 'Denudation rates in northeast Papua from potassium-argon dating of lavas', *Am. J. Sci.* 265, 545–61

SCHMID, J. (1955) *Der Bodenfrost als morphologischer Faktor*

SCHUMM, S. A. (1956) 'Evolution of drainage systems and slopes in badlands at Perth Amboy, New Jersey', *Bull. geol. Soc. Am.* 67, 597–646

SCHUMM, S. A. (1962) 'Erosion on miniature pediments in Badlands National Monument, South Dakota', *Bull. geol. Soc. Am.* 73, 719–24

SCHUMM, S. A. (1964) 'Seasonal variations of erosion rates and processes on hillslopes in western Colorado', *Z. Geomorph. Suppl.* 5, 215–38

SCHUMM, S. A. (1967) 'Rates of surficial rock creep on hillslopes in western Colorado', *Science, N.Y.* 55, 560–1

SCHUMM, S. A. and R. J. CHORLEY (1966) 'Talus weathering and scarp recession in the Colorado Plateaus', *Z. Geomorph.* 10, 11–36

SIMONETT, D. S. (1967) 'Landslide distribution and earthquakes in the Bewani and Torricelli Mountains, New Guinea' in J. N. JENNINGS and J. A.MABBUTT (eds) *Landform studies from Australia and New Guinea*, 64–84

SLAYMAKER, H. O. (1972) 'Patterns of present sub-aerial erosion and landforms in mid-Wales', *Trans. Inst. Br. Geogr.* 55, 47–68

SMITH, R. M. and W. L. STAMEY (1965) 'Determining the range of tolerable erosion', *Soil Sci.* 100, 414–24

SOONS, J. M. (1971) 'Factors involved in soil erosion in the Southern Alps, New Zealand', *Z. Geomorph.* 15, 460–70

SOONS, J. M. and J. N. RAYNER (1968) 'Microclimate and erosion processes in the Southern Alps', *Geogr. Annlr* 50A, 1–15

STARKEL, L. (1962) 'Stan badan nad wspolczesnymi procesami morfogenetycznymi w Karpatach', *Czas. geogr.* 33, 459–73

STEERS, J. A. (1953) *The sea coast*

STOUT, M. L. (1969) 'Radiocarbon dating of landslides in southern California and engineering geology implications', *Spec. Pap. geol. Soc. Am.* 123, 167–79

STRAKHOV, N. M. (1967) *Principles of lithogenesis*

SWEETING, M. M. (1972) *Karst landforms*

TERZAGHI, K. (1913) 'Hydrographie und Morphologie des Kroatischen Karstes', *Mitt. Jb. K. ung. geol. Anst.* 20, 255–369

TERZAGHI, K. (1958) 'Landforms and subsurface drainage in the Gačka region in Yugoslavia', *Z. Geomorph.* 2, 76–100

THORNBURY, W. D. (1954) *Principles of geomorphology*

TOWNSHEND, J. R. G. (1970) 'Geology, slope form and process and their relation to the occurrence of laterite in the Mato Grosso', *Geogrl J.* 136, 392–9

URSIC, S. J. (1963) 'Sediment yields from small watersheds under various land uses and forest covers', *Misc. Publs U.S. Dep. Agric.* 970, 47–52

WASHBURN, A. L. (1967) 'Instrumental observations of mass-wasting in the Mesters Vig District, northeast Greenland', *Medd. om Grønland* 166

WASHBURN, A. L. (1970) 'Instrumental observations of mass-wasting in an arctic climate', *Z. Geomorph. Suppl.* 9, 102–18

WHIPKEY, R. Z. (1965) 'Subsurface storm flow from forested slopes', *Bull. int. Ass. scient. Hydrol.* 10, 74–85

WILLIAMS, M. A. J. (1968) 'Termites and soil development near Brocks Creek, Northern Territory', *Austr. J. Sci.* 31, 153–4

WILLIAMS, M. A. J. (1973) 'The efficacy of creep and slopewash in tropical and temperate Australia', *Austr. geogrl. Stud.* 11, 62–78

WILLIAMS, P. J. (1957) 'The direct recording of solifluction movements', *Am. J. Sci.* 255, 705–15

WILLIAMS, P. J. (1966) 'Downslope soil movement at a sub-arctic location with regard to variations with depth', *Can. geotech. Jnl* 3, 191–203

WILLIAMS, W. W. (1956) 'An east coast survey: some recent changes in the coast of East Anglia', *Geogrl J.* 122, 317–34

YOUNG, A. (1960) 'Soil movement by denudational processes on slopes', *Nature, Lond.* 188, 120–2

YOUNG, A. (1963a) 'Soil movement on slopes', *Nature, Lond.* 200, 129–30

YOUNG, A. (1963b) 'Deductive models of slope evolution', in *Neue Beiträge zur internationalen Hangforschung (Nachrichten der Akademie der Wissenschaften in Göttingen, II Math.-Phys. Klasse 5)* 45–66

YOUNG, A. (1969) 'Present rate of land erosion', *Nature, Lond.* 224, 851–2

YOUNG, A. (1972) *Slopes*

RÉSUMÉ. *La vitesse du recul des versants.* Ce travail rassemble les résultats de 109 études de mouvements mesurés de creep, de solifluxion, de ruissellement, de dissolution, de glissement de terrain, de retrait de falaise, de recul de pentes et de dénudation totale des versants. Le creep déplace plus de sol que le ruissellement sous climats tempérés humides; le ruissellement est le processus dominant dans les climats semi-arides et de savane tandis que les creep et le ruissellement sont tous deux modérément rapides sous la forêt humide. La dissolution est un processus important sous tous les climats humides. L'importance relative de la dénudation périglaciaire sur le recul Quaternaire des versants n'a pas été établie, pas plus que l'importance relative occupée par les processus catastrophiques et continus sous toutes les zones climatiques. Pour faire progresser les recherches sur les processus actifs en surface, il serait très utile que soit réalisé, dans les sites étudiés, un échantillonage stratifié établi d'après les formes de terrain.

ZUSAMMENFASSUNG. *Die Rate der Hangabtragung.* Die Ergebnisse von 109 Arbeiten über beobachtete Raten von Abtragungsvorgängen (Bodenkriechen, Solifluktion, Abspülung, Lösung, Rutschbewegungen, Rückverlegung von Steilwänden und Hängen, sowie Gesamtabtragung) werden zusammengefasst. Im feucht-gemässigten Klima wird mehr Material durch Bodenkriechen hangabwärts verlagert als durch Abspülung; im semiariden Klima und im Savannenklima herrscht Abspülung vor, während im Regenwald sowohl Kriech- als auch Abspülungsvorgänge mässig rasch arbeiten. Die relative Bedeutung periglazialer Denudation für die quartäre Hangabtragung in der gemässigten Zone ist noch nicht bestimmt worden; dasselbe gilt für die relative Leistung von katastrophalen im Vergleich zu kontinuierlichen Vorgängen. Für die zukünftige Arbeit über Oberflächenvorgänge ist ein räumlich geordnetes System für die Verteilung einzurichtender Beobachtungsstationen nötig.

The degradation of a coastal slope, Dorset, England

DENYS BRUNSDEN

Lecturer in Geography, University of London, King's College

Revised MS received 5 July 1973

ABSTRACT. An undercliff in the centre of a coastal landslide near Charmouth, Dorset has been monitored for five years in an attempt to establish general models of evolution and rates of erosion by large- and small-scale processes. The cliff is 21 m high at a mean angle of 35°, is unvegetated and suffers basal removal of debris by mudslides. The slope was studied from 1966–71 with the aid of a sequence of aerial photographs from 1946–69, field mapping, estimation of landslide volumes, and erosion pins. The rate of erosion is summarized in maps, tables and isoline maps to show the temporal, spatial and magnitude variability of landslide processes and a three-dimensional pattern of slope evolution. It is suggested that short-term measurements of erosion rates should not be used for long-term calculations unless the spatial and temporal variability of the process is known.

IN *The Everlasting Hills*, D. L. Linton (1957) described the fundamental geomorphological problem of establishing the rates at which land surfaces are denuded. Linton relied on stratigraphical and denudation chronology techniques to estimate the amount of ground lowering in terms of the thickness of rock removed from an area in long periods of geological time. Measured rates of erosion by currently active processes and variations in erosion rates were not considered. It was the grand view, the framework within which later studies would fit the details of how 'prodigious denudation' was achieved.

Earlier some attempts had been made to establish precise rates of denudation using the sediment and dissolved solids outputs from large drainage basins. Such measurements, however, only provide composite figures which incorporate the work of many processes acting over wide areas. For this reason they have been complemented by a rapidly increasing number of studies which are yielding rates of operation of specific processes. In particular the processes of soil creep, mass movement, surface wash and throughflow have recently received close attention (M. A. Carson and M. J. Kirkby, 1972).

It continues to be important that such measurements are obtained, for the number of processes and environments so far studied is small. The time periods rarely exceed five years and very little is known about the variability of processes over short time periods or even small distances. This paper attempts to extend this work by reporting the nature of erosion and erosion rates on selected coastal slopes and, in particular, draws attention to the time and space variability which is usually hidden in large-scale calculations of total denudation.

THE STUDY AREA

The slopes studied form an undercliff known as Fairy Dell (M. A. Arber, 1941) in the centre of a large landslide complex on the coastal side of Stonebarrow Down, Dorset (Fig. 1). Above the undercliff is a series of large landslides from a 36·6 m cliff of chert, Upper Greensand and Gault. The undercliff forms the back-slope of several mudslides and rotational landslides which move across a bench to the top of a 45·7 m sea cliff. The surveyed slopes are approximately 21·3 m high and are composed of three elements: an upper, inaccessible portion developed on the eroded toes of the large Upper Greensand landslides, a central free face of Middle Lias and a lower talus

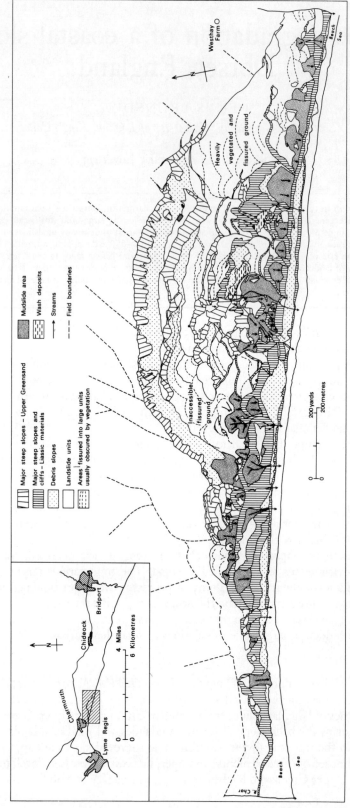

FIGURE 1. Geomorphological map of the coastal landslide complex of Stonebarrow Hill (1969)

slope of mixed silt, sand and clay. The mudslides form amphitheatre-shaped hollows which eat landward into the undercliff. They are separated either by knife-edged ridges or wider divides which are surmounted by degraded remnants of rotational slides. The undercliff, in plan, is therefore a series of concave embayments and sharp spurs.

The Upper Greensand is a yellow-orange, fine sand capped by thick chert beds. Together with a thin, sandy Gault layer it rests unconformably on the Middle Lias deposits (V. Wilson *et al.*, 1958). The sand weathers into individual grains and will readily flow in wet conditions. In local construction work it is known as running sand. It has a grain size of predominantly 0·06–0·2 mm, an average liquid limit of 36·8 per cent, plastic limit of 21·9 per cent and a plasticity index of 12·9.

The Middle Lias deposits are overconsolidated, closely fissured silts and clays with irregular limestone bands known as the Three Tiers. Following erosion of the slope, lateral and vertical unloading leads to expansion of joints and fissures. The surface softens or fragments if frost occurs and quickly weathers in a shallow 0·2 m layer which is very unstable. These deposits have a grain size characteristic of medium to fine silts and clays, 0·001–0·02 mm, an average liquid limit of 48·2 per cent, plastic limit of 24·3 per cent and plasticity index of 23·9.

The top of the slope, on the Upper Greensand, is covered by a thirty-year growth of gorse, bramble and broom which prevents access to the landslides. Small patches also occur between the mudslides and on the talus slopes. The main part of the undercliff is, however, very sparsely vegetated and bare ground occupies as much as 80 per cent of the surface.

MEASUREMENT PROGRAMME

Detailed measurement of the slope erosion began in 1966 as part of a wider study of mass movement processes (D. Brunsden, 1968, 1973; D. Brunsden and D. K. C. Jones, 1972). The slopes were studied in the following ways: first, vertical aerial photographs for 1946, 1948, 1958, 1969, oblique air photographs of 1956 and 1966 and 1:2500 Ordnance Survey maps of 1877 (revision 1901, 1921) and 1953 (revision, 1964) were used with geomorphological field mapping to construct detailed maps of the major changes in morphology (Figs 1–4). Using these maps it was then possible to construct general models (Figs 5 and 7) of slope development and obtain approximate figures of slope erosion.

Secondly, a small portion (100 m) of the undercliff was surveyed using a theodolite with Ewing-Stadialtimeter (J. C. Pugh and D. Brunsden, 1964) plane table and telescopic alidade to serve as a base map for repetitive surveys of landslides, volume estimation and general description (Figs 8 and 9). The occurrence of each small-scale mass movement and areas of frost action were mapped at monthly intervals from September 1966–September 1969 and measurements of landslide width, depth and length were taken with a steel tape. Although efforts were made to keep this as accurate as possible, access to wet clay slopes of up to 60° angle and the desirability of keeping disturbance of the slope to a minimum meant that measurements should be regarded as approximate only.

Thirdly, six slope profiles were randomly located and measured by laying a short levelling board on the ground surface and determining the slope angle of the generalized surface with an Abney level. Readings were taken to the nearest one-half degree (Figs 10 and 11). At five foot intervals on each profile steel pins, 0·64 cm diameter, 0·61 m long, were driven into the slope normal to the surface. The pins initially protruded 2·54 cm. Each year in September the profiles were resurveyed, pin exposure noted and missing pins replaced. After the first year new pins were inserted on the talus slope and to overcome the problem of pin loss by burial were left 15·2 cm proud. Measurement of ground length between pins was repeated each year to establish downslope pin movement.

TABLE I

Geometry of slopes with active landslides and mudslides

Type	HE m	VI m	VI HE	Length m	Angle °
Successive	35·1	33·5	0·96	47·2	45·0
rotational	109·7	42·7	0·39	112·8	25·0
	68·6	48·8	0·71	74·7	31·5
	53·3	30·5	0·57	62·5	29·0
	51·8	18·3	0·35	54·9	20·5
	30·5	12·2	0·40	33·5	21·0
Mudslide	22·9	9·1	0·40	24·4	23·5
	48·8	18·3	0·38	53·3	21·0
	112·8	24·4	0·22	114·3	12·5
	32·0	10·7	0·33	33·5	19·5
	56·4	16·8	0·30	57·9	15·0
	109·7	23·5	0·21	111·6	12·0

Where H.E. = horizontal extent of slope
V.I. = vertical height of slope

In addition samples were taken for determination of soil properties, continuous recording of precipitation was made with a Kahlsico remote location rain gauge and the amount of talus subsidence at the slope base was measured.

LARGE-SCALE PROCESSES

The aerial photographs, historical maps and geomorphological surveys reveal two general ways in which the undercliff develops.

Rotational landslides

Some of the ridges between the mudslides are surmounted by successive rotational landslides which move sporadically seaward at rates as high as 81·3 m per annum. This activity is maintained by the retreat of the underlying sea cliff which is eroded by wave action and rockfalls at approximately 1–5 m per year. The toes of the landslides are eroded by shallow (1–2 m) mudslides and deep gullies which undermine them and cause continued instability. The slope angles of these ridges vary from 21–45° and no stable slides have been identified (Table I) (Fig. 6). The seaward movement of the degraded slides removes lateral support from the base of the landslide scar and eventually a further large slide occurs.

The available records (W. D. Lang, 1942–44) show that on 14 May 1942 a very complex movement took place which affected the whole landslide complex. This not only affected the undercliff area but also caused two huge slides on the uppermost cliff, involved the whole Upper Greensand slope and destroyed a Radiolocation (Radar) Station which had been placed on the hill-top.

On the undercliff four large rotational slides began to move (Fig. 2A, Sites A, B, C, E). Aerial photographs of 1946, 1948, 1958 and 1969 (Figs 3 and 4) show how these blocks are broken into smaller units as they are eroded from below and move toward the sea cliff. At the same time new tension cracks are forming on the slope above. By 1958 one of the original blocks had been removed completely, the others were low down on the mudslide slopes and a new slide had begun to descend (Fig. 2C–D, Site D). By 1969 (Fig. 2D) few remnants of the original slides existed, slide D was already being destroyed and a further succession had begun with a major collapse on 8 January 1968 (Fig. 2D, Site F). These records suggest a time period of approxi-

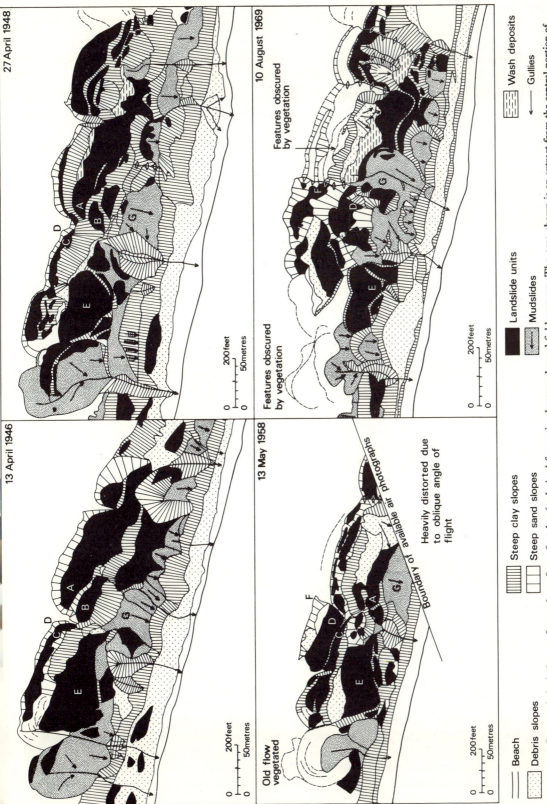

FIGURE 2. Geomorphological maps for 1946, 1948, 1958 and 1969 derived from air photographs and field surveys. The area shown is an extract from the central portion of Figure 1 and includes the undercliff, mudslide areas and sea cliff discussed in this paper

FIGURE 3. A. Oblique air photograph 1956　　　　　B. Oblique air photograph 1966
Courtesy J. K. St Joseph
The undercliff studied is in the centre of the photograph above the sea cliffs. The landslides of 1942 can be distinguished clearly as can the growth of the study mudslide between photo-periods

mately twenty-five years from the initial failure to destruction of a landslide unit although this will vary, being shorter for smaller slides (Site B—approximately twenty years) and longer for big units especially if these are in relatively inactive locations (Site E—30 years).

The units are not only destroyed by erosion of the toe but also by lateral flows and slides which originate on the side slopes and in the linear depressions behind each slide. This removes preferentially the Upper Greensand and chert from the top of the slide which thus reduces in thickness as the slide moves toward the sea (Fig. 5). Small-scale mass movement also occurs on the slide scar, reduces the scar angle and creates a small talus slope. The talus is pushed forward and over the back-slope of the slides by each successive new slide.

Headward erosion by mudslide

The slopes immediately above the sea cliff are composed of a complex series of mudslides which either eat headward into the rotational landslides on the ridges or form arcuate hollows backed by steep clay cliffs in the undercliff itself. The largest hollows are usually drained by intermittent streams which may form gullies at the cliff edge. The slope angles of these areas range from 12·00–23·50° with a mean of 16·31° (Table I).

The pattern of erosion by mudslides is complex and will therefore be described for one site (Fig. 2, Site G) starting with the large failures of 1942 (Fig. 7). At that time the slide undercut the Middle Lias undercliff and two failures took place. These broke up as they descended slowly but formed an effective barrier across the base of the cliff for nearly twenty years. During this time the hollow behind the slides had become infilled with thinly laminated wash deposits of alternating sand and clay. In addition a talus slope had built up and the slide scar had reduced in angle by small scale mass movements from near vertical to an average of 30–35° (Table III).

By 1958 the last remnants of the 1942 slides were being destroyed and during 1960–70 the

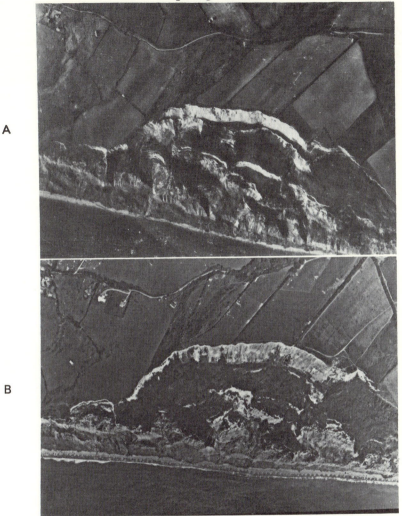

FIGURE 4. A. Section of vertical air photograph 1946 (R.A.F.)
B. Section of vertical air photograph 1969 (Fairey Surveys)
The changing pattern of rotational slides and mudslides is clearly shown and should be compared with Fig. 2

mudslide began to erode the wash deposits. This process also took place at depth so that a water-filled depression developed into which the talus slope began to subside.

At this time the frequency and scale of small failures on the undercliff increased as did the slope height, due to basal subsidence, and slope angle, by basal undercutting. In the near future the mudslide will evacuate all the slide debris, wash deposits and talus from its hollow and it is therefore anticipated that steepening of the undercliff will proceed rapidly leading to a further large-scale failure and a repetition of the cycle (Fig. 7). So far the process has run for 30 years and it is therefore suggested that a cycle length of thirty to forty years is possible.

The two models of large-scale slope evolution are representative of all parts of the undercliff (Fig. 1) and indicate the variability of erosion patterns and the extremely complex spatial distributions that exist in rapidly eroding areas. It is possible to calculate that since 1942 the crest of the slope has retreated 45–60 m in the ridge areas and up to 45–50 m in mudslide areas yielding average annual rates 1·5–2·0 m and 1·5–1·7 m respectively. This, however, only serves to empha-

D.BRUNSDEN

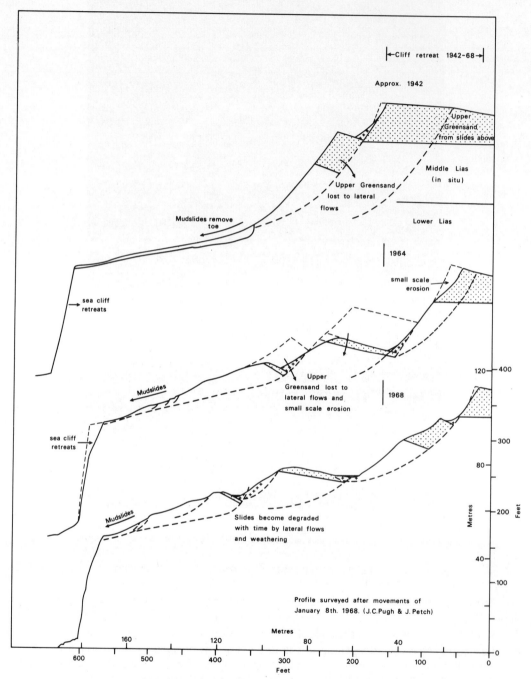

FIGURE 5. Diagram of a portion of the cliff affected by rotational landslides. Derived from air photographs and field surveys

FIGURE 6. View of the area of rotational landslides to the west of the mudslide area. The mudslide occupies the foreground. The successive nature of the landslides is clearly seen. The landslide of 8 January 1968 lies on the right

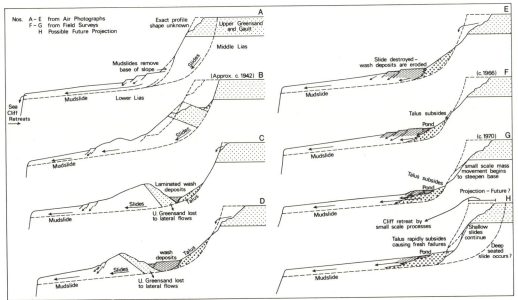

FIGURE 7. Schematic diagram of a portion of the cliff at the head of a mudslide amphitheatre. Derived in part from air photographs and field surveys. The model is generalized and no scale is shown

size how much temporal and spatial variability is hidden by average or general statements of erosion.

Repetitive survey and volume measurements

The variability of erosion becomes even more apparent when surveys are made of the very small-scale failures (0·1–100 m³) which occur on the undercliff during the time interval between large-scale movements. Such processes were mapped in 1966/7, 1967/8, 1968/9 (Fig. 8) and volumes estimated from measurements of slide geometry (Fig. 9).

The surveys show a steady increase in number of failures from twenty-six in 1966/7 to sixty-six in 1968/9 (Table II). The volume moved was lower in 1967/8 than the previous year but during 1967/8 there was continuous subsidence at the slope base which steepened the slope

88

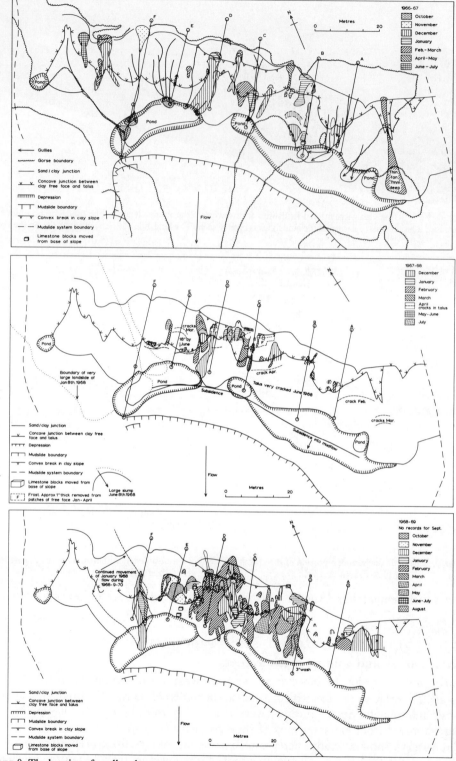

FIGURE 8. The location of small-scale mass movement
(a) 1966/7   (b) 1967/8   (c) 1968/9

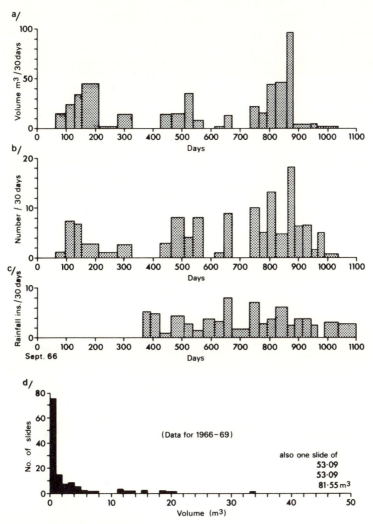

FIGURE 9. A. Volume of landslides 1966–69
B. Number of landslides 1966–69
C. Precipitation
D. Frequency of landslides by volume
The volumes and number of slides are corrected for the unequal measurement periods

and led to a markedly increased volume in 1968/9. The slope thus seems to be very sensitive to boundary conditions at the slope base with a relaxation time of less than a year.

The maps and volume estimates reveal the extent of the temporal, magnitude and spatial variability. There is a strong seasonal component in the frequency of slides with maximum activity in January, February and March and a smaller peak in July (Fig. 9). It was at first thought that this was associated with seasonal precipitation (Fig. 9) but although some correspondence of peaks is evident no relationship in a statistical sense was discovered. The relationship between precipitation and small-scale failure is obviously a complex one which probably requires a knowledge of antecedent moisture and erosion, evaporation, pore pressure and strength reduction due to weathering. Unfortunately such data are not available for the study site.

TABLE II

Volume and number of small-scale mass movements from the free face 1966–69 in m³

Date 1966/7	Sept./Oct.	Nov.	Dec.	Jan.	Feb./Mar.	Apr./May	June/July	Aug./Sept.
Volume of individual slides from free face m³	Wash from gullies only	15·30	18·12 0·23 3·44 0·07 1·42 0·03 0·43 0·57	3·40 0·99 0·51 20·39 0·43 5·38	0·28 81·55 0·85 4·53 2·95	0·09 3·27	19·03 2·72 0·96 1·33	Wash from gullies only

(joint monthly readings)   Year

	Sept.	Oct.	Nov.	Dec.	Jan.	Feb./Mar.	Apr./May	June/July	Aug.	Sept.	Year
Total vol.	0	0	15·30	24·31	31·10	90·16	3·36	24·04	0	0	188·27
No. slides	0	0	1	8	6	5	2	4	0	0	26·00
Mean vol.	0	0	—	3·04	5·18	18·03	—	6·01	0	0	7·24

Date 1967/8	Sept./Oct.	Nov.	Dec.	Jan.	Feb.	Mar.	Apr.	May/June	July	Aug./Sept.
Volume from free face m³	Wash only	Frost erosion	1·70 2·55 13·59 0·14	11·89 0·51 0·32 0·26 0·85 0·57 0·09 1·27 0·85 1·70 3·74 0·45	33·98 0·26 0·07 1·10	1·02 0·89 0·07 0·43 1·61 0·06 4·25 0·57 0·34	Frost erosion	0·37	0·23 0·85 0·23 0·34 2·18 7·14 0·45 0·28	Wash only

I————————————I
Talus subsidence

joint reading   Year

	Sept.	Oct.	Nov.	Dec.	Jan.	Feb.	Mar.	Apr.	May/June	July	Aug.	Sept.	Year
Total vol.	0	0	0	17·98	22·50	35·41	9·24	0	0·37	11·70	0	0	97·20
No. slides	0	0	0	4	12	4	9	0	1	8	0	0	38·00
Mean vol.	0	0	0	4·50	1·88	8·85	1·03	0	—	1·46	0	0	2·56

Date 1968/9	Sept.	Oct.	Nov.	Dec.	Jan.	Feb.	Mar.	Apr.	May	June/July	Aug.	Sept.
Volume from free face m³	Wash only	0·58 15·30 1·75 3·06 0·20 0·26 0·25 0·19 0·11 0·09	0·94 11·21 0·30 0·51 1·91	0·99 0·04 5·35 12·04 3·40 11·78 0·11 0·06 0·17 0·82 0·10 4·53	3·82 4·53 53·09 0·26 0·04 0·07	2·69 4·08 6·16 53·09 0·85 1·59 1·53 0·34 2·27 0·04 0·51 0·53 0·61 0·14	0·58 0·35 0·32 0·23 1·13 0·85	0·68 0·34 0·46 2·48 0·07 0·04 0·06	0·02	0·43 0·68 0·02 0·02	2·66	Wash only

I————————— Talus subsidence —————————I

Year

	Sept.	Oct.	Nov.	Dec.	Jan.	Feb.	Mar.	Apr.	May	June/July	Aug.	Sept.	Year
Total vol.	0	21·79	14·87	39·39	61·81	74·43	3·46	4·13	0·02	1·15	2·66	0	223·71
No. slides	0	10	5	12	6	14	6	7	1	4	1	0	66
Mean vol.	0	2·18	2·97	3·28	10·30	5·32	0·58	0·59	—	0·29	—	0	3·39

TABLE III

Slope geometry of profiles A–F

Profile	Elevation m	Mean angle °	Mean angle–sand °	Mean angle–clay °	Mean angle–talus °
A	16	35	35	47	29
B	18	32·6	32·8	39	24
C	20	37·3	40	42	26
D	20	35·5	38	42	26
E	21	41·5	40	47	29
F	21	32·6	37	44	22

The variability of mass movement size is also considerable. The distribution (Fig. 9) is strongly skewed with the majority of slides being smaller than 1 m³. The mean size is 4·1 m³ with a range of 0·02–81·6 m³. Both monthly and annual volumes show considerable variation (Table II). It is also of interest that the mean volume of 4·1 m³ has a recurrence interval of only 1·6 months, thus smaller slides may occur many times in any month.

The spatial variability is apparent on both an annual and a monthly basis (Fig. 8). In 1966/7 the failures are distributed randomly across the slope surface and no areas show preferential development. However, in 1967/8 a pattern of erosion concentration began to develop. In the vicinity of profiles C, D and E talus subsidence was very marked (0·5 m) and superimposed on the random failure pattern was an area of intense erosion. This pattern was accentuated in 1968/9 and although most areas of the slope experienced some mass movement the area between profiles C–E was again intensely eroded. Not only the number but also the volume of slides increased in this zone. Profile E, in particular, was affected by larger and deeper slides. Although monthly recording ceased in September 1969, annual observations show that this pattern has continued with one exceptional movement on profile E and an extension of the active zone as far as profile B.

The normal distribution of slides which has occurred in each year can perhaps be explained in terms of the weathering of the whole face which probably takes place at an approximately equal rate. The concentrated erosion is due to the basal boundary conditions. The apex of the mudslide occurs between profiles B–E and it is here that maximum removal of support, increase of slope height and basal erosion are felt.

Surveyed profiles and erosion pins

In an attempt to determine the overall rates of erosion on the undercliff by all small-scale processes including landslides, gullies, frost and solution, erosion pins were inserted at 1·5 m intervals on the six surveyed profiles (Fig. 10). The geometrical properties of these profiles are given in Table III. (Erosion-accumulation figures for each pin are available from the author.)

Profile A located at the eastern end of the sample slope was the shortest slope measured and the survey detected no subsidence at the slope base. Even though the mean angle of the clay free face was high the erosion on the profile was low in comparison to the rest of the area. Nevertheless the total pin ground loss on the lower part of the face of 21·6 cm in five years is a very rapid rate of erosion.

Profile B was composed in 1966 of three concave sections separated by two thin limestone bands. The structural influence of these layers on the micro-relief of the slope was strong and there was a concentration of erosion just below the lower band which undermined the limestone and allowed a large block to slide down the talus slope. The profile was severely affected by

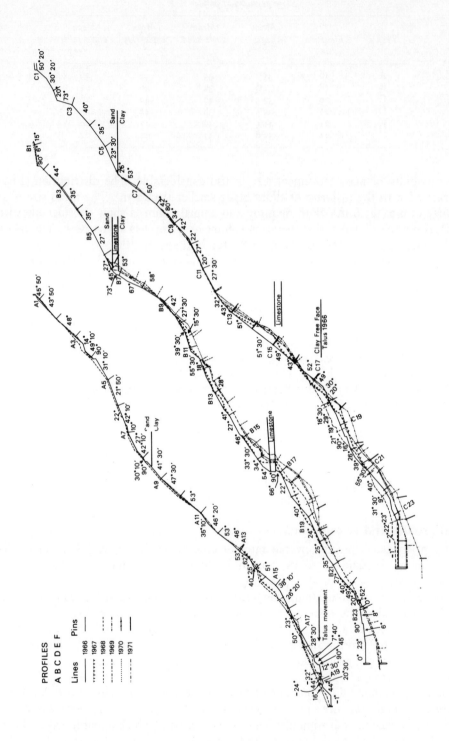

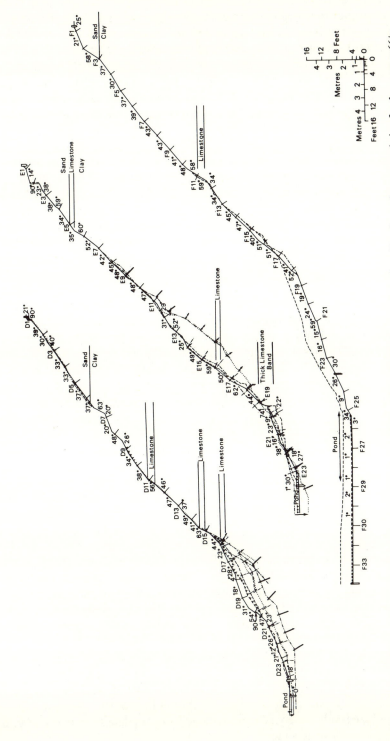

FIGURE 10. Slope profiles and erosion pins on part of the coastal slope. To demonstrate the effect of small-scale erosion-accumulation for the years 1966/7, 1967/8, 1968/9, 1969/70, 1970/1

subsidence in 1969–71 (0·5 m) and the rate of erosion increased. The central part of the profile showed parallel retreat but the erosion was mainly accomplished by a slump which occurred to the west of the line and not by processes acting directly downslope. As with profile A the sand portion was little eroded and most activity occurred just above the talus.

Profile C was higher and separated into three concave sections at the sand–clay junction and in midslope. A limestone band was present but highly jointed and had little effect on slope form. Talus subsidence was very high at 1·94 m and the lower clay face retreated in a parallel manner in all years of observation. Changes in the upper part of the slope were also low so that the dominant changes were an increase in height and angle due to the basal conditions. The slope is becoming increasingly convex and unstable.

Profile D was initially a straight slope with a slight basal concavity into a pond. A total subsidence of only 0·48 m was recorded but this is an underestimate due to rapid deposition in the pond from an unmeasured lateral landslide. The head of the talus was lowered by a total of 1·22 m in five years which is probably closer to the true subsidence figure. The clay face of this profile seemed to bear a charmed life with failures on either side of it but only the thinnest erosion on the profile itself. Even in February 1969 when in the centre of intense erosion (Fig. 5 C) it was only affected by debris from a failure above and to the side of the line which passed across the surface with a minimum of erosion.

Profile E was also a straight slope with a short talus slope leading into the centre of the subsiding pond. Three limestone bands formed only very small breaks in form. The profile showed little change in 1966/7 but thereafter it suffered the maximum recorded erosion which culminated in 1969/70 in a slump, 1·5 m deep, from the lowest part of the clay face. This was associated with rapid talus subsidence during 1967/8 and again in 1969/70. This subsidence revealed a 1 m thick limestone band which broke away in large blocks and sank into the talus presumably to be incorporated in the subsurface movement. The overall effect of this failure was to steepen the base of the profile while the upper part remained unchanged to produce a convex, unstable form. In 1970/1 and continuing into 1972 further failures have extended the failure scar upslope.

Profile F, a straight slope with a small break caused by a limestone outcrop and a short talus slope leading to a flat wash slope, suffered only slight erosion until 1968 when on 8–9 January the whole of the talus slope, the wash slope and the base of the clay face was buried by the toe of a massive new landslide on the adjacent ridge. The profile was therefore abandoned for further measurement since the dominant process was acting from outside the study area. Casual observations, however, suggest that basal conditions have fluctuated as the landslide settled further down-slope.

These descriptions show that small-scale failures on the slope face are closely related to the activity of the mudslide at the base of the profiles. The profile data confirms that the maximum ground loss was concentrated immediately above the talus, was greatest near the apex of the mudslide and increased rather erratically from 1966–71. All the profiles became more convex towards the base and exhibited parallel retreat on the portion of the slope which was actively eroding (Fig. 10).

It is clear that there is again considerable variation in location, magnitude and time of change. This is further demonstrated if isolines of erosion-accumulation are drawn for the successive years (Fig. 11). Before discussion, however, attention should be drawn to the fact that the isolines give a distorted picture in that they emphasize horizontal links rather than vertical movement and are based only on erosion pin data and not the landslide maps (Fig. 8) which were of a much lower order of accuracy.

Although the maps are clearly very different in detail it is possible to suggest some general

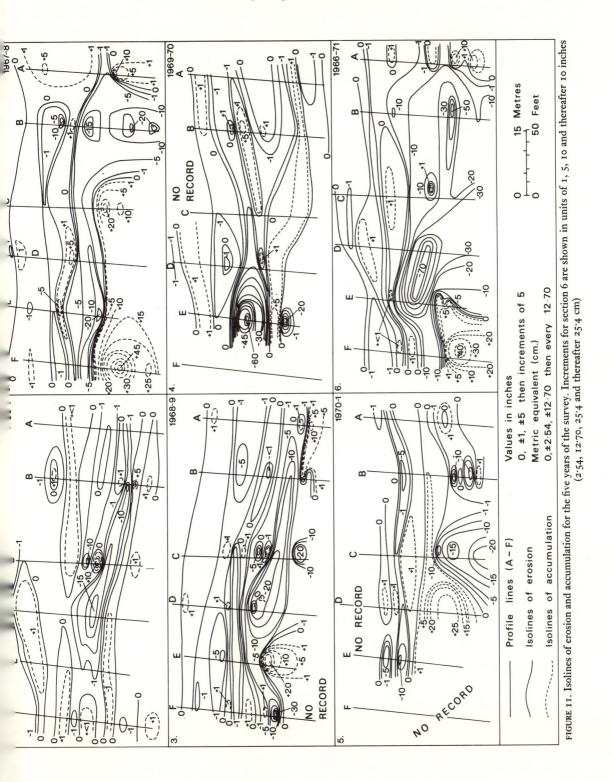

FIGURE II. Isolines of erosion and accumulation for the five years of the survey. Increments for section 6 are shown in units of 1, 5, 10 and thereafter 10 inches
(2·54, 12·70, 25·4 and thereafter 25·4 cm)

trends. The maximum variability occurs at the base of the slope where the talus alternately builds up and subsides. Above this a belt of maximum erosion occurs and the upper slope shows local changes at much lower rates.

More significant is the pattern of erosion and accumulation concentration at different loci each year. This may be partly due to the regularly spaced data points with no intervening information but also reflects the effect of larger individual failures, centres of frost action and areas of subsidence or deposition.

The isolines for the whole survey period reflect the two accumulation centres at each end of the slope; on the west where deposition from 8–9 January 1968 landslide is important and in the east where no subsidence was experienced. The centre of erosion associated with the large slump on profile E is also dominant. Elsewhere the isolines become further apart and show that the sum totals for each year amount to a more even loss over the whole slope face.

It is therefore conjectured that although erosion is intermittent at any one place the effect over a longer time period might be a comparatively even removal and an overall parallel retreat. This idea is supported by the relative straightness of the slope both in plan and profile.

CONCLUSION

It has long been recognized in geomorphology that generalized figures of erosion rates must be treated with extreme caution and that there is a very large gap in our knowledge between rates derived in studies of perhaps 1–25 years duration and gross figures of large-scale erosion on a geological time-scale. There is an obvious temptation to use short-term figures to obtain long-term effects but this study demonstrates that, in areas of rapid erosion, short-term variability may be so great that such calculations are meaningless.

It is true that the figures obtained are for very rapid erosion and that studies of slower processes such as soil creep may show less variation and therefore may be more amenable to long-term extension. In such cases, however, we do not yet know the time period required for temporal variability to affect the record and we have even less idea of the spatial variability.

This survey has shown that in one selected site of rapid erosion the magnitude of ground loss varies from 0·02 m³ to many thousands of cubic metres in both a short time and distance. The frequency of events varies from twenty to forty years for the largest slides to a recurrence interval of only 1·6 months for the mean small-scale event. The spatial pattern is equally complex especially if individual events are considered and although it is often convenient to summarize the figures into overall figures of evolution, it is important that we understand how a given geomorphological change is achieved if we are to place reliance on our gross figures or if we are to predict future events.

ACKNOWLEDGEMENTS

Many colleagues and students of the Joint School of Geography have assisted in the monthly field work necessary for this study. Particular thanks are due to C. Brooks, D. K. C. Jones, D. Maugham, P. Parkinson, J. Petch, J. Pittam, E. Relph and D. Weyman. Professor and Mrs J. C. Pugh assisted with surveying and constant encouragement. The financial support of King's College and the Natural Environment Research Council who awarded a generous research grant for the investigation of landslides on the Dorset coast is gratefully acknowledged. Photographic assistance was received from the Captain, H.M.S. *Osprey* and invaluable technical assistance from Mr C. R. Polglaze and the Cartographers of the Department of Geography, King's College, London. The author is very grateful to Mrs Gisella Banbury, Mrs Helen Muir and Mrs Karen Wach for the translations of Abstracts and captions.

NOTE

For reasons of space the data for erosion-accumulation on the six, surveyed, erosion pin sections are not included. These data which comprise six tables and cover the period 1966–71 can be obtained on request from the author.

REFERENCES

ARBER, M. A. (1941) 'The coastal landslides of West Dorset', *Proc. Geol. Ass.* 52, 273–82

BRUNSDEN, D. (1968) 'A preliminary report on an investigation of landslides near Charmouth, Dorset', *Coastal Engineering Meeting, Engineering Geol. Group* Geological Society (preprint), p. 10

BRUNSDEN, D. (1973) 'The application of systems theory to the study of mass movement', *Proc. I.R.P.I. Conference on Natural Slopes, Stability and Conservation*, Naples–Cosenza (October 1971) (in press)

BRUNSDEN, D. and D. K. C. JONES (1972) 'The morphology of degraded landslide slopes in south-west Dorset', *Q. Jl Eng Geol.* 5, 1–18

CARSON, M. A. and M. J. KIRKBY (1972) *Hillslope form and process.* This provides summaries of the recent work in establishing short- and long-term erosion rates

LANG, W. D. (1942) Geological Notes, 1941–42 *Proc. Dorset nat. Hist. archaeol. Soc.* 64, 129–130 also in *Geological Notes* 1944, 66, 129

LINTON, D. L. (1957) 'The everlasting hills', *Advmt Sci.* 54, 58–67

PUGH, J. C. and D. BRUNSDEN (1965) 'Geographical applications of the Ewing Stadialtimeter', *Trans. Inst. Br. Geogr.* 37, 157–67

WILSON, V. and F. B. A. WELCH, J. A. ROBBIE, G. M. GREEN (1958) 'Geology of the country around Bridport and Yeovil' (Sheets 327 and 312), *Mem. geol. Surv. U.K.* 238

RÉSUMÉ. *La dégradation d'une pente à la côte de Dorset, Angleterre.* Pendant cinq années un talus au centre d'un glissement à la côte près de Charmouth, Dorset, a été enregistré afin d'établir des modèles généraux d'évolution, de meme qu'estimer combien les processus en grand et en petit l'aurait érodé. La falaise, dont l'hauteur est de 21 mètres sous un angle moyen de 35°, ne présente pas de végétation, et le détritus au pied a été enlevé par des éboulis ordonnés.

Une succession de photographies aériennes de 1946–69, les cartes dressées au champ, le volume estimé des glissements, et les jalons posés pour mesurer l'érosion, tous ont aidé l'étude de la pente. Le degré auquel la pente s'érodait s'est résumé en cartes, en tables et en cartes isolines qui montraient combien se variaient les processus de glissement par rapport au temps, à l'espace et à la grandeur; on s'y servait aussi d'une modèle à trois dimensions de l'évolution de la pente. La proposition se fait que peut-être les mesures de l'érosion à court terme ne doivent pas être employées pour faire les calculs à long term au moins qu'on ne sache la variabilité spatiale et temporelle du processus.

FIG. 1. Carte géomorphologique du tout du glissement le long de la côte de Stonebarrow Hill (1969)

FIG. 2. Cartes géomorphologiques pour 1946, 1948, 1958 et 1969, tirées de photographies aériennes et de levés faits au champ. Le terrain ici représenté est un morceau choisi du parti central de la Figure No. 1 où on peut voir le talus les éboulis ordonnés et la falaise dont il s'agit en cette étude

FIG. 3. (Gravures A–B) A. Photographie aérienne oblique 1956
  B. Photographie aérienne oblique 1966
  à titre gracieux de J. K. St Joseph.
Le talus dont il s'agit se trouve au centre de la photographie au-dessus des falaises. On y peut distinguer les glissements de 1942 et les photographies montrent combien se sont augmentés les éboulis ordonnés pendant des intervalles.

FIG. 4. (Gravures A–B) A. Section de photographie aérienne verticale 1946 (R.A.F.)
  B. Section de photographie aérienne verticale 1969 (Fairey Surveys)
On y voit clairement combien se change la forme des glissements suivant une surface courbe, et des éboulis ordonnés; cette forme doit être comparée au terrain chez la figure No. 2

FIG. 5. Diagramme d'une section de la falaise altérée par des glissements suivant une surface courbe, tirée de photographies aeriennes et de levés au champ

FIG. 6. (Gravure) Vue du territoire où se produisent les glissements suivant une surface courbe à l'ouest de l'éboulis ordonné qui occupe le premier plan. Le caractère successif des glissements s'y voit clairement. Le glissement du 8 janvier 1968 se trouve à droit

FIG. 7. Diagramme schématique d'une section de la falaise à la tête d'un amphitheatre d'un éboulis ordonné. Tirée en partie de photographies aériennes et de levés au champ. C'est un modèle generalisé sans aucune échelle

FIG. 8. La situation du mouvement de la masse en grand
  (a) 1966/7 (b) 1967/8 (c) 1968/9

FIG. 9. A. Volume de glissements 1966–69
  B. Nombre de glissements 1966–69
  C. Précipitation
  D. La fréquence des glissements relativement au volume
Les volumes et le nombre de glissements sont corrigés pour les périodes de mesure inégale

FIG. 10. A–B. Les profils de la falaise, et les jalons situés sur une partie de la falaise pour expliquer les effets de l'accumulation en petit de l'erosion pendant les années 1966/7, 1967/8, 1968/9, 1969/70, 1970/1

FIG. 11. Isolines d'érosion et d'accumulation pour les 5 années de l'étude. Incréments pour la section No. 6 s'y montrent en unités de 1, 5, 10 et en suite de 10 pouces (2·54, 12·70, 25·4 et en suite de 25·4 cm)

ZUSAMMENFASSUNG. *Die Abtragung einer Küstenhanger in Dorset, England.* Über einen Zeitraum von 5 Jahren wurde ein Kliff in der Mitte eines Küstenerdrutsches in der Nähe von Charmouth untersucht, um auf diesem Wege ein allgemeingültiges Beispiel für die Evolution und den Erosionsgrad bei grösseren und kleineren Vorgängen zu schaffen. Das Kliff ist 21 m hoch bei einem Mittelwertswinkel von 35 Grad, es ist unbewachsen und hat durch Schlammrutsche Geröllabtragungsschäden an der Basis.

Der Hang wurde mit Hilfe von einer Serie von Luftaufnahmen (1946–69), Landvermessungen, Volumenschätzungen von Erdrutschen und Verwitterungsmessungen (1966–71) erforscht. Der Erosionsgrad wurde in Karten, Tabellen und Konturenkarten summiert, um die zeitlichen, räumlichen und grössenmässigen Schwankungen von Erdruschen und ein dreidimensionales Bild von einer Hangevolution zu zeigen. Es wird vorgeschlagen, dass kurzfristige Messungen der Erosionsgrade nicht für langfristige Berechnungen verwendet werden sollen, es sei denn, die räumlichen und zeitlichen Schwankungen des Vorganges sind bekannt.

ABB. 1. Geomorphologische Karte des kompletten Küstenerdrutsches von Stonebarrow Hill

ABB. 2. Geomorphologische Karte von 1946, 1948, 1958 und 1969 basiert auf Luftaufnahmen und Feldvermessungen. Das hier gezeigte Gebiet ist ein Auszug des Mittelteils von Bild 1 und schliesst das in dieser Veröffentlichung besprochene, Kliff, das Schlammrutschgebiet und Klippe ein

ABB. 3. (Bildtafel A–B) A. Schräge Luftaufnahme 1956
B. Schräge Luftaufnahme 1966
Mit freundlicher Genehmigung von J. K. St Joseph

Das beobachtete Kliff befindet gich in der mitte der Photographie über der Klippe. Die Erdrutsche von 1942 können deutlich erkannt werden, ebenso das Wachstum des beobachteten Schlammrutsches zwischen den Aufnahmezeit punkten.

ABB. 4. (Bildtafel A–B) A. Teilansicht der vertikalen Luftaufnahme 1946 (R.A.F.)
B. Teilansicht der vertikalen Luftaufnahme 1969 (Fairey Surveys Ltd)

Der Wechsel des Bergsturze mit rückwärtige kippung und Schlammrutsche ist deutlich gezeigt und ist mit Abbildung 2 zu vergleichen

ABB. 5. Diagramm der von dem Bergsturze mit rückwärtige kippung betroffenen Kliffabschnittes. Basiert auf Luftaufnahmen und Feldvermessungen

ABB. 6. Ansicht des Bergsturze mit rückwärtige kippung gebietes westlich des Schlammrutschgebietes. Der Schlammrutsch ist im Vordergrund. Die vielfältiqe Eigenschaft der Erdrutsche ist deutlich zu erkennen. Der Erdrutsch am 8 Jan. 1968 liegt rechts

ABB. 7. Schematisches Diagramm des Kliff abschnittes ober halb des von dem Schlammrutsch gebildeten Amphitheaters Teilweise basiert auf Luftaufnahmen und Feldvermessungen. Das Beispiel ist verallgemeinert und kein Masstab ist gezeigt.

ABB. 8. Stellen kleinerer Massenbewegungen

(a) 1966/7   (b) 1967/8   (c) 1968/9

ABB. 9. A. Erdrutschvolumen 1966–69
B. Anzahl der Erdrutsche 1966–69
C. Niederschlag
D. Häufigkeit der Erdrutsche nach volumen.

Volumen und Anzahl der Erdrutsche sind korrigiert, um die unterschiedlichen Messperioden auszugleichen.

ABB. 10. (Bildhafel A–B) Hangprofile und abtragungpflocke in einem Abschnitt des Küstenhanges, um den Effekt kleinerer Erosionsansammlungen über die Jahre 1966/7, 1967/8, 1968/9, 1969/70, 1970/1 zu zeigen.

ABB. 11. Isolinen der Erosion und Ansammlung über die 5 Jahre der Untersuchungsperiode. Werte für Abschnitt 6 sind in Gruppen von 1, 5, 10 und danach 10 inches angegeben (2·54, 12·70, 25·4 und danach 25·4 cm).

The ages of land surfaces, with special reference to New Zealand

MAXWELL GAGE

*Professor of Geology, University of Canterbury**

AND

JANE M. SOONS

Professor of Geography, University of Canterbury

MS received 17 October 1972

ABSTRACT. For entire land surfaces, parts of surfaces or particular assemblages of landforms, a complete statement of age which is not always possible, should define the earliest date from which development of the present surface could have begun as well as the time since when modification has been negligible. For both erosional and constructional surfaces there may be uncertainty as to how much they are younger than the material beneath. Older surfaces may be dated from the time when their present general form was attained, or it may be preferable to date from subsequent modifications such as those due to Pleistocene climatic changes. Within tectonically active regions, eroding surfaces that have been subject to prolonged, rapid degradation and continual renewal, and likewise depositional surfaces continually built up, may be regarded as of virtually zero age, or alternatively they could date from the inception of the present dynamic situation. The relevant aspects of New Zealand tectonic history are outlined and applied to the problem of dating landforms. Surfaces initiated prior to the Late Pleistocene can be recognized only outside the chief belt affected strongly by the Late Cenozoic 'Kaikoura Movements'. Even in more stable regions the possibility of present surfaces having evolved continuously since before Pliocene times is very restricted, but there are important tracts of resurrected Cretaceous erosion surface in southern parts of the country. The history of certain major river systems is reviewed. Widespread Pleistocene volcanic activity in the North Island has complicated the picture, but progress towards a complete tephrachronology is now assisting with the documentation of the later geomorphic history in central North Island districts.

THIS essay attempts some analysis of the meaning of 'age' as applied to land surfaces, some clarification of the concepts involved and an application of these to New Zealand landforms. Relative age in the sense of the Davisian sequential models is not considered here. New Zealand is chosen to illustrate many of the ideas developed because land surfaces of different ages occur within a limited area and also because the tempo of processes, ranging from moderate to exceptionally high, raises general questions of age not commonly encountered in regions that have had less extremely—or less recently—diversified geologic histories. These questions are nevertheless vital if any acceptable body of theory on the age of landforms is to be established.

AGES OF LANDFORMS

Defining age limits

It is a safe assumption that no portion of the present surface of the earth has existed unchanged through the several thousand million years since the planet attained its present condition. For every landscape there is consequently a base-line to which present landforms may be referred, a time at which the land surface differed significantly from its present aspect. There may also be an upper limit of age. The generation of landforms or of landform assemblages must not only have had a beginning, but must normally have occupied a finite span of time. Establishment of landform age is thus a question of deciding first whether a clear starting point for its evolution

*Dr Gage retired on 31st January 1974

99

can be distinguished finitely or within determinable limits of error; next, whether development has virtually ceased, and if so, how long ago. The dates may be expressed in geologic chronology or in the secular time-scale or both.

Dating the beginning

The face of the earth fortunately differs from the grin on the Cheshire Cat in an important respect, namely, that some substance never fails to be present under a landform. If the geologic or secular age of the youngest underlying materials is known, so also is the latest possible date for initiation of the landform. Surfaces resulting from terrestrial construction processes, fluvial, volcanic or other, having suffered only negligible subsequent modification, date from the cessation of the constructional process. If this is sharply defined the limits of error may be simply those inherent in the dating method. The problem usually is to confirm that the surface is in fact original and not significantly degraded.

In the case of erosional landforms the age of the youngest underlying materials may differ little from that of the surface, as for example where a surface of glacial outwash was trimmed quickly by meltwater from receding ice to form a flight of terraces. In a terrain of marine sedimentary rocks the maximum possible age must be somewhat less than that of the rocks, and there are usually several sources of uncertainty. The geologic age of the strata may not be determinable precisely and the time of final emergence of the area as dry land may or may not be known. This emergence may justifiably be taken as beginning an uninterrupted process of erosional modelling from which the present landscape resulted, but there may be other possibilities. The geology of the surrounding region will, however, generally offer some basis for establishing maximum and minimum limits. In general, the older the rocks, the less relevant their age to that of the landscape.

Dating the end

Statements about the age of surfaces are seldom explicit, but generally they can be taken to refer not to the beginning but to the end of the generation process. This implies that it is possible to demonstrate when geomorphic development effectively ceased or slowed down to a negligible rate. In fact, this is usually possible only in rather vague terms and almost invariably involves subjective judgements as to how much subsequent modification may justifiably be neglected, or indeed recognized. A case in point is the belief held by many Australian geologists and geomorphologists that the landscape evolution of that continent effectively ceased prior to the Quaternary Period (e.g. Öpik, 1958).

Minimum ages for erosional surfaces may be suggested where these are partly mantled by dateable surficial materials, e.g. alluvium, colluvium, loess, tephra, soil. The most favourable case arises where it can be demonstrated that erosion surfaces pass beneath surface mantlings without change of form or of depth of weathering. In many cases, however, there will be no evidence as to whether the underlying erosional surface attained its present form a long or a short time prior to the formation of the oldest surficial materials.

Datings of surficial limiting cover may be obtained palaeontologically, more directly by C_{14} or other isotope-ratio methods, or less directly from relative advancement of the pedogenic process on the covering materials compared with that on the underlying surface. The case of more or less deeply buried ancient erosion surfaces that have been re-exposed is an extension of the above, but will be considered separately later.

Age limits

A complete statement of age should ideally indicate both ends of the time-span, but this is not always a practical possibility. Much depends on the scale at which an investigation is being undertaken. Thus it may be possible to assign an individual landform, such as a spit, a moraine or a landslide, to a very precise time interval, whereas the dating of an extensive and varied surface may be placed only within the framework of major tectonic events. An interesting problem is also posed by the curious and rather irrational perspective of geologic time, where a span of uncertainty tolerable in the case of a supposed, early Cenozoic surface would be considered excessive in a late Pleistocene example.

Climatic events, with all the uncertainties to which definitions of these are subject, may serve to fix limits to the development of certain landforms. Thus in higher latitudes and altitudes the dominant features of a landscape may be attributable to the combined effects of various periglacial processes operating during the latest episode of Pleistocene cold upon a surface due to previous glacial sculpture. The significant dates may be the onset and end of periglacial conditions rather than of the preceding glaciation or the pre-glacial origin of the surface. More complex situations arise in areas subject to a succession of Pleistocene glaciations and intervening warm periods or pluvials and interpluvials. The original, pre-glacial, forms may be hardly or not at all discernible and successive modifications reflect the changing climatic environments of the area.

In such situations, the problems of scale, whether of time or magnitude, are important in determining the desirable or practicable accuracy with which age limits may be established. At one level of refinement, it may be desirable to establish the position of each element in a hierarchy of ages; at another, it may be neither practicable nor necessary to say more than that the whole assemblage of forms originated during the Pleistocene climatic oscillations. At this level, the difficulty of distinguishing minor differences with increasing overall age is accepted in terms of time-perspective.

The tectonic factor

The well-established principle that surface relief will eventually be reduced to a minimum unless erosion is somehow compensated has a corollary which states that strong terrestrial relief implies recent or current tectonic uplift. In dynamic terms, terrestrial relief expresses a momentary algebraic sum of elevation minus erosion, which for the time being has a positive sign. Extending this corollary, long-enduring strong regional relief, indicated for example by the ages, amount and texture of detrital products, is compatible only with a history of sustained tectonic activity. The global association of strong relief with evidence of crustal mobility is well known. Conversely, prolonged freedom from tectonic activity is obviously an essential condition if erosional surfaces are to survive without modification from the remote past.

Tectonically stable provinces of the main continents provide examples of surfaces for which Cenozoic or older ages have been claimed, and which are probably acceptable provided that we ignore minor modifications of surface detail and soils attributable to Pleistocene climatic fluctuations. In contrast, a recently tectonically active area may show few traces of early land surfaces. Thus it is difficult to find surfaces of possible Pliocene or greater age in New Zealand except, perhaps, in the relatively stable northern peninsula of the North Island and in the south-east of the South Island. Within the tectonically active regions of both islands vigorously eroding mountain slopes and very young constructional features, both sedimentary and volcanic, present surfaces which are changing so rapidly that their geological age is virtually zero. This would have been true of these surfaces at any time since the inception of the present high tempo of tectonism. The vertical rather than the horizontal components of tectonic displacement are chiefly involved, and these, in New Zealand, have been highly differential in amount and direction. Suggate (1965a) reported uplift rates, based on the deformation of late Pleistocene deposits near the Alpine Fault in Westland, ranging from imperceptible to as much as 10 m/1000 years over the past 13,400 years. Evidence of a different kind confirms that displacement of this magnitude accumulated through the Pleistocene to give a maximum vertical movement of about 14·5 km. The maximum present elevation is, however, only 4000 m, and there is no evidence that the alpine chain was formerly more lofty. Elsewhere in the mountain region, where the recent rate of uplift is only 1·25 m/1000 years, the summits are nevertheless almost half as high as those in the region of greater uplift. It is clear that the relationship between absolute tectonic uplift and surviving erosional relief is far from simple. 'The first thousand feet gain of height may be easy against the forces of erosion, but the tenth thousand feet may require ten times the rate of uplift' (Suggate, 1965a).

Elevation alone undoubtedly adds to the potential energy of the erosional process, though its effects are less immediate than those from tilting and warping which tend to increase gradients. Evidence from warped and tilted surfaces and stream profiles indicates trends of deformation similar to those which accompanied the main, late-Cenozoic period of mountain building in New Zealand and have been represented by a model of elongated crustal segments rotating between 4° and 10°/10⁶ years about longitudinal axes, with more intense deformation occurring locally. Sediment studies amply confirm that a fast tempo of exogenous change has prevailed generally in the most active regions throughout the period, erosion surfaces being continually maintained in a steady-state condition. In contrast, areas least affected by late Cenozoic tectonism include features which may date from pre-Pleistocene periods.

Fossil landscapes

This discussion would be incomplete without mention of resurrected ancient erosional surfaces, with or without fossil weathering profiles, buried under substantial thicknesses of younger rock which have been more or less stripped away. Where the rocks below the unconformity are notably more resistant to erosion than those above, re-emergence can occur with little dissection of the older surface, although some modification must normally accompany both burial and exhumation. The New Zealand 'Late Cretaceous Peneplain' (Benson, 1935) is now widely exposed in the southern part of the country. Its associated deep weathering zone, complete with soil and vegetal remains, survives in the basal parts of the overlying transitional coal beds. Such remains have usually been lost from the completely exhumed areas, but local faulting and the varying depth of Cretaceous weathering has preserved odd pockets. These have been related to the modern occurrence of tors (McCraw, 1965).

While the value of such exhumed surfaces in the reconstruction of the later geological history of appropriate areas may be considerable, their significance in relation to the history of the present landforms may be low, and it may be advisable to regard a resurrected surface as a new feature, originating at the time of exhumation, unless a clear causative link with modern forms can be established.

LANDFORM AGES IN NEW ZEALAND
Tectonic history

Before attempting to apply the principles set out in the first part of this paper to the challenging problem of dating landforms in New Zealand, it is necessary to summarize those aspects of tectonic, sedimentary and volcanic history most relevant to a historical geomorphologic study. The geologic record dates from Cambrian and possibly earlier times; but for the purpose of studying the present landform development, the story may be taken up conveniently from the end of the Rangitata Orogeny, which followed a late-Carboniferous-to-late-Jurassic geosynclinal episode and culminated during early Cretaceous times. Paralleling the history of other parts of the world, the later Cretaceous was marked by profound erosion of the orogenic highlands, followed by a gradual marine flooding across an erosional surface of low relief. Interrupted locally by mild tectonism, the transgression continued until in late Oligocene time a landmass of sub-continental dimensions had dwindled to scattered islands (Fleming, 1962). Fleming has depicted the mobility of the Pacific margin thereafter as 'a kind of writhing' such that a constantly changing island archipelago must be envisaged during the preliminary differential phase of the Kaikoura Orogeny which commenced in the Miocene and accelerated to a climax early in the Pleistocene (Suggate, 1965a).

The Kaikoura Orogeny was mainly responsible for creating the broad framework of present New Zealand relief, which expresses differential uplift and faulting varying in intensity in different regions. The cover of younger Cretaceous and Tertiary strata was broken, folded and uplifted along with the underlying more highly indurated sedimentary rocks and metamorphics, and it was largely stripped away from the rising blocks where the cover was thin and relatively unresistant. Remnants are preserved in downfaulted areas and on the flanks of ranges which now form the backbone of the landmass. The situation was most ably summarized by Cotton (1916) with the phrase 'a concourse of earth blocks'. Detritus eroded from the rising blocks accumulated in adjoining depressions, and the sea was finally expelled. With increasing relief the detritus tended to become coarser, and locally it attained thicknesses of 1000 m or more. Erosional remnants of the conglomerates occur up to higher altitudes and are more deeply weathered than the later Pleistocene deposits formed after the culmination of the orogeny. The earliest glacial deposits were involved along with late Tertiary strata in the orogenic movements, and the nature of the land surface which supported the glaciers is therefore entirely obscure.

Progress of the Kaikoura Orogeny towards its mid-Pleistocene climax was accompanied by climatic changes inferred from the sedimentary record from mid-Tertiary onwards. Erosion at the time of climax was too intense for evidence of glaciations to have survived, but four late Pleistocene glaciations have been claimed from the evidence of post-orogenic deposits and features (Suggate, 1965b).

Glaciation in the North Island was very restricted, and there is no evidence for the existence of glaciers prior to the last (Otira) glaciation. Continual addition of volcanic products from Pliocene time onwards has made it difficult to decipher the history of land surfaces in central districts, though recent years have seen much progress towards a complete tephrachronology for the region.

The present coastal outline of both the larger islands is of comparatively recent date. Certain coastal districts of what is now the North Island were still submerged and accumulating marine sediments until a relatively late date in the Pleistocene. These deposits have afforded valuable faunal evidence of climatic fluctuations. Nearly all coasts show terraces reflecting sealevel fluctuations in the late Pleistocene, but interpretation of these is difficult in the tectonically active areas.

Age of materials

The accompanying maps (Figs 1 and 2) generalize information about the age of materials underlying the land surfaces of New Zealand, indicating broadly the earliest date from which their development could have begun in different regions. They are highly generalized because of scale limitations, showing only the age of the materials bearing the dominant landforms and omitting smaller areas of younger or older rock which would be significant in a more detailed study. Also shown are the boundaries of the chief zone of Late Cenozoic and continuing vigorous tectonism (reflected in geologically recent elevation, active faults, warped and tilted Late Pleistocene surfaces, steeply dipping Late Cenozoic strata and current seismicity) and the area containing

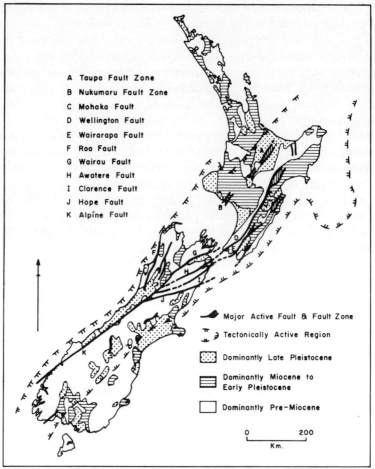

FIGURE 1. Generalized map showing (a) age of materials, distinguishing areas dominated at the surface by pre-Miocene rocks, Miocene to Early Pleistocene rocks (co-eval with the crescendo of the Kaikoura Orogenic Phase), and Late Pleistocene and younger rocks (post-dating the Kaikoura climax); (b) outline of tectonically active region, encompassing areas of steeply-dipping Miocene–Early Pleistocene rocks strongly affected by Kaikoura deformation, tectonically active areas with many recently active fault-traces, or warped and tilted Late Pleistocene surfaces, and major active faults and fault-zones

extensive highland above 1000 m altitude together with the general extent of Late Pleistocene glaciation. This information is sufficient to determine where surfaces must generally be significantly younger than the underlying materials, or of negligible age. On a more detailed scale other factors locally affecting the vigour of degrading processes could be brought in.

The age categories in Figure 1 were chosen for the following reasons. The beginning of the Miocene Period was a turning point in the history of the region because it coincided with the inception of the orogenic phase of Kaikoura diastrophism. Seas which until then covered virtually all the present land area, apart from discontinuous small islands, began to withdraw. Land areas increased, though the present outline was not assumed until quite recently. As the land emerged, sedimentary rocks and volcanics continued to accumulate locally to thicknesses of thousands of metres, but within a framework of depositional troughs determined by relative elevation and depression of crustal segments as the Kaikoura movements gained momentum. Culmination of tectonism in mid-Pleistocene times has been confirmed (Suggate, 1965a) so that within the active

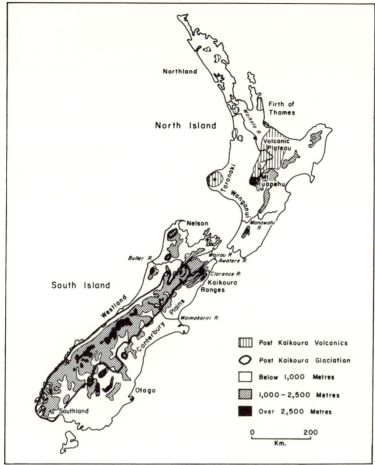

FIGURE 2. Locality map showing relief and major drainage pattern; limits of Late Pleistocene (post-Kaikoura) valley glaciation; larger areas of post-Kaikoura volcanism

tectonic zone it is only in the areas dominated by Late Pleistocene deposition that we find substantial areas of constructional surface comparable in age to the underlying materials.

Age of surfaces

Outside the active tectonic zone, survival of Miocene or older surfaces is possible within the limits of pre-Miocene rocks, but since so little of this region can have escaped the Early Cenozoic marine transgression these cannot be extensive. Pre-Miocene terrains cannot be shown on the present maps, but include areas of exhumed Cretaceous erosion surface, extensive in parts of the South Island. Possible examples of original constructional surfaces of at least Pliocene age may be expected to have survived, more or less undissected, in parts of Otago and Southland marginal to active regions during the Kaikoura crescendo, where considerable thicknesses of terrestrial gravels and sands accumulated. Rocks of corresponding age in the stable northern part of the North Island are thinner, more easily eroded, and apart from some basaltic lava areas, land surfaces are probably everywhere erosional. Various authors have suggested early Pleistocene or greater age for extensive upland surfaces there (e.g. Turner and Bartrum, 1929; Brothers,

1954), while Schofield (1958) is among those emphasizing the crustal quiescence of the region since the Miocene.

Younger surfaces in Northland occur on Late Pleistocene and younger deposits, but apart from some fine examples of little dissected volcanic mountains and basalt lava fields, the deposits are generally unconsolidated and easily eroded, and elucidation of landscape history correspondingly difficult.

Dating of surfaces within the active zone encounters diverse and complex problems, of the kind discussed in the first part of this paper. It is unlikely that any parts can be ancient, unless there are scraps of recently uncovered Cretaceous surfaces surviving in areas of favourable lithology. Over much of this zone there may be no correlation between age of surface and age of underlying rock, and consequently the dating of superficial deposits, such as loess, and of soils assumes greater importance in establishing a minimum age for the surfaces.

It deserves some emphasis that the generalizations in the previous paragraphs apply to roughly 27 per cent of the present land area. That in which surfaces as old as Miocene are theoretically possible amounts to only 14 per cent, while the actual proportion is likely to be even smaller by an unknown amount because of the high probability that the mid-Cenozoic marine transgression extended beyond the areas in which sedimentary evidence has survived.

Within the active zone, geologically recent flood-plain and low-level terraces composed dominantly of gravels penetrate along river valleys far into the mountains and occupy the floors of intermontane basins. Marine terraces of middle or late Pleistocene age flank the central mountains of both islands, as in the Wanganui district (Fleming, 1953) and on the west coast of the South Island, where their relationships with suites of glacial outwash gravels have been worked out by Suggate (1965b). The largest continuous area of flat land in New Zealand, the Canterbury Plains, is largely a constructional surface built up during successive advances of glaciers down the major mountain valleys during the Late Pleistocene. Other lowland areas are physiographically similar, and although their histories differ in detail, the climatic factor has been important in all cases. Surfaces are predominantly constructional and contemporaneous with fluvial deposition of the underlying gravels. Local areas of erosional surface reflect degradational interludes between successive periods of gravel deposition. Precise dating of these surfaces is not always possible, but loess, tephra and soil sequences are proving valuable for this purpose. Early Pleistocene conglomerates, e.g. the Moutere Gravels of the Nelson district, are truncated by older surfaces that have undergone considerable dissection but still retain traces of their original form. A deeply dissected upland surface cutting across tilted Pliocene-to-Early-Pleistocene marine strata throughout Taranaki and the Wanganui district has been described as a peneplain by several authors. Its mid-Pleistocene age and its erosional history are discussed by Fleming (1953).

Initial surface

It will be clear from the foregoing that the ideal concept of 'initial surface' from which modern erosional forms have been developed is inappropriate for New Zealand geomorphology, unless one can accept as such the extensive surface of low relief resulting from Cretaceous erosion. However, in most parts of the country this surface remains either as a plane of unconformable contact within rock sequences, or as a 'ghost' surface to be dotted-in far above the ground profile in geologic cross-section diagrams. Mention has already been made of upland surfaces of easy relief, variously deformed and more or less dissected, and mainly to be found in interior parts of Otago and Southland. In places their aspect is that of rolling downland, but they may stand 1000 m or more above the general level of modern river valleys. It is generally accepted that these surfaces are uplifted, exhumed remnants of fossil peneplain. Outliers of formerly continuous Tertiary cover rest on deeply weathered greywacke or schist basement. Widespread development of tors points to a history of prolonged weathering and stripping of the ancient regolith (McCraw, 1965). Since the date and duration of the period of exhumation is usually uncertain, so also is the length of time since the resurrected Cretaceous landscape began to experience the erosional regimes of later geologic times. These surfaces are thus unsuitable for reference purposes for other reasons beside their absence from many areas, and unfortunately no alternatives older than the post-Kaikoura constructional surfaces of Late Pleistocene alluvium, volcanics, etc., are available.

The problem of identifying and reconstructing an initial reference surface is acute in the more northerly mountains of the South Island and the southern end of the North Island. Remnants of the Cretaceous peneplain are rarely obvious in these regions because the covering strata were generally thick, and scarcely less resistant than the undermass rock. A hint of some former enveloping surface is given by small flattish areas on mountain summits, and particularly by a strong visual impression of summit height accordance throughout the Southern Alps.

This impression was given some substance by summit-height contouring (Wellman and Willett, 1943), while Wellman (1948) demonstrated intriguing relationships between summit-profile contours and the major river pattern. Benson (1935) and others from time to time postulated a Late Tertiary erosion surface overstretching the entire mountain region, but on the whole the suggestion has received little support. Since strongly deformed Late Tertiary and Early Pleistocene strata would be truncated by the supposed erosion surface along with older, harder rocks, little time is available for the necessary peneplanation of early Kaikoura mountains before the final uparching, which in turn must have been well advanced before the beginning of the earliest of several Late Pleistocene glaciations.

History of river patterns

Younger rock cover undoubtedly was more extensive when most of the major river courses established themselves, in consequent fashion, more or less contemporaneously with the movements that blocked out the present mountain ranges. Only in the relatively stable regions well outside the active tectonic zone could pre-orogenic drainage patterns conceivably have survived, and no certain examples are known. In some areas, denudation of Tertiary cover was accompanied by the imposition of structural control, expressing the attitudes and variable resistance to erosion of strata. This led to deflection of river courses, capture, increasing emphasis of subsequent elements, and the development of relief features due to differential erosion (e.g. Jobberns, 1937; Gage, 1950).

The consequences of this phase include the initiation of superposed gorges which still conduct major rivers through and across the axes of mountain ranges, e.g. the Manawatu Gorge (Lillie, 1953), Buller Gorge, and probably several others, although in some cases there is evidence that the axes have continued to rise, so that in part these gorges are antecedent. On a smaller scale, there are many examples of superposition through unusually thick Late Pleistocene glacial outwash.

The valleys of the Wairau and Awatere rivers, which conduct drainage from the very mountainous interior of the northern part of the South Island, are aligned chiefly along active faults, while the Clarence follows a remarkably tortuous path through fault-angle depressions amid a jumble of young tectonic blocks. The history of the present pattern is thus closely linked with a vigorous tectonic history, and accordingly it cannot be traced back very far. Upper and middle reaches of the Waimakariri and other rivers farther south must also have developed through linkage of a series of intermontane depressions, but the present pattern reflects considerable simplification by Late Pleistocene glacial processes.

Besides the Manawatu river, North Island river systems in general present more confused patterns than those of the South. A major part of the reason for this is evident from inspection of Figure 1, which shows a less well-ordered distribution of contrasting rock terrains, especially in the northern half of the North Island. A well-defined axial watershed determined by young tectonic ranges is present only in the southern half. Generally unresistant Late Tertiary formations covering wide areas on the flanks of these greywacke ranges and surrounding the central Volcanic Plateau provide weaker structural control of stream patterns than is the case in similar terrains in the South Island. Further complications result from the extensive Late Pleistocene ignimbrite and ashfall deposits covering much of the central highlands and beyond, resting in places upon erosional land surfaces of various ages and in others upon upwarped surfaces of marine planation dating from the Late Pleistocene. Late Pleistocene volcanic mountains dominate the landscape from Mt Ruapehu northwards into the North Auckland peninsula. Curious problems of drainage evolution have resulted, among them the comparatively recent diversion of the Waikato river towards the west coast from a former course to the Firth of Thames. At the same time, repeated ash showers continuing through the period of Polynesian legend into the colonial era (Tarawera eruption, 1886) are receiving much attention, permitting progress towards the sound tephrachronology required for a well-documented account of the later geomorphic evolution of the central North Island region.

ACKNOWLEDGEMENTS

The information generalized in Figs 1 and 2 has been drawn from many sources, including *Geological Map of New Zealand 1 : 2,000,000* (N.Z. Geological Survey, 1958), *N.Z. 1 : 2,000,000* (NZMS84, Lands and Survey Department) and published papers listed in the References. The figures were redrawn for production by Mrs P. Oliver. This assistance is gratefully acknowledged by the authors.

REFERENCES

BENSON, W. N. (1935) 'Some landforms in southern New Zealand', *Aust. Geogr.* 2 (7), 3–23

BROTHERS, R. N. (1954) 'The relative Pleistocene chronology of the South Kaipara District, New Zealand', *Trans. R. Soc. N.Z.* 82 (3), 677–94

COTTON, C. A. (1916) 'The structure and later geological history of New Zealand', *Geol. Mag.* 3, 243–9 and 314–20

DICKENSON, G. E. and R. D. ADAMS (1967) 'A statistical survey of earthquakes in the main seismic region of New Zealand. Part 3—Geographical distribution', *N.Z. Jl Geol. Geophys.* 10 (4), 1040–50

EIBY, C. A. (1964) 'The New Zealand Sub-crustal rift', *N.Z. Jl Geol. Geophys.* 7 (1), 109–33

FLEMING, C. A. (1953) 'The geology of Wanganui Subdivision', *Bull. geol. Surv. N.Z.* 52

FLEMING, C. A. (1962) 'New Zealand Biogeography: a palaeontologist's approach', *Tuatara* 10, 53–108

GAGE, M. (1950) 'Stream patterns in the Greymouth District', *Trans. R. Soc. N.Z.* 78 (4), 418–25

JOBBERNS, G. (1937) 'The Lower Waipara Gorge', *Trans. R. Soc. N.Z.* 67 (2), 125–32

LILLIE, A. R. (1953) 'The geology of the Dannevirke Subdivision', *Bull. geol. Surv. N.Z.* 46

McCRAW, J. D. (1965) 'Landscapes of Central Otago' in R. G. LISTER and R. P. HARGREAVES (eds) *Central Otago*, New Zealand Geographical Society, Christchurch, N.Z.

ÖPIK, A. A. (1958) 'The geology of Canberra City District', *Bull. Bur. Miner. Resour. Geol. Geophys. Aust.* 32

SCHOFIELD, J. C. (1958) 'Pliocene shell beds south of the Manakau Harbour', *N.Z. Jl Geol. Geophys.* 1 (2), 247–55

SUGGATE, R. P. (1963) 'The Alpine Fault', *Trans. R. Soc. N.Z. (Geology)* 2 (7), 105–29

SUGGATE, R. P. (1965a) 'The tempo of events in New Zealand geological history', *N.Z. Jl Geol. Geophys.* 8 (6), 1139–48

SUGGATE, R. P. (1965b) 'Late Pleistocene geology of the northern part of the South Island, New Zealand', *Bull. geol. Surv. N.Z.* 77

TURNER, F. J. and J. A. BARTRUM (1929) 'The geology of the Takapuna-Silverdale District, Waitemata County, Auckland', *Trans. N.Z. Inst.* 59 (4), 864–902

WELLMAN, H. W. and R. W. WILLETT (1942) 'The geology of the West Coast from Abut Head to Milford Sound—Part 1', *Trans. R. Soc. N.Z.* 71, 282–306

WELLMAN, H. W. (1948) 'Tararua Range summit height accordance', *N.Z. Jl Sci. Technol.* 30 (22B), 123–7

RÉSUMÉ. *Les âges des surfaces du terrain ainsi qu'on les trouve à la Nouvelle-Zélande.* Pour des entières surfaces du terrain, des parties des surfaces ou des assemblages particuliers des formes du terrain, un exposé complet d'âge, ce qui n'est pas toujours possible, devrait définir la date la plus tôt, à partir de laquelle le développement de la surface actuelle pouvait avoir commencé, ainsi que le moment depuis lequel modification est négligeable. Pour les surfaces d'érosion et de construction il peut y avoir d'incertitude quant à l'âge du matériel en dessous—par combien les surfaces sont les plus jeunes. On peut dater les surfaces plus âgées du moment auquel la forme générale actuelle a été atteinte ou il peut être préférable de les dater par le moyen des modifications subséquentes, telles que celles dûes aux changes climatiques du Pléistocène. En dedans des régions tectoniquement actives, les surfaces érondantes soumises à dégradation prolongée et rapide, et à renouvellement continuel, et également les surfaces de dépôt continuellement en construction, peuvent être considérées comme d'un âge en pratique de zéro, ou alternativement elles pouvaient dater du commencement de la situation dynamique actuelle. On expose à grands traits les aspects utiles de l'histoire tectonique de la Nouvelle-Zélande et on les applique au problème d'assinger des dates aux formes du terrain. On ne peut reconnaître les surfaces amorcées avant le Pléistocène Ultérieur qu'en dehors de la bande principale fortement affectée par les 'Mouvements de Kaikoura' du Cénozoïque Ultérieur. Même dans les régions plus stables, la possibilité que les surfaces actuelles eussent développé sans interruption depuis avant les époques du Pliocène est tres limitée, mais il y en a des étendues importantes de la surface cretacée d'érosion qui ont été ressuscitées dans le sud du pays. On passe l'histoire de certains grands systèmes fluvials en revue. L'activité volcanique étendue du Pléistocène dans l'Ile Nord a compliqué le tableau mais l'avancement vers und tephrachronologie complète aide-t-on maintenant avec la documentation de la suite de l'histoire géomorphique dans les régions de l'Ile Nord centrale.

FIG. 1. Karte généralisée qui indique (a) l'âge des matériaux, en distinguant les zones dominées à la surface par roches du pré-Miocène, roches Miocènes jusqu'à la première époque du Pléistocène (contemporain du crescendo de la Phase Kaikoura Orogénique) et roches des dernières époques du Pléistocène et celles plus jeunes (datées après le comble Kaikoura); (b) profil de la région tectoniquement active qui entoure des zones de roches du Miocène jusqu'à la première époque du Pléistocène raidement plongeantes et fortement affectées par la déformation Kaikoura, des zones tectoniquement actives à plusieurs traces de faille récemment actives ou des surfaces des dernières époques du Pléistocène déformées et inclinées, et des majeures failles actives et zones de failles

FIG. 2. Karte d'emplacement indiquant le relief et le majeur réseau hydrographique; limites de la glaciation des dernières époques du Pléistocène (post-Kaikoura) des vallées; zones plus grandes du volcanisme post-Kaikoura

---

ZUSAMMENFASSUNG. *Die Landoberflächenalter, besonders bezüglich Neuseelands.* Für ganze Landoberflächen, Teile der Oberflächen oder besondere Versammlungen von Landormen, sollte eine vollständige Altersangabe, welche nicht immer möglich ist, das früheste Datum, an dem die Entwicklung der gegenwärtigen Oberfläche begonnen haben könnte, festlegen, wie auch die Zeit seit der Modifikation geringfügig ist. Es könnte unsicher sein, weiviel beide Abtragungs- und Aufbauflächen jünger sind als das Material darunter. Man kann ältere Oberflächen von der Zeit, als sie die gegenwärtige allgemeine Form erreichten, datieren, oder es möge besser sein, von späteren Modifikationen zu datieren, wie solche infolge klimatischer Veränderungen des Pleistozäns. Innerhalb tektonish tätiger Gebiete mag man die Abtragungsflächen, die der anhaltenden schnellen Verwitterung und dauernder Erneuerung unterworfen gewesen sind, und gleichfalls die immer wieder aufgebauten Ablagerungsflächen, als eigentliches Nullalter betrachten, oder könnten sie andernfalls vom Anfang des gegenwärtigen dynamischen Zustandes datieren. Die sachdienlichen Seiten der tektonischen Geschichte Neuseelands werden umgerissen und angewandt auf das Problem der Landform-datierung. Man kann die Oberflächen, die früher als das späte Pleistozän begannen, nur ausserhalb des Hauptgürtels erkennen, der von den späten 'Cenozoic' 'Kaikoura Bewegungen' stark betroffen wurde. Sogar in stabileren Gebieten ist die Möglichkeit, dass die gegenwärtigen Oberflächen sich seit vor der Pliozänzeit unaufhörlich entwickelt haben, sehr beschränkt; es gibt aber bedeutende Strecken einer kreideartigen Ablagerungsfläche in südlichen Teilen des Landes. Die Geschichte bestimmter Hauptfluss-systeme ist nachgeprüft. Weit verbreitete vulkanische Tätigkeit in der Nordinsel hat das Bild kompliziert, aber jetzt hilft der Fortschritt zu einer vollständigen Tephrachronologie dem dokumentarischen Nachweis der späteren geomorphologischen Geschichte in zentralen Nordinsel Gegenden.

ABB. 1. Eine verallgemeinerte Karte zeigt (a) Materialalter, Fonen unterscheidend, die an der Oberfläche von Vor-Miozänfelsen dominiert sind; von Miozäm- bis früh-Pleistozänfelsen (gleichzeitig mit dem Crescendo der Kaikoura Orogenishcen Phase) dominiert; und von spät-Pleistozan und Kaikoura Orogenischen Phase) dominiert; und von spät-Pleistozan und jüngeren Felsen (später also den Kaikoura Höhepunkt datierend) dominiert; (b) Umriss des tektonisch tätigen Gebiets, welches Fonen der steil neigenden Miozän- bis früh-Pleistozänfelsen einschliesst, die von der Kaikoura Verformung stark eingewirkt wurden, tektonisch tätige Fonen mit vielen kürzlich tätigen Verwerfungszeichen oder gekrümmten und gekippten spät-Pleistozän Oberflächen, und bedeutend tätige Verwerfungen und Verwerfungszonen

ABB. 2. Gebietskarte, die Relief und Hauptgewässernetz; Grenzen der Talvergletscherung des spät-Pleistozäns (seit-Kaikoura); grössere Zonen des Post-Kaikoura Vulkanismus; zeigt

III. PALAEOGEOMORPHOLOGY

Introduction

The title of David L. Linton's M.Sc. thesis in the University of London (1930) was 'Contribution to the knowledge of the drainage, morphology and structure of the Wessex region'. The study of the inter-relationships between structure, surface and drainage ran through his research for the whole of his professional life (1932). His interest in drainage evolution was extended to Scotland when he moved there in 1929 through investigations of the rivers Tweed (1933), Clyde (1934), Earn and Tay (1940). When, in 1945, he moved south to Sheffield, characteristically he started to investigate Midlands drainage (1951), and later resumed his study of the drainage of south-east England, especially the Sussex rivers (1956). In later years, his interest in rivers led him to investigate some of their hydrological characteristics (1959). His early research on the land surface of Wessex (1930) was extended, in conjunction with his friend and teacher, S. W. Wooldridge, to the whole of south-east England, and culminated in their seminal study of its structure, surface and drainage (1939). His move to Edinburgh enabled him to pursue a similar theme in Scotland, the problems of whose scenery he discussed in 1951. In his years at Sheffield, this particular interest found expression in his regional studies of landforms in Lincolnshire (1954) and the Sheffield region (1956). Perhaps it is not too fanciful to see his study of geomorphology by the light of the moon (1966 and 1969) as a continuation of the same interest.

The three papers which follow in Part III relate very closely to themes which Linton himself pioneered. First, Professor George from Scotland reviews the broad outlines of the geomorphological evolution of Britain in the light of evidence, much of which was not available at the time of Linton's death.

From London, Christopher Green returns to Wessex to re-examine the validity of Wooldridge and Linton's model, and David Jones re-examines the drainage evolution of south-east England and the influence of the Calabrian transgression upon it.

REFERENCES

D. L. LINTON

1930 'Contribution to the knowledge of the drainage morphology and structure of the Wessex region', Unpubl. M.Sc. thesis, Univ. of London
1932 'The origin of the Wessex rivers', *Scott. geogr. Mag.* 48 (3), 146–66
1933 'The origin of the Tweed drainage system', *Scott. geogr. Mag.* 49 (3), 162–75
1934 'On the former connection between the Clyde and the Tweed', *Scott. geogr. Mag.* 50 (2), 82–92
1939 (with S. W. WOOLDRIDGE) 'Structure, surface and drainage in south-east England', *Trans. Inst. Br. Geogr.* 10, 124 pp.
1940 'Some aspects of the evolution of the rivers Earn and Tay', *Scott. geogr. Mag.* 56 (1), 1–11, and (2) 69–79
1951a 'Midland drainage: some considerations bearing on its origin', *Advmt Sci.* 7 (28), 449–56
1951b 'Problems of Scottish scenery', *Scott. geogr. Mag.* 67 (2) 65–85
1954 'The landforms of Lincolnshire', *Geography* 39 (2), 67–78
1956a 'Geomorphology' (of the Sheffield region), Chap. 2 in *Sheffield and its region* (ed. DAVID L. LINTON), 24–43
1956b 'The Sussex rivers', *Geography* 41 (4), 233–47
1959 'River flow in Great Britain, 1955–56', *Nature, Lond.* 183 (4663), 714–16
1966 'Geomorphology by the light of the moon', *Tijdschr. K. ned. aardrijksk. Genoot.* 83 (3), 249–65
1969 'Lunar landscapes', *Geogrl J.* 136 (3), 344–64

Prologue to a geomorphology of Britain

T. NEVILLE GEORGE

Professor of Geology, University of Glasgow

Revised MS received 24 May 1973

ABSTRACT. The distribution and structure of residual Chalk make the probability remote that present-day landform in Britain is to be explained by reference to a former Cretaceous cover. Eocene and Oligocene rocks and structures are signs of strongly pulsed tectonism during Palaeogene times before the advent of the Miocene 'Alpine' movements. The complementary relations of mainland Britain and flanking submarine troughs promote a Cenozoic denudation chronology that emphasizes a contrast between repeatedly renewed corrugations of landscape and the smoothed Neogene profile. Stepped marine platforms, not greatly deformed, are the characteristic products of Neogene agents: their degradation by rivers provides supplementary evidence of the geologically youthful regional landscape of Britain.

LINTON was a geomorphologist of wide sympathies. To him a geomorphological analysis of the present-day physique of Britain combined two main activities: an interpretation of landform as a product of the sculpturing and moulding of the geological foundation by a variety of processes, and an interpretation as history of the sequence and stages of morphogenetic events leading to the terminus of Holocene times. His own work well illustrates the two aspects of the method. Thus in his discussions of glacial erosion in upland country, notably in the Scottish Highlands, he persuasively argued for the influence of transfluent and diffluent ice in the gouging of discordant trenches, like the Lomond trench, across terrain evolved mainly under fluvial controls; while in his earlier work with Wooldridge and in his later work on drainage and landform he traced the development of much of Britain in a synthesis of stratigraphy, tectonics, and palaeogeography that placed the events of Cenozoic times in a chronological sequence, mainly of denudation, and gave a new evolutionary coherence to an understanding of how present terrain is to be explained by reference to the terms of a long inheritance. Linton's work is exemplary, and is an appropriate base on which to build.

A major premise to be found in Linton's regional synthesis, partly derived from earlier theory, is the central place in an integrative geomorphology accorded to the blanket of Upper Cretaceous rocks that formerly covered Britain. The premise is provisionally to be accepted. The Albian-Cenomanian transgression in north-west Europe was an outstanding event in the stratigraphical obliteration of an earlier physique, that in magnitude and geomorphological importance, as a sign of contemporary erosion and tectonics, compares with the Permo-Triassic and Devonian transgressions of earlier geological periods.

THE UPPER CRETACEOUS DATUM

The ancient foundation (as seen in present land outcrop) on which the greensands and the Chalk rest was widely variegated. In the south-east of England there were only comparatively minor breaks in sedimentation from Jurassic into Upper Cretaceous times, deep basins of Mesozoic rocks, preserved in thicknesses exceeding 2000 m, sagging into the Palaeozoic floor; but elsewhere progressive Cretaceous overstep extinguished all older rocks, in Devon down to low Permian (almost to Carboniferous), on the East-Anglian platform down to Silurian or older, in west Scotland and in Ulster down to Dalradian and Moinian. Supplementary evidence of Upper Cretaceous distribution, provided by recent offshore surveys in the North Sea, in the Western

Approaches and the English Channel, and in the Celtic Sea, is strongly corroborative: it extends the range of submarine outcrops into basins flanking the landmass, whose generalized structural pattern, with added hints in the flint-rich Neogene deposits of Orleigh and Buchan, intensifies an impression that a proto-Britain was surrounded by and drowned beneath the sea, and covered by Chalk, in later Cretaceous times.

The impression is not wholly to be discounted by variant rock-sequence in the outcrops that remain, but there should be some reservation in assuming the original cover to have been unbroken when it is recognized that even in southern England there are strong signs of internal overlap as the Upper Cretaceous rocks are followed westwards into Dorset and Devon (J. M. Hancock, 1969): and that nonsequences, unconformities, and rapid changes in facies of the Chalk in Scotland and Ulster reveal a restless and irregular floor implying contemporary land repeatedly emergent, or oldland remnants standing up as residual islands, especially above a Turonian sea (H. E. Wilson, 1973, Fig. 22). Even so, thick Senonian deposits in Antrim, and a conjectural overlay of Maestrichtian (of which only very small outliers are still preserved in Britain), give some reassurance to a hypothesis of a final drowning and a deposition of a complete Chalk cover in latest-Cretaceous times.

It has been a persistent theme for many decades, strongly supported by Linton, that the Cretaceous cover (or with its progressive removal the exhumed sub-Cretaceous floor) was the prime determinant of geomorphological evolution in Cenozoic times, indeed that British geomorphology finds its zero chronobase, its initial datum, in the interface between Mesozoic and Palaeogene rocks; and that whatever landforms have since evolved in the sub-Cretaceous oldland have in major part been inherited by superimposition from the Cretaceous cover (D. L. Linton, 1951, p. 69; Linton, 1964). The reasons for attaching such importance to a selected geological event are not far to seek when no comparable deposition of thick and widespread sediments is known to have taken place in Cenozoic times, when mid-Cenozoic ('Alpine') deformation has hitherto appeared to be minor except in local belts and centres, and when circumstantial evidence (even if cyclical in its deductive application) of regional plateau form and of repeated drainage pattern, mostly discordant except in secondary adjustment to the geological foundation, appears both to find its explanation in and to reinforce the inference of controls exerted by a Chalk cover, complete or not.

The hypothesis of a dominant place of Cretaceous events in British landform evolution must, nevertheless, be abandoned, despite its traditional respectability and its attractively plausible content—not because Britain lacked an extensive cover of Chalk, but because contrary evidence, even in ground where Chalk is still preserved, is now sufficiently weighty in cumulative consensus to demonstrate otherwise. The events of Palaeogene times, leading to a culmination in 'Miocene' deformation, were far more complex than has generally been supposed: successive surfaces of Palaeogene denudation, and a corresponding denudation chronology, relegate Cretaceous erosion to an incidental place, rarely to be identified, in a sequence of phases of erosion each of which tended to obliterate more or less completely relics of any earlier phase, and which together reflected tectonic pulses no less effective in defacing inherited surfaces of erosion and deposition. Already in very early Cenozoic times the Chalk cover had become greatly mutilated, and its geomorphological influence become diminished to insignificance.

THE EOCENE BASE

The geomorphic effects of a crustal instability seen in the transition from Cretaceous to Eocene times need no emphasis: the Eocene rocks of southern England may be regarded as in fluctuating offlap from an emergent dome of Chalk on which were stencilled the first lines of Palaeogene denudation. The rocks are terrigenes, rhythmically alternating sediments utterly in contrast to the Chalk, restricted to the salient of the Paris basin, and giving the strongest signs, especially in the west and north-west of the Hampshire basin, of lateral passage from marine to estuarine and fluviatile sediments as the emergent dome is approached.

This orthodox version of an emergent landscape is, however, much oversimplified. The tectonism that produced the palaeogeographical contrasts was multiple. It had repeatedly affected Mesozoic sedimentation, and it affected the Upper Cretaceous rocks not only in the discontinuities of the Antrim sequence but also in the warps, locally with accompanying faulting, displayed in the variations and the interrupted sequence of all three divisions of the Chalk in southern England (Hancock, 1969; W. J. Kennedy, 1970, Fig. 30; E. A. Edmonds *et al.*, 1969, p. 73). It reached greater intensity during Danian and Montian interludes when the Chalk, uplifted above the zone of sedimentation, first became widely degraded. Overlap and overstep at the base of the Eocene are spectacular: almost the whole of the Senonian is missing on the crest of the Hertford swell between Maestrichtian outcrops in Norfolk and Dorset (S. W. Wooldridge and D. L. Linton, 1938b, p. 281; D. Curry 1965, Fig. 2, p. 163), and the complementary syncline to the west is rapidly transgressed in Devon, where in residual outliers all the Chalk is extinguished as the Haldon gravels (of mid-Eocene age) descend onto Albian greensands—and very nearly descend onto Carboniferous rocks.

The sub-Eocene folds in the Chalk have a cumulative amplitude exceeding 400 m in the London and Hampshire basins. The accompanying overstep resulted in the complete obliteration of post-Albian rocks in oldland Devon before local Eocene sedimentation began, and it may well have done so on the rising crest of the Weald anticline. In notional extension of structures north and east it is likely to have been greater, perhaps much greater, in oldland Wales and northern England, where the surface in Palaeozoic rocks was presumably exposed to prolonged subaerial (including fluviatile) denudation throughout Eocene times, and was only briefly accommodated, perhaps not at all significantly, to the form of a stripped sub-Cretaceous floor from which almost all vestiges of the Chalk had been removed.

The Eocene rocks of Scotland and Ulster, igneous basalts and intrusives concentrated in half a dozen centres of an independent structural province, were very different in their rough physique of mountainous and high-plateau form, and in their great variability, from the sediments of the gentler environments of southern England, but they followed the Chalk with equal discordance. Thus Maestrichtian beds are preserved in northern Antrim, but elsewhere all the Chalk (to not less than 150 m at minimum) and the greensands were overstepped across folds of corresponding amplitude, the basalts descending onto Lower Palaeozoic rocks (including Dalradian) in Ulster, onto Precambrian Moinian in Scotland (Wilson, 1972, Fig. 23; T. N. George, 1966, Figs 2 and 3; 1967, Fig. 3). To see the ghost of an end-Cretaceous surface in the present landscape is to ignore the plain evidence of these stratigraphical relations and determinedly to impose expectation on inference.

OLIGOCENE TRANSGRESSION

A denudation chronology of Palaeogene times that runs from Eocene into Oligocene events amply confirms the widespread extinction of a Cretaceous heritage. The apparent continuity of Sannoisian deposits on Bartonian (Ludian) in the Hampshire basin, and their being regarded as a sixth or seventh cyclothem in the Palaeogene sequence, have tended to diminish a recognition of the effects of contemporary movement on landscape and foundation; but the Creechbarrow outlier of Bembridge Beds nonsequentially resting on Eocene in Dorset has allusion to the Bovey Tracey deposits that places the whole region of southern England in a tectonic setting of recurrent deformation from latest-Cretaceous to Miocene times.

The importance of the Bovey Beds (taken to be of Oligocene age), and analogously of the

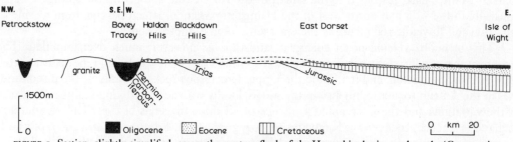

FIGURE 1. Section, slightly simplified, across the western flank of the Hampshire basin: to show the 'Cenomanian transgression' across Mesozoic and Permian rocks; overstep of the diachronous Eocene base across the Upper Cretaceous rocks; and the gross discordance of the basins of Oligocene rocks in Devon, Eocene and Cretaceous rocks extinguished as an indication of locally acute mid-Palaeogene movement

similar lignitic clays and fine sands of the Marland Beds in the Petrockstow basin, is two-fold. First, they are stratigraphically isolated, resting with gross unconformity on Palaeozoic rocks, without sign of Eocene or Chalk beneath. Yet in their nearest outcrops they lie little more than a mile from the Haldon Hills, where Albian Upper Greensand is followed by Lower Chalk (that originally continued upwards into Upper Chalk before Eocene overstep by the Haldon gravels). A palaeogeographical concept of a unitary Palaeogene cuvette in Wessex is therefore inadequate: it may be valid for the Thames–Hampshire terrain, even if it needs to be modified by reference to a growing Weald arch; but it is not to be extended much beyond the Haldon outcrops—the hinterland of the Cornubian uplands appears not to have received sediments of the unitary cuvette in any notable thickness. The hinterland was also completely denuded, by both pre-Eocene and pre-Oligocene erosion, of whatever Cretaceous rocks it may once have carried. The direct evidence thus discourages any hypothesis of the influence of a Chalk cover in the evolution of the Cornubian oldland during mid and later Cenozoic times.

Secondly, the unconformity at the base of the Bovey and Marland beds was not the simple product of a gentle westward rise of the floor on which the Hampshire Tertiaries were deposited, for although there was such a rise, with internal overlap of the lower Eocene beds and with lateral facies changes from marine in the east to estuarine and fluviatile in the west, the contrasts in thickness between the Bovey Beds, which probably exceed 750 m (M. J. R. Fasham, 1971), and their assumed counterparts in the Isle of Wight, at less than 100 m, must be ascribed to Oligocene warping and perhaps faulting. Even in the small Petrockstow pocket, 45 km from the Haldon Hills, the Marland Beds are known to exceed 630 m. The basins in which the sediments accumulated, depressions in an environment of rough physique (as the interbedded gravels and sands reveal), are no doubt in part a product of 'subaerial' and fluviatile erosion, but only in small part: at such an amplitude they compare in order of deformation with the pre-Cenomanian folds of the East-Anglian platform and the Purbeck–Weald cuvette (George, 1962, pp. 200ff.) (Fig. 1).

The residual Oligocene outcrops in southern England are small, and their geomorphological significance has been neglected, but what they indicate is strongly confirmed by the comparable rocks of the Lough Neagh basin in Ulster, where unconformable overstep by the Lough Neagh Clays from horizons high in the Eocene basalt sequence across thin Chalk onto Trias is as revealing as the attitude of the Bovey Beds, and demonstrates even more fully the intensity of pre-Oligocene deformation and erosion (George, 1967, p. 424). As in Devon, the provenance of the detritus in the sediments also provides lithological evidence of origin in a hinterland of Palaeozoic rocks lacking a protective younger cover, the absence of flints being notable.

Palaeogene palaeogeography is yet further amplified by the sequence of rocks and structures proved in the Mochras borehole of Merioneth, which disclosed a wholly unexpected occurrence of Oligocene-Miocene sediments of 'normal' kind, 550 m thick in residual preservation, resting with great discordance on Lias, without sign of younger Mesozoic sediments, particularly without sign of Chalk. In Tremadoc Bay the Mochras beds overstep the Lias north-westwards and descend onto New Red Sandstone over the flank of an anticline of amplitude (in the Mesozoic rocks) of 1200 m. The interpretation put upon the geophysical data got in Cardigan Bay and the southern Irish Sea, although only partly confirmed by drilling, is concordant if not so precise: Cretaceous rocks appear over the greater part to be absent from the Mesozoic sequence beneath unconformable Palaeogene (which may include Eocene and may continue upwards into Miocene); and only as the Irish Sea merges southwards through St. George's Channel into the Celtic Sea does the Chalk reappear in thickness (A. Woodland, 1971; D. J. Blundell *et al.*, 1971, Fig. 14; M. R. Dobson *et al.*, 1973; George, 1974, Figs 95 and 97).

The minute pocket of pipe-clays at Flimston in Pembrokeshire, and the numerous similar pockets not in original position in North Wales (P. Walsh and E. H. Brown, 1971), contain rocks assumed to be Oligocene (although the dating is not certain) that find an appropriate setting as signs of comparably transgressive Palaeogene and perhaps early Neogene sedimentation in oldland terrain. They rest with gross unconformity on Palaeozoic rocks, and point to deep erosion of a degraded post-Mesozoic surface from which, no doubt in pulses, Triassic and Jurassic rocks, and probably Upper Cretaceous, were removed in early Cenozoic times; and they exclude from later geomorphic evolution the influence of Mesozoic events.

MIOCENE DEFORMATION

To lay special emphasis on Miocene earth-movement is to be highly selective in the Cenozoic time-span. The folds and faults of the south of England are impressive in the Cretaceous and Palaeogene rocks, and the generally conformable pattern of structures in the London and Hampshire basins and in the Weald have encouraged an assumption of a Miocene culmination of movement in Britain. The comparable structures recently proved in the floor of the English Channel strengthen the assumption (see Curry *et al.*, 1970; R. G. Dingwall, 1971; D. T. Donovan, 1972). Nevertheless, it is now clear that where in the Oligocene outliers west of the Dorset–Hampshire 'Alpine' folds an intermittent stratigraphy can be determined, pre-Eocene deformation of an order of magnitude measured in hundreds of metres of fold height or fault throw can be shown to be followed by even more intense pre-Oligocene deformation. Moreover, if stratigraphy is carried back into the Mesozoic succession, post-Hercynian restlessness, of which the Cenomanian transgression was perhaps the most spectacular sign, punctuated sedimentation in discontinuities that now make overstep seen in present-day outcrop to be cumulative and long-sustained, and to be most obvious in Miocene elements only through accidents of preservation.

The best-known instances of pulsed tectonism are the East-Anglian platform and the Market Weighton structure in the Mesozoic rocks, the Weald in the Palaeogene; but geophysical survey during the past decade, vastly extending knowledge onto the sea floor around Britain, has widely demonstrated a complexity of tectonic relations of equivalent grade and adds to the undermining of an assumption of a few strong briefly concentrated pulses in orogenic periodicity. In a general sense, a British landmass, partly defined by present-day coastal limits, is a persistent positive block, vaguely horst-like, that is surrounded by deeply subsiding basins in which Mesozoic and Cenozoic rocks have accumulated in relatively great thickness. Thus in the North Sea basin from the Dutch coast to the Shetlands, Cretaceous rocks lie beneath Cenozoic at depths of 2000 to 3000 m, perhaps 4000 m, in slowly growing synclines of corresponding amplitude; comparable troughs in the Minch and the Sea of the Hebrides are also of the order of 2000 m deep in Mesozoic rocks; the trough of the southern Irish Sea and Cardigan Bay may descend

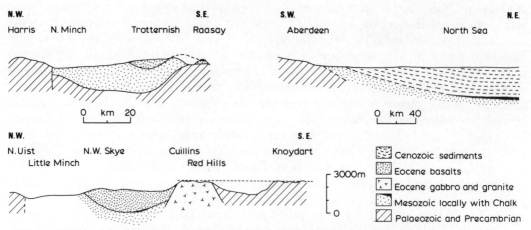

FIGURE 2. Sections, simplified and partly reconstructed, across northern Scotland: to show the relations of oldland terrain to shelf basins; the signs of repeated Cenozoic tectonism; the degradation of Palaeogene landforms; and the congruence of summit levels in residual hills of Eocene, Dalradian, and Precambrian rocks

to 7000 m in post-Carboniferous rocks, of which locally not less than 600 m are Cenozoic. There is only partial indication of the stratigraphical range of sediments in these enormous troughs, and of their onshore links; but in reconstucted profiles of marker beds (including the Chalk) it is possible to see oldland cores in Scotland and Wales as anticlinal warps onto which the Mesozoic and Cenozoic rocks overlapped or overstepped (P. E. Kent, 1967; M. H. P. Bott, 1968; Kent and P. J. Walmsley, 1970, Figs 2 and 10; Bott and A. B. Watts, 1970; R. A. Eden et al., 1971; Blundell et al., 1971; D. K. Smythe et al., 1972; J. Hall and Smythe, 1974; J. A. Chesher et al., 1972) (Fig. 2).

At the same time, as Kent (1949) showed in his synthesis of the sub-Permian floor, troughs like those offshore are delineated inland by the New Red Sandstone and later rocks in England: the eastern Irish Sea basin to 2000 m extending into Lancashire; the Solway basin (in New Red Sandstone and thin Lias only) to 1350 m; the Cheshire basin (also in New Red Sandstone and thin Lias only) to more than 2000 m; the Severn basin to the east of the Malvern line to 1000 m; the extension of the North Sea basin into eastern Yorkshire to 2000 m; the Hampshire basin to 3500 m. Post-Liassic Mesozoic and Cenozoic sediments are absent from these troughs except in the marginal Hampshire basin and the east-Yorkshire sag: if younger rocks are notionally added —it may be supposed that they have been eroded to some considerable thickness—a distinction between structures in the oldland and structures offshore is not absolute, especially when it is recalled that Palaeozoic rocks are just below the Upper Cretaceous surface in East Anglia and extend at shallow depth beneath the North Sea, and crop out on the sea floor between Anglesey and the Isle of Man where they separate the basins of the eastern Irish Sea (with New Red Sandstone to perhaps 5000 m) and the southern Irish Sea. In being cumulative, staggered in their sagging, the basins imply persistent stress over long periods: the energy source for such sustained crustal deformation may well (in terms of fashionable theory) be plate migration—a general easterly drift of the east-Atlantic plate, a northerly drift of the north-European cis-Alpine plate (see A. Hallam, 1972).

The geomorphic implication of the crustal deformation is continual warping —and 'eustatic' becomes a descriptive term applicable to relative sea level only approximately and temporarily, over comparatively small areas. Moreover, the basins in cumulative thickness give the impression of persistence, and the 'oldland' appears to have long been 'old'—detritus constantly pouring

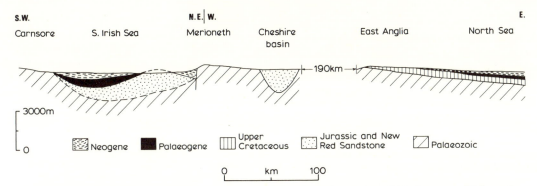

S.W.
N.E.| W.
E.

Carnsore   S. Irish Sea   Merioneth   Cheshire   East Anglia   North Sea
                                            basin

190km

3000m

0

☒ Neogene   ■ Palaeogene   ▥ Upper Cretaceous   ⬚ Jurassic and New Red Sandstone   ▱ Palaeozoic

0   km   100

FIGURE 3. Section, slightly simplified and partly conjectural, from the southern Irish Sea across Wales and England to the North Sea: to show the effects of pulsed subsidence of the offshore basins in relation to the positive mainland mass; the landforms arising from repeated marine overstep; and the cumulative product of tectonic waves on an emergent Neogene landform

into the troughs, and denudation, as constantly revived, paring down the land. The appearance, however, is deceptive in being over-simplified: basin floors repeatedly emerged, uplands sank, in alternations of movement. There is ample evidence in basinal rock-succession of non-sequence, unconformity, upfaulting, and overstep. Terrestrial Eocene basalts in the heart of the Minch depression are corrugated by post-Eocene tectonism that took the rocks 1500 m below sea level in synclines contiguous with anticlines of amplitude exceeding 2500 m. The 'Miocene' Llanbedr fault on the landward flank of the Cardigan Bay basin has a local throw of more than 800 m in Oligocene-Miocene (George, 1974). Unconformable Miocene sediments in the Western Approaches, Oligocene absent, rest on Eocene and probably overstep onto Cretaceous (Curry *et al.*, 1970, Figs 1 and 4). The 'internal' Cheshire and Severn basins retain no rocks younger than Jurassic, perhaps only because they are not so deeply troughed into the Palaeozoic floor, or lie at higher altitudes and deeper erosion levels than the marine basins; and the narrow faulted syncline of the Bristol Channel, without Cretaceous or Cenozoic rocks, adds indirectly to the evidence (M. Brooks and M. S. Thompson, 1973, Figs 8 and 11; A. J. Lloyd *et al.*, 1973).

Miocene earth-movement in Britain takes its place in the structural series as no more than a rather spectacular episode in southern England, where its effects are commonly assumed to be marginal elements in 'Alpine' orogeny (the migrating north-European plate having its energies rapidly dissipated beyond the Hampshire basin and the eastern English Channel). The very gentle dips along the north-western limits of the Chalk, in the Chiltern and neighbouring escarpments, are correspondingly 'explained', and the inference supported of insignificant dips in the former Chalk cover that has been removed from ground farther north. The inference cannot be carried very far, however, for other energy sources, perhaps in the drifting Atlantic plate, were responsible for the large-scale faulting of the broken Tremadoc Bay syncline, and for the greater deformation of the Lough Neagh basin the amplitude of which is about 1000 m in the Oligocene rocks, about 1500 m in the Cretaceous (George, 1967, Fig. 6). The Ulster folds compare in magnitude with those of the Hampshire basin; the faults are larger.

The signs are as yet few and far apart, indirect where Oligocene rocks do not form a datum or are not surely dated, uncertain where (as in the North Sea basin) a thick Cenozoic succession as yet lacks stratigraphical refinement; but the signs are impressive in suggesting allusively the occurrence of similar concentrated folding in the 'inland' Cheshire, Solway, and perhaps Midland Valley basins; and they leave little doubt of the highly irregular physique of much of Britain, far north of the south-English 'Alpine front', that resulted from 'Miocene' deformation (Fig. 3).

MIOCENE-PLIOCENE TRANSITION

There are no known land deposits of Miocene age in Britain, except perhaps in the Brassington Formation and in the upper part of the Mochras clays; but submarine Miocene in a synclinal pocket has been identified in the Western Approaches, and it probably forms the lower part of the thick Neogene sequence in the southern Irish Sea, as it does of the thicker Neogene sequence in the North Sea. Pliocene overstep is known to obliterate Miocene rocks on the Norfolk offshore shelf (Kent and Walmsley, 1970, Fig. 2), and their absence from mainland southern Britain may equally be attributed to Pliocene transgression, perhaps through combined overstep and overlap. That Stampian rocks are the youngest preserved in the Hampshire basin is no proof that formerly they were not followed by Aquitanian and Burdigalian. Conjecture does not lead far, but it is cautionary in suggesting that a prolonged interval of subaerial erosion, during which the 'Miocene summit plain' was eroded at (present) heights between 230 and 280 m, is not to be placed too confidently between the closure of the 'Alpine' episode of folding and the first deposits of the Pliocene sequence.

The evidence of the Miocene-Pliocene Brassington Formation in Derbyshire (M. C. Boulter *et al.*, 1971; Walsh *et al.*, 1972; T. D. Ford, 1972) leads to the inference of a mid-Cenozoic surface in Carboniferous terrain, now lying at heights (projected from sands resting in solution cavities below the level of original deposition) up to 480 m. Uplift of equivalent magnitude, exceeding the uplift of the southern 'Miocene summit plain', and the proximity of a contemporary sea, are to be inferred. The non-marine sediments of the Formation have analogy with those of Merioneth and Powys—sediments that (on the floral dating) together suggest like environments of deposition continuing from Oligocene into Pliocene times. The mid-Cenozoic surface on which the sediments accumulated, perhaps peneplaned perhaps wave-cut but evolving, may have extended over much of Britain: conceivably it included remnants of planed surfaces at comparable heights elsewhere, notably the 'Middle Peneplain' identified by Brown in Wales, but an altimetric correlation of such early geomorphic elements is not easily accommodated in a tectonic history that includes strong mid-Cenozoic deformation.

Thick Pliocene rocks almost certainly occur in the Neogene succession of the southern Irish Sea. They are known to be of the order of 300 m off the Norfolk coast. In general relations they follow the pattern of rock-accumulation, persistent from Mesozoic times, and of pulsed subsidence of sedimentary basins, of which the margins displayed wide lateral oscillations as the pulses alternated between deepening and emergence. The oldest of the Norfolk crags, the Gedgravian Coralline Crag, is very late Pliocene in age; its overstep from London Clay onto Chalk is a local sign of the restlessness in Neogene times of the North Sea margin (B. M. Funnell, 1961). If it is accepted as being younger than the Diestian Lenham Beds (and comparable deposits of the London Basin), it may be regarded as overlapping northwards from a continuing London embayment; and derivatively a diachronous Pliocene base may be mapped as descending beneath the floor of the southern North Sea to depths of not less than 600 m (Wooldridge and Linton, 1955, Fig. 11; Kent and Walmsley, 1970, Fig. 2; Walsh *et al.*, 1972, Fig. 6; R. G. West, 1972, Fig. 3). The known range of altitude of the Pliocene rocks in south-eastern England onshore and offshore is thus of the order of 800 m. Comparably, Neogene rocks, some almost certainly Pliocene, descend to depths exceeding 200 m in the southern Irish Sea. Conversely, they appear not to be present in the eastern Irish Sea Basin, or the Bristol Channel, or to any great thickness in the English Channel. The persistent deformation characteristic of the whole British area during Cenozoic times, in warps and swells where it was not more violent, thus continued from Miocene into Pliocene and (as seen in the post-Gedgravian crags) Pleistocene times, and a tectogenetic contrast between Palaeogene and Neogene events is to be seen only in relative terms.

NEOGENE GEOMORPHY

The effects of continual movement over the greater part of 'inland' Britain during Pliocene times are not to be identified by reference to a recognizable stratigraphy but are to be inferred from stages of geomorphic evolution. The one guide is the highly instructive synthesis of landforms and minute residual outliers that allowed Wooldridge and Linton (1938a; 1955, pp. 45ff.) to plot in southern England a shoreline and marine shelf of Pliocene age, biting into the Chalk dip-slopes and indifferent to underlying structure, and thus to 'explain' the drainage pattern as one discordant to structure superimposed from the shelf onto the Miocene folds. The altitude to which the shelf rises is about 220 m: above it the long-maturing 'Miocene summit plain' is to be discerned.

The synthetic method, with less certainty, can be applied to other areas of Britain, where remnant Pliocene outliers are not to be found. Rivers are equally spectacular in their indifference to structure almost everywhere, and at almost any altitude, in 'Alpine' or Hercynian or Caledonian Britain, in a simplicity of superimposition that matches the superimposition of the Weald and Hampshire rivers across Chalk and Palaeogene. Only on the supposition that 'inland' Britain has suffered insignificant deformation during Palaeogene and 'Alpine' times—a supposition belied where circumstantial tests are possible—can it be inferred that the simplicity of the superimposition is pre-Neogene. The late date of secondary adjustment of drainage pattern to structure in upland country is also a positive sign of Neogene, in later stages of Pleistocene, evolution (George, 1974, p. 368).

The composite Pliocene platforms of the south-east of England may be followed, little deformed, westwards into Devon and Cornwall, across the Channel into France, into Wales, and perhaps into Cumbria. Correlated with these indications of pulsed uplift the valley profiles of the linked rivers show repeated evidence of rejuvenation in knickpoints and stepped terraces, some of the major marks of ancient base-levels (at conventionally emphasized '200-foot', '400-foot' and '600-foot' stands), suggesting uplift with little warping and being traceable for some hundreds of kilometres (Wooldridge, 1950; B. W. Sparks, 1953; Wooldridge, 1954; Brown, 1960; C. Kidson, 1961; Linton, 1964).

Higher platforms, typified for instance by Brown's 'Low Peneplain', 'Middle Peneplain', and 'High Plateau' that rise to or above 600 m in much of upland Wales, are less crisply preserved, correlation between their remnants is less certain, and their mode and date of origin are less surely known. In Scotland high-level elements of comparable platforms give every appearance of extending from Precambrian terrain across the truncated mountain-tops of unroofed Eocene intrusions in the Hebridean islands, and then to be without doubt post-Eocene. At intermediate levels fragmented platforms in Ulster interpolated across the depression of the Lough Neagh syncline link heights in Eocene basalts and Dalradian rocks, and therefore are post-'Alpine'; and similar relations of geomorphic elements span Tremadoc Bay from Lleyn to Merioneth. A widespread integration of the many fragments leads with a high degree of confidence to the conclusion that the plateau form of upland Britain, to heights above 950–1000 m (above 1200 m in Scotland), is in its major geomorphic constituents a product of Neogene erosion, young in a geological sense, and that its composite elements descend in unbroken geomorphic sequence to the 'coastal plateau' (up to about 200 m), recognizably Pliocene in southern England. The higher reaches of the rivers, superimposed on the upland plateau, strengthen the impression of continually pulsed regrading in their long profiles, interrupted by knickpoints, as they descend to lower reaches (Brown, 1960; J. T. Parry, 1960; George, 1966, Figs 6 and 7; 1967, pp. 43ff.).

The origin of the stepped upland plateau (the margins of which are locally steep coastal cliffs) is a source of controversy. Because of the complexity of intervening events, the planed form can no longer be given a distant ancestry in Cretaceous times; but if in traditional terms it remained Cretaceous, there would be no hesitation in ascribing it to marine planation across discordantly underlying geological structure. A reluctance to accept a marine origin in post-'Alpine' times is thus prejudice, partly reflecting a discomfort in imagining high sea levels during the brief 'Pliocene' interval. The alternative, of postulating subaerial 'peneplanation' (unspecified in detail as a process), implies a sequence of scarp retreats, uncontrolled by geological structure, to explain the recognizable steps in the composite upland plateau—whereas the same kinds of steps at lower altitudes are readily understood as a sign of pulsed emergence of the land from beneath the sea during 'Pliocene' times. Moreover, stepped subaerial peneplains, contained not as upland 'basins' but only by the coast, imply corresponding changes in 'base level' scarcely different as measures of changing relative altitude from stepped marine platforms (Brown, 1957; 1960).

The scale of deformation of Pliocene deposits in the troughs flanking mainland Britain—in amplitudes exceeding 800 m in the southern North Sea, in probably greater amplitudes in the northern North Sea, in uncertain but comparable amplitudes in the southern Irish Sea—is commensurate with a Neogene uplift of the upland plains in Britain; and there thus does not appear to be an intrinsic tectonic reason for invalidating a general theory, embracing a multiplicity of geomorphic features, that would for the greater part attribute landscape in Britain to a Neogene origin, its profile carved in a geological foundation revealing a complexity of events in Palaeogene and 'Alpine' stages on a Cretaceous base.

The Pliocene sediments deep in the troughs discourage any postulate that the Neogene emergence of upland Britain, as subaerial peneplains or marine platforms, was a product of eustasy; and although there may have been block uplift (as the Mochras and perhaps the west-Scottish structures indicate), it is likely that there was some warping also, which would contribute to a general difficulty in identifying specific denudation levels at higher altitudes. Such warping would add doubt to the altimetric correlation of platforms except over comparatively small areas and would remove all confidence in extending British scales across the Atlantic or to the Mediterranean or even to continental north-west Europe. Nevertheless, the repeated evidence of concordant summit levels in a variety of kinds of geological terrain suggests that the scale of the warping was not so great as to obliterate altogether identifiable platform-units; and in longitudinal belts of no great width, as along the west coast of Scotland, and in more extensive inland tracts, as over much of Wales, an approximate correlation of some of the platform-units is convincing (Brown, 1961).

As latest-Cenozoic stages in the deformation of an evolving terrain, the Quaternary changes in sea level are not to be isolated from the effects of ice-load. Whether the lowest members of the composite 'coastal plateau' are technically Pleistocene or not, they appear everywhere in Britain to be pre-glacial (in a local application of that term). The 'raised beaches' and their submarine counterparts, on the other hand, clearly are not, and in time-sequence with the crags are signs of inter-glacial movement. Their correlation, and the measure they give of relative sea level, are still sources of controversy (J. B. Sissons, 1967, pp. 147ff.; West, 1968, pp. 246ff.; G. F. Mitchell, 1972, p. 182), not least because of a local lack of stratigraphical evidence; but the tilting of some of the highest beaches in Scotland, in a fall from heights exceeding 35 m, reflects differential late-glacial rebound and offers some explanation of stepped coastal form; and the post-glacial 'Neolithic' beach in northern Britain may be similarly explained. The uniformity in level of some of the inter-glacial beaches in southern Britain from South Wales to Sussex, and their derived correlation as Hoxnian or Ipswichian, are less easily understood, except on the assumption of a factor, presumably tectonic, operative beyond the limits of glaciation—some of them appear to have their counterparts in northern France. Conversely, the pervasive post-glacial Flandrian transgression with its submerged peats is an orthodox example of eustasy, directly attributable in greatest part to a world-wide rise in sea level (West, 1972, Figs 1 and 6). Together, Quaternary events, though relatively minor, show themselves in their geomorphic effects not to be distinguished in kind from earlier movements, but to point to a continuity of the phases of deformation (including some faulting) recognized in Pliocene and earlier Cenozoic evolution.

A CENOZOIC GEOMORPHOLOGY

A history of the main post-Cretaceous geomorphic events—generalized, for there is much variation in local detail—may be integrated in the following sequence:

(1) A widespread erosion of the Chalk cover, in places completely, before the accumulation of the unconformably overlying Eocene rocks; a transformation of the geomorphological profile, notably in Scotland and Ulster but also in North Wales, by volcanic and intrusive igneous action; and the evolution of strongly corrugated landforms.

(2) Mid-Palaeogene folding and faulting, in amplitude and displacement of the order of hundreds of metres, with accompanying deep erosion, demonstrated by the transgressive interface between Oligocene and older rocks in Devon, in Merioneth, and in Ulster.

(3) Widespread Oligocene deposition marine in the south in conformity with the Eocene rocks beneath, non-marine elsewhere unconformably on a varied floor, perhaps in localized basins, locally continuing intermittently into Miocene and Pliocene, the floor perhaps still being recognizable as remnant outliers at appropriate heights.

(4) Post-Oligocene mid-Cenozoic ('Alpine') movements, continuing in the same order of magnitude those of Palaeogene times, illustrated in the widespread deformation of the Oligocene rocks (as in Tremadoc Bay and along the Llanbedr fault, and in the Lough Neagh basin), and implying variegated 'Miocene' geomorphological surfaces mostly not to be recognized in the present landscape.

(5) Neogene (possibly late-Pliocene) marine planing of the inherited rough relief, and the pulsed epeirogenic emergence of a little-deformed series of stepped platforms that, dissected in more or less degree, persist as the major element of the present regional landform; and equally late Neogene drainage evolution, accordant in its superimposed and its rejuvenated elements.

(6) A complementary relationship of fluctuating relative altitude between a positive mainland massif with large gaps in rock sequence, and flanking submarine basins differentially subsiding to receive great but not wholly unbroken thicknesses of sediment during most or much of Mesozoic and Cenozoic times.

REFERENCES

BLUNDELL, D. J., F. J. DAVEY and L. J. GRAVES (1971) 'Geophysical surveys over the south Irish Sea and Nymphe Bank', *Q. Jl. geol. Soc. Lond.* 127, 339–75

BOTT, M. H. P. (1968) 'The geological structure of the Irish Sea basin' in *Geology of shelf seas*, (ed. D. T. DONOVAN) 93–115

BOTT, M. H. P., and A. B. WATTS (1970) 'Deep sedimentary basins proved in the Shetland-Hebridean continental shelf', *Nature* 225, 265–8

BOULTER, M. C. *et al.* (1971) 'Brassington Formation: a newly recognized Tertiary formation in the southern Pennines', *Nature phys. Sci.* 231, 134–6

BROOKS, M. and M. S. THOMPSON (1973) 'The geological interpretation of a gravity survey of the Bristol Channel', *J. geol. Soc.* 129, 245–74

BROWN, E. H. (1957) 'The physique of Wales', *Geogrl J.* 126, 318–34

BROWN, E. H. (1960) *The relief and drainage of Wales*

BROWN, E. H. (1961) 'Britain and Appalachia: a study of the correlation and dating of planation surfaces', *Trans. Inst. Br. Geogr.* 29, 91–100

CHESHER, J. A. *et al.* (1972) 'IGS marine drilling with m. v. Whitehorn in Scottish waters 1970–71', *Rep. Inst. geol. Sci.* 72/10

CURRY, D. (1965) 'The Palaeogene beds of south-east England', *Proc. Geol. Ass.* 76, 151

CURRY, D., D. HAMILTON and A. J. SMITH (1970) 'Geological and shallow subsurface geophysical investigations in the Western Approaches to the English Channel', *Rep. Inst. geol. Sci.* 70/3

DINGWALL, R. G. (1971) 'The structural and stratigraphical geology of a portion of the eastern English Channel', *Rep. Inst. geol. Sci.* 71/8

DOBSON, M. R., W. E. EVANS and R. WHITTINGTON (1973) 'The geology of the southern Irish Sea', *Rep. Inst. geol. Sci.* 73/11

DONOVAN, D. T. (1972) 'Geology of the central English Channel', *Mem. Bur. geol. min.* 79, 215–20

EDEN, R. A., J. E. WRIGHT and W. BULLERWELL (1971) 'The solid geology of the east Atlantic continental margin adjacent to the British Isles', *Rep. Inst. geol. Sci.* 70/14, 111–28

EDMONDS, E. A., M. C. McKEOWN and M. WILLIAMS (1969) 'British regional geology: south-west England', *Inst. geol. Sci.*

FASHAM, M. J. R. (1971) 'A gravity survey of the Bovey Tracey basin', *Geol. Mag.* 108, 119–30

FORD, T. D. (1972) 'Evidence of early stages in the evolution of the Derbyshire karst', *Trans. Cave Res. Gp Gt Br.* 14, 73–7

FUNNELL, B. M. (1961) 'The Palaeogene and early Pleistocene of Norfolk', *Trans. Norfolk Norwich Nat. Soc.* 19, 340–91

GEORGE, T. N. (1962) 'Tectonics and palaeography in southern England' *Advmt Sci.* 50, 192–217

GEORGE, T. N. (1966) 'Geomorphic evolution in Hebridean Scotland', *Scott. J. Geol.* 2, 1–34

GEORGE, T. N. (1967) 'Landform and structure in Ulster', *Scott. J. Geol.* 3, 413–48

GEORGE, T. N. (1974) 'The Cenozoic evolution of Wales' in '*The upper Palaeozoic and post-Palaeozoic rocks of Wales*' (ed. T. R. OWEN)

HALL, J. and D. K. SMYTHE (1974) 'Discussion of the relation of Palaeogene ridge and basin structures of Britain to the north Atlantic', *Earth. planet. Sci. Lett.* 19, 54–60

HALLAM, A. (1972) 'Relation of Palaeogene ridge and basin structures and vulcanicity in the Hebrides and Irish Sea regions of the British Isles to the opening of the north Atlantic', *Earth planet. Sci. Lett.* 16, 171–7

HANCOCK, J. M. (1969) 'The transgression of the Upper Cretaceous sea in south-west England', *Proc. Ussher Soc.* 2, 61–83

KENNEDY, W. J. (1970) 'A correlation of the uppermost Albian and the Cenomanian of south-west England', *Proc. Geol. Ass.* 81, 613–77

KENT, P. E. (1949) 'A structure contour map of the surface of the buried pre-Permian rocks of England and Wales', *Proc. Geol. Ass.* 60, 87–104

KENT, P. E. (1967) 'Outline geology of the southern North Sea basin', *Proc. Yorks. geol. Soc.* 36, 1–22

KENT, P. E. and P. J. WALMSLEY (1970) 'North Sea progress', *Bull. Am. Ass. Petrol. Geol.* 54, 168–81

KIDSON, C. (1962) 'The denudation chronology of the River Exe', *Trans. Inst. Br. Geogr.* 31, 43–66

LINTON, D. L. (1951) 'Midland drainage, some considerations bearing on its origin', *Advmt Sci.* 7, 449–56

LINTON, D. L. (1964) 'Tertiary landscape evolution', in *The British Isles, a systematic geography* (eds. J. W. WATSON and J. B SISSONS), 110–30

LLOYD, A. J., R. J. G. SAVAGE, A. H. STRIDE and D. T. DONOVAN (1973) 'The geology of the Bristol Channel floor', *Phil. Trans. R. Soc.* A, 274, 595–626

MITCHELL, G. F. (1972) 'The Pleistocene history of the Irish Sea: second approximation', *Sci. Proc. R. Dublin Soc.* A, 4, 181–99

PARRY, J. T. (1960) 'The erosion surfaces of the south-western Lake District', *Trans. Inst. Br. Geogr.* 28, 39–54

SISSONS, J. B. (1967) *The evolution of Scotland's scenery*

SMYTHE, D. K., W. T. C. SOWERBUTTS, M. BACON and R. McQUILLIN (1972) 'Deep sedimentary basin below northern Skye and the Little Minch', *Nature phy. Sci.* 236, 87–9

SPARKS, B. W. (1953) 'Erosion surfaces around Dieppe', *Proc. Geol. Ass.* 64, 105–17

WALSH, P. T. and E. H. BROWN (1971) 'Solution subsidence outliers containing probable Tertiary sediment in north-east Wales', *Geol. J.* 7, 299–320

WALSH, P. T., M. C. BOULTER, M. ITTABA and D. M. URBANI (1972) 'The preservation of the Neogene Brassington Formation of the southern Pennines and its bearing on the evolution of upland Britain', *Q. Jl geol. Soc. Lond.* 128, 519–59

WEST, R. G. (1968) *Pleistocene geology and biology*

WEST, R. G. (1972) 'Relative land-sea-level changes in south-eastern England during the Pleistocene', *Phil. Trans. R. Soc.* A, 272, 87–98

WILSON, H. E. (1972) 'Regional geology of Northern Ireland', *Geol. Surv. Northern Ireland*

WOODLAND, A. (ed.) (1971) 'The Llanbedr (Mochras Farm) borehole', *Rep. Inst. geol. Sci.* 71/18

WOOLDRIDGE, S. W. (1950) 'The upland plains of Britain', *Advmt Sci.* 7, 162–75

WOOLDRIDGE, S. W. (1954) 'The physique of the South-West', *Geography*, 186, 231–42

WOOLDRIDGE, S. W. and D. L. LINTON (1938a) 'Influences of Pliocene transgression on the geomorphology of south-east England', *J. Geomorph.* 1, 40–54

WOOLDRIDGE, S. W. and D. L. LINTON (1938b) 'Some episodes in the structural evolution of south-east England', *Proc. Geol. Ass.* 49, 264–91

WOOLDRIDGE, S. W. and D. L. LINTON (1955) *Structure, surface and drainage in south-east England*

RÉSUMÉ. *Prologue à une géomorphologie de Grande-Bretagne.* D'après la distribution et la structure de la Craie résiduelle il est peu probable que la physique actuelle de la Grande Bretagne puisse s'expliquer par une couverture crétacée ancienne. Les roches et les structures éocènes et oligocènes font supposer de fortes pulsations tectoniques pendent les temps paléogènes, avant la venue des mouvements miocènes « alpins ». Les relations complémentaires du continent britannique et les dépressions sous-marines limnotrophes entraînent une chronolgie de la dénudation cénozoique qui souligne le contraste entre des plissements souvent renouvelés du terrain et le profil néogène palané. Des plateformes marines terrassées et peu déformées sont le produit caracteristique des agents néogènes: leur dégradation par des fleuves constitue une évidence supplémentaire de la jeunesse géologique des contrées regionales de la Grande Bretagne.

FIG. 1. Section, légèrement simplifiée, pratiquée à travers du flanc ouest du bassin du Hampshire: pour montrer la « transgression cénomène » à travers les roches mésozoiques et permiennes; le dépassage de la base diachronique éocène à travers les roches haute-crétacées et la grande discordance des bassins des roches oligocènes dans le Devon, ou les roches éocènes et crétacées eteintes sont le signe d'un mouvement paléogène localement aigu.
FIG. 2. Des sections, légèrement simplifiées et en partie conjecturales, pratiquées à travers de l'Ecosse septentrionale: pour montrer les relations entre le terrain ancien et les bassins de la plateforme; les signes du tectonisme cénozoique répeté; l'érosion de la physique paléogène; et l'accord des points de passage dans les montagnes résiduelles de roches éocènes, dalradiennes, et précambriennes
FIG. 3. Section, légèrement simplifiée et en partie conjecturale, depuis la Mer d'Irlande du sud à travers le Pays de Galles et L'Angleterre jusqu'à la Mer du Nord: pour montrer les effets d'un affaissement à coups de pulsations dans les bassins sous-marins par rapport au massif du continent; la physique qui resulte d'une transgression marine répetée; et le produit cumulatif de vagues tectoniques sur une physique néogène émergente.

ZUSAMMENFASSUNG. *Prolog zu einer Geomorphologie von Britannien.* Die Verteilung und Struktur der restlichen Kreide lässt die Möglichkeit höchst unwahrscheinlich erscheinen, dass die heutige Geomorphologie Grossbritanniens mit Bezug auf eine frühere Kreidedecke zu erklären sei. Der Fels und der Aufbau des Eozäns und Oligozäns geben Anzeichen starker tektonischer Tätigkeit während der Zeiten des Paläogens vor den Beginn der miozänen 'alpinen' Bewegungen. Die Verzahnung von dem britischen Festland und den flankierenden submarinen Trögen lässt auf eine Abtragungschronologie des Känozoikums schliessen, die den Kontrast zwischen wiederholt erneuerter Zerklüftung der Landschaft und dem abgerundeten Profil der Neogens hervorhebt. Gestufte, nur in geringen Mass veränderte Abrasionsplattformen sind das charakteristische Ergebnis neogener Kräfte: ihre Abtrangung durch Ströme zeugen zusätzlich für die morphologische Jugend der Landschaft Grossbritanniens.

ABB. 1. Querschnitt, etwas vereinfacht, durch die Westseite des Talbeckens von Hampshire: um die "Transgression des Cenomans" über des Mesozoikum und Permformationen darzustellen; ferner das Übergreifen des Unterbaus des diachronen Eozäns über die Oberkreide; und die klare Diskordanz mit den Felsenbecken des Oligozäns in Devon, wo Eozän und Kreideformationen abgetragen wurden also Anzeichen der örtlich heftigen Bewegung des Mittelpaläogens
ABB. 2. Querschnitte, vereinfacht und zum Teil rekonstruiert, durch Nordschottland: um das Verhältnis den Altlandterrains zu Üferbecken; die Merkmale des wiederholten Tektonismus; die Verwitterung der Landformen des Paläogens; und die Übereinstimmung der Gipfelflur bei den übriggebliebenen Felsenhügeln des Eozäns, Dalradischen, und Präkambriums darzustellen
ABB. 3. Querschnitt, etwas vereinfacht und zum Teil vermutet, von der sudlichen Irischen See durch Wales und England bis zur Nordsee: um die Auswirkungen der intermittierenden Absenkung küstenabgewandter Becken vergleichen mit der eigentlichen Festlandmasse; die Lanformen also Folge der wiederholten marinnen Transgression; und die Summe der Auswirkungen tektonischer Wellen auf eine aufsteigende neogene Landform darzustellen.

The summit surface on the Wessex Chalk

CHRISTOPHER P. GREEN

Lecturer in Geography, Bedford College, University of London

Revised MS received 28 June 1973

ABSTRACT. The suggested type locality for the Mio-Pliocene peneplain of Wooldridge and Linton (1955) is re-examined. The superficial geology of the Chalk summits and the relief and drainage of the area are described. As a result the summit surface is separated into two elements. The earlier is the product of an almost complete planation which left only isolated residuals, capped with relics of Palaeogene sediments, on the main watersheds. The later was formed by slope retreat from drainage lines similar to those of the present, but was less extensive and left substantial residuals of the earlier surface. These form the main watersheds in the present landscape. Both surfaces are of Neogene age. The later surface is uplifted over the Wardour anticline and the present drainage pattern in the Vale of Wardour arose in response to this uplift. The status of the Calabrian marine transgression in the area is considered, and no evidence that the late Tertiary landforms were significantly modified by it can be found.

IN the early years of the present century, British geographers, reading the works of Davis (1895), Mackinder (1902) and Bury (1910), were familiar with the idea that lowland Britain had suffered planation in the latter part of the Tertiary period. The question still at issue when Wooldridge was writing in 1927 was not the reality of the plain but whether it was of sub-aerial or marine origin. Wooldridge himself, examining the summits of the Chalk around the London Basin, recognized 'a (sub-aerial) surface of low relief, representing an approximation to a late Miocene base level' (Wooldridge, 1927).

In 1932 Linton identified this sub-aerial peneplain on the Wessex Chalk, and in 1939 Wooldridge and Linton, describing the surface as the Mio-Pliocene peneplain, suggested that an area in Wessex, on the southern border of Wiltshire, might be regarded as its type locality. The main challenge to the work of Wooldridge and Linton in southern England has come from Pinchemel (1954) who considered the summit relief to approximate closely to a polycyclic surface of early Tertiary age.

Wooldridge and Linton also distinguished in Wessex, mainly on morphological grounds, the 'Pliocene marine plain' which Wooldridge (1927) had originally described in the London Basin. This feature has been recognized in many parts of the British Isles at levels up to 210 m O.D., and is now generally related to a marine transgression in the Lower Pleistocene Calabrian.

The area considered in this account includes the suggested type locality for the Mio-Pliocene peneplain. The relief of the area shows a significant measure of structural control. The principal structures are fold axes and faults trending east to west. These are examples of a large group of structures in southern England, attributed to mid-Tertiary (Alpine) earth movements. The Vale of Wardour (Fig. 1) is an anticlinal vale in which the Chalk has been denuded to expose earlier strata down to the Kimmeridge Clay.

To the north of the Vale, east of Rook Hill, the Chalk rises gradually without any well-marked escarpment to a synclinal upland (the Great Ridge) forming the watershed between the Rivers Nadder and Wylye. Farther west, the Mere fault cuts out strata between the Chalk and the Kimmeridge Clay, and a bold Chalk escarpment rises to the Mere Downs. To the south of the Vale, west of Berwick Saint John, a Chalk escarpment is again present and forms a watershed between drainage in the Vale of Wardour and drainage southward to the Hampshire Avon.

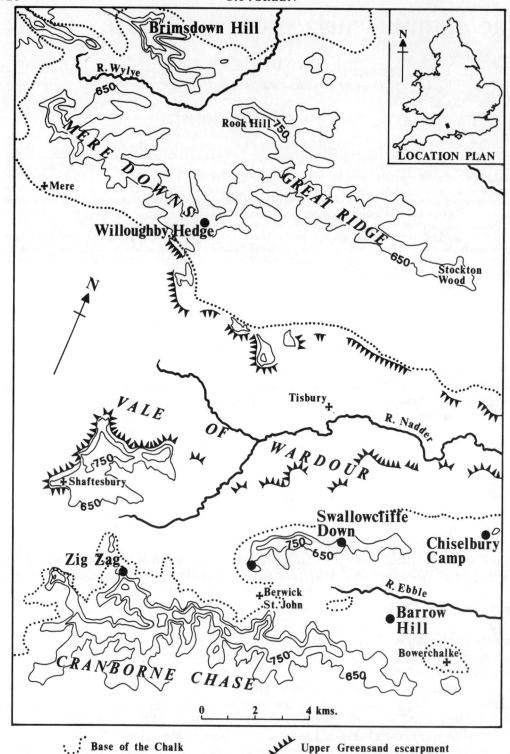

FIGURE 1. The Vale of Wardour and the neighbouring Chalk uplands in southern Wiltshire. Contours at 650, 750 and 850 feet O.D. Heavy dots: chert-bearing gravel relics

Farther east, the Chalk escarpment forms a watershed between the rivers Nadder and Ebble. The Ebble drains a synclinal tract separating the Wardour fold from the smaller, partially denuded, Bowerchalke anticline.

Wooldridge and Linton appear to have regarded all the summits here, above approximately 200 m O.D., as remnants of their Mio-Pliocene peneplain. The eastern parts of the principal watersheds, which fall below this level, they distinguish as remnants of the 'Pliocene marine plain'.

THE RELIEF AND SUPERFICIAL GEOLOGY OF THE CHALK UPLANDS

A re-examination of the Chalk uplands shows that in terms of relief and superficial geology the summits fall into three main groups.

Watershed residuals with relics of quartzose gravels

Brimsdown Hill (284 m) and neighbouring summits rise conspicuously above the summit peneplain (Fig. 3 A, B, C). Relics of quartzose gravels, possibly of early Tertiary age, are found on these summits, and may indicate proximity to the level of an early Tertiary surface (Green, 1969a).

The summit peneplain: Lower Surface

In Figure 3, the separation of the summit peneplain into two elements, termed here the Higher and the Lower Surfaces, is apparent. The Lower Surface is extensively preserved between Rook Hill and Willoughby Hedge (Fig. 3B). Rook Hill and neighbouring summits rise above the Lower Surface to the level of the Higher Surface. The Lower Surface here is flat and relatively undissected, and a gentle northward gradient is evident. Farther east (Fig. 3A), the Lower Surface has been mapped by Gifford (1957), as her Higher Plain. In the field it is represented by well-defined benches on the spurs extending northward and southward from the Great Ridge. The Higher Surface survives on the summit of the Great Ridge. In the west, on the Mere Downs (Fig. 3C), the Lower Surface completely replaces the Higher Surface, and the watershed residuals around the upper Wylye rise abruptly from it.

In the south of the Vale of Wardour the identification of two surfaces is scarcely warranted on the basis of topographic evidence alone. In the west (Fig. 3D) the Lower Surface will have been developed not on the Upper Chalk, as it is on the northern side of the Vale, but on the relatively weaker Middle and Lower Chalk, brought up here in the Bowerchalke anticline. The Lower Surface can be tentatively identified only at the Zig Zag and on the Upper Greensand near Shaftesbury. The higher summits of Cranborne Chase form part of the Higher Surface. Farther east (Fig. 3E and F) the junction between the Higher and Lower Surfaces will have been near the axis of the Bowerchalke anticline and this structural weakness has been exploited by the River Ebble and its tributaries. The Lower Surface is well preserved on the Nadder–Ebble watershed; its southward gradient and the break of slope separating it from the Higher Surface to the south are tentatively reconstructed.

The Lower Surface is notably free from superficial deposits. Traces of reddish clay are occasionally present, but chalk is almost always seen in the plough layer. However, at a number of points extraneous debris is encountered in the chalky soil. These debris appear to be, in part, the 'relics of old gravel with pieces of Upper Greensand chert' described by Andrews (1892). Wooldridge and Linton accept the view proposed originally by Andrews that the high-level gravel relics in the Vale of Wardour are of the same age as the Clay-with-Flints on the summit of the Great Ridge, and that all these deposits rest on the same planation surface, the summit

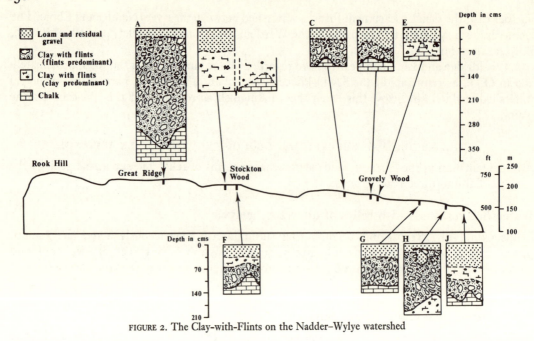

FIGURE 2. The Clay-with-Flints on the Nadder–Wylye watershed

peneplain of Wooldridge and Linton. The present investigation shows, however, that these relics are present only on the Lower Surface.

On the Mere Downs near Willoughby Hedge, a reddish clay is present to a depth of 1–2 m; a shallow surface layer (10–20 cm) includes Upper Greensand chert and sandstone, pieces of ironstone, a few pebbles of quartz and a quantity of unidentified cherts and sandstones (? Jurassic). This material appears to be the remains of a coarse gravel. Upper Greensand cherts and sandstones probably form at least 90 per cent of the gravel fraction (>6 mm). On Swallowcliffe Down, on the Nadder–Ebble watershed a similar accumulation is present. Inspection shows that although Upper Greensand material is predominant (possibly 99 per cent), Jurassic material is also present. Here and on an un-named summit above Berwick Saint John, fragments of a chalk breccia are found in association with the other debris. This breccia is of chalk fragments up to 3 cm long in an incomplete matrix of crystalline calcite. Its character suggests formation at or near the water table and comparison with a breccia described by Prestwich (1891) in the Darent valley. Greensand debris is also found on the small scarp face bench at the Zig Zag (213 m) and on the summit of Barrow Hill (185 m); at the latter site fragments of chalk breccia are again found.

The summit peneplain: Higher Surface

The remnants of the Higher Surface occupy the major watersheds to the north and south of the Vale of Wardour, namely, the Rook Hill–Great Ridge–Grovely Wood and Cranborne Chase uplands. Unlike the Lower Surface, this ground is occupied by deep superficial deposits. A section on the Great Ridge (Fig. 2A) shows 2·1 m of superficial material, comprising a shallow loamy horizon resting on very flinty clay. The flint is broken but rarely abraded. Infrequent abraded flints have a brown patina, whereas the angular material is grey or white. The interstices between the larger fragments are filled with a gritty buff coloured clay. Apart from flint, only a few small ferruginous concretions are observed even in the fine gravel fraction (2–6 mm), and

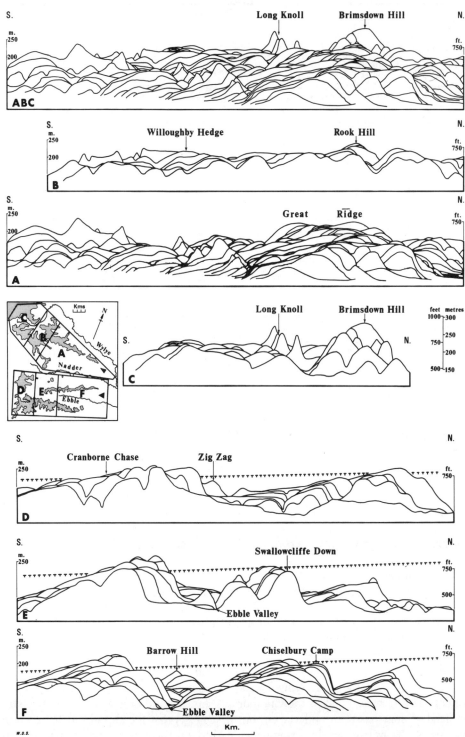

FIGURE 3. Profiles across the Chalk uplands north and south of the Vale of Wardour. The plane of the Lower Surface is indicated

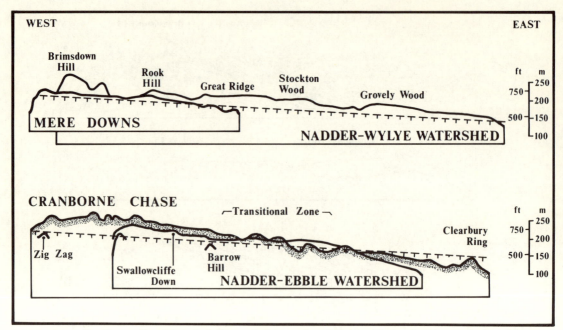

FIGURE 4. Crestline profiles of the principal watersheds in southern Wiltshire. The break of slope at the foot of the escarpment separating the Higher and Lower Surfaces is indicated

one small pebble of quartz was collected from the exposure. Further exposures and surface indications suggest that this section is typical of the Mio-Pliocene peneplain of Wooldridge and Linton on these uplands, and that the Clay-with-Flints here is generally uniform in character. In the medium and coarse gravel fractions, at some sites only angular flint is present, at other sites abraded 'mahogany' flint occurs, in proportions ranging up to 11 per cent. At most sites a thin layer (1–5 cm) of much less gritty reddish clay is present at the base of the Clay-with-Flints, and a layer of unbroken flint nodules may also occur at this level.

On the Cranborne Chase upland exposures are infrequent. White (1923) described a superficial deposit of 'loose gravel composed of unworn flints'. This deposit, regarded by White as of residual origin, was sufficiently deep to be dug as gravel in the nineteenth century. Present-day surface indications confirm White's account, nothing but angular flint is observed in the plough layer.

Both the Nadder–Wylye watershed and the Cranborne Chase upland have been generally regarded at their eastern ends as portions of the Calabrian marine bench (Wooldridge and Linton, 1955; Pinchemel, 1954; Small, 1964). On the upland occupied by Stockton and Grovely Woods eight sections (Fig. 2 B–J) were dug down to the Chalk (Green, 1965). The Clay-with-Flints here is, in most important respects, indistinguishable from the deposit on the Rook Hill–Great Ridge upland. In its composition and arrangement there is no hint of a marine transgression. Significant in this respect is the absence from the gravel fraction of material derived from sources other than the underlying Chalk, or of material bearing evidence of a marine environment. At the eastern end of the Cranborne Chase upland only a reconnaisance study of the superficial geology has been undertaken, but it suggests that the deposits at the eastern and western ends of the upland are indistinguishable, and no trace of a Calabrian marine transgression has been observed.

Thus, on the Higher Surface the superficial deposits comprise a 'mantle of deeply weathered rock waste' (Wooldridge and Linton, 1955). There is no evidence to support the view proposed by Pinchemel (1954) and Small (1964) that the summit surface here is of sub-Eocene origin. Early Tertiary deposits are found only on the watershed residuals. We can therefore accept the conclusion proposed by Wooldridge and Linton (1955) that the Higher Surface is of sub-aerial origin and Neogene age. The abraded 'mahogany' flint may be the product of the planation processes which originally formed the surface. Nevertheless it seems clear that the surface has been substantially modified by solution of the Chalk; the residue of this solution forms the bulk of the Clay-with-Flints described in this account. The Lower Surface is evidently of more recent age, lacks a deep mantle of residual debris, and may approximate more closely to the original erosional terrain.

Profiles along the main watersheds (Fig. 4) clarify the principal relief features. The general eastward inclination of all the watersheds is apparent and although the crestlines are uneven, they are nowhere interrupted by benches or bluffs which can be referred convincingly to marine erosion. The Nadder–Wylye watershed can be seen throughout its length as a partially dissected remnant of the Higher Surface, falling from over 215 m in the west to about 165 m in the east. On the Cranborne Chase upland the Higher Surface can be identified in the west, but it appears to have been replaced farther east by the Lower Surface. There is a transitional zone between the two surfaces in which residual outliers of the Higher Surface can be seen, suggesting (cf. Brown, 1960) that both the surfaces are of sub-aerial origin. The Lower Surface can be tentatively reconstructed at the eastern extremity of this profile, where it falls to the level of undoubted Pleistocene gravels in the Avon valley at Clearbury Hill (Green, 1969a). The Nadder–Ebble watershed and the Mere Downs can be seen as remnants of the Lower Surface.

THE DRAINAGE PATTERN IN THE VALE OF WARDOUR

The drainage pattern in the Vale of Wardour is relatively well adjusted to structure (Fig. 5). The westerly part of the Vale is drained by the Dorset Stour, flowing generally southward. The principal stream within the Vale, the Nadder, also rises in the western part of the Vale on the Kimmeridge Clay, and flows eastwards on a course which coincides closely with the structural axis of the Wardour anticline. At the eastern end of the Vale, the Nadder crosses onto the Chalk and continues eastward to join the Hampshire Avon. The main right bank tributaries of the Nadder rise on the Middle or Lower Chalk and follow uncomplicated courses from south to north. Tributary to them, in the vale separating the Upper and Lower Chalk escarpments, are subsequents in an early stage of development (Small, 1961). The main left bank tributaries of the Nadder now rise near the base of the Chalk and follow uncomplicated courses from north to south. Above their present sources, however, there is an extensive system of dry valleys on the outcrop of the Upper Chalk. Small (1961) has suggested that the pattern here is inherited from obsequent elements on the summit surface. Easterly trending components are conspicuous in this dry valley system.

The drainage pattern throughout the Vale of Wardour was thought by Wooldridge and Linton (1955) to have arisen by rejuvenation of a pattern already established on the summit surface. The evidence in this account shows that the summit peneplain can be separated in the Vale of Wardour into a Higher and a Lower Surface, and that the Lower Surface generally slopes away from the anticlinal axis. This is clear from an inspection of the relief (Fig. 3) and is also suggested by the distribution of Upper Greensand debris on the Chalk summits. The evidence seems to indicate drainage on the Lower Surface away from the anticlinal axis and therefore unlike the present pattern.

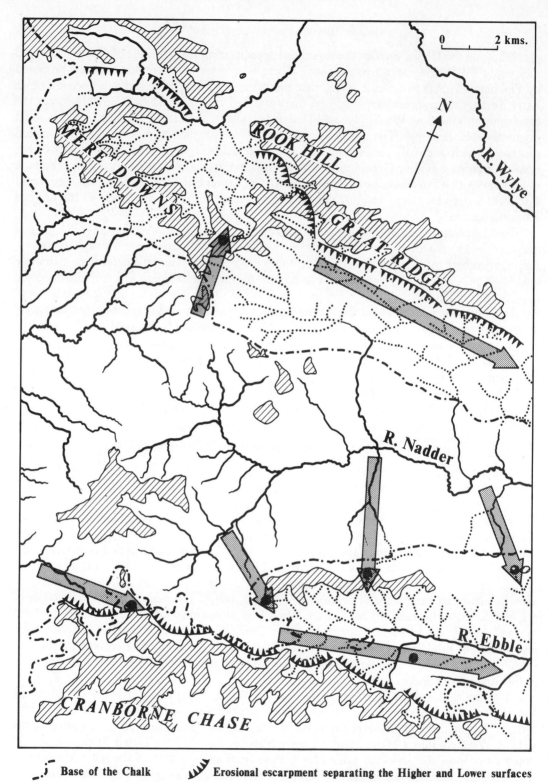

FIGURE 5. Drainage and dry valley pattern on the southern border of Wiltshire. Arrows indicate initial drainage lines on the uplifted Lower Surface. Land above 600 feet shaded. Heavy dots as in Figure 1

DISCUSSION

In the following paragraphs the origin of the summit relief and of the drainage pattern are re-examined. In general terms certain comparisons are possible with the scheme of landscape evolution recognized in the Paris basin (e.g. Cholley, 1957). In the Paris basin, following the development of a polycyclic land surface in the Palaeogene, planation occurred during the Miocene period, adjusted initially to an Upper Oligocene base level. Planation was gradually terminated by deformation of the Miocene surface. Planation of Mesozoic and Cenozoic rocks was almost complete, only a few residual summits remaining on certain watersheds. In the Vale of Wardour the Higher Surface appears to be the product of this Miocene cycle; Brimsdown Hill and neighbouring summits around the upper Wylye form a small group of residuals, capped with the remnants of Palaeogene sediments, and occupying the main watershed between the English and Bristol Channels.

In this scheme of landscape evolution the Pliocene period is occupied by the development of a less extensive planation surface, equated in the Vale of Wardour with the Lower Surface. The latter appears to fall eastward to the level of the highest Pleistocene gravels in the valley of the Hampshire Avon, and can be regarded with some confidence as a late Tertiary feature. In the Paris basin, as in Wessex, evidence of a Calabrian marine transgression is deficient and its identification south and west of the Artois axis is disputed (Pinchemel, 1954; Cholley, 1957).

The apparently scarp-like junction between the Higher and Lower Surfaces suggests that it was the product of back wearing rather than down wearing. King (1962) has indicated that in the evolution of the pedimented landscape 'cyclic scarps have in nearly all instances transgressed from weak onto resistant formations'. In the present case it can be suggested that the Lower Surface in the Vale of Wardour developed initially on relatively weak Jurassic rocks already exposed on the Higher Surface near the axis of the anticline, and extended later, northward and southward, onto the more resistant Upper Chalk.

Nevertheless the relics of the Lower Surface now slope away from and not towards the axis of the anticline, and near the axis rise to, and locally above, the level of the Higher Surface. Thus, the Lower Surface has, since its formation, been upwarped in the axial part of the Vale of Wardour. This movement can be tentatively referred to the late Pliocene–early Pleistocene interval.

Evidence of a drainage pattern developed on the newly upwarped Lower Surface can be detected in the Vale of Wardour. First, sparse relics of chert-bearing gravels are found on the Lower Surface, and their distribution suggests consequent drainage elements, relatively poorly adjusted to structure, and unlike the present drainage pattern. Linton (1969) has suggested that this evidence indicates surviving remnants of drainage consequent on folds of Oligo-Miocene age. If however the Lower Surface originated as a pediment, and formerly sloped towards the axis of the Vale, then the survival of any Oligo-Miocene consequents is difficult to accept. Secondly, in at least two cases the high-level gravel relics are aligned with major tributaries of the present Nadder (Fig. 5). These streams appear to mark the original courses of major consequents on the upwarped Lower Surface. Thirdly, the eastward course of the Ebble, and the easterly trending dry valleys on the Upper Chalk in the north of the Vale suggest that drainage arising on the uplifted Lower Surface gathered at the foot of the escarpments separating the Higher and Lower Surfaces and flowed away to the east. The integration of such eastward-flowing drainage may have been favoured by the pre-existence of marginal depressions on the pediment surface (Pugh, 1956). A proto-Ebble flowing from the Upper Greensand outcrop onto the Upper Chalk is confirmed by the survival of Upper Greensand debris at Barrow Hill in the present valley of the Ebble. In the north of the Vale, easterly trending components in the dry valley pattern are interpreted as relics of a proto-Nadder system. This drainage pattern, consequent on the uplift of the Lower Surface, was evidently dismembered rather rapidly throughout the Vale of Wardour and the present Ebble valley by the revival of a subsequent pattern—a readjustment favoured by negative movements of base level during the Pleistocene and by the slight relief of the Lower Surface.

The foregoing account of drainage evolution in the Vale of Wardour suggests that substantial adjustment of the drainage pattern to geological structure has taken place during the present cycle of erosion, since the warping of the Lower Surface and probably within the Pleistocene period. This conclusion is important in the reappraisal of the role of the Calabrian marine transgression in the evolution of the Vale of Wardour. In Wessex the extent of the Calabrian transgression has been defined in terms of drainage pattern, following Wooldridge and Linton who 'concluded in 1938 that marine transgression had occurred throughout the area in which epigenesis could be established' (Linton, 1964). In the Vale of Wardour the drainage pattern appears, on the basis of Davisian criteria, to be in a second cycle of erosion and the Vale is accordingly considered to lie beyond the area of epigenesis. The present account suggests, however, that the drainage pattern is the product of a single cycle of erosion. Such a conclusion undermines the fundamental concept upon which the recognition of the Calabrian transgression in Wessex rests. The earlier parts of this account have confirmed that in the Vale of Wardour, neither superficial geology nor relief provide evidence of the transgression. It is difficult to prove that no transgression occurred in the Calabrian period but the conclusions reached in this account suggest that in any case its effects in the Vale of Wardour left no significant impression on the late Tertiary landscape. Kidson (1968) indicates that this may generally have been the case in Britain.

CONCLUSIONS

By reference to the foregoing account and to recent work in this area (Green, 1969a), it now seems possible to relate the main features of the summit relief to a sequence of geomorphological episodes extending back into the Palaeogene.

The earliest landscape of which any part is preserved in the present terrain is a land surface of Palaeogene age. Small fragments of this survive on the summits of residual elevations which rise from the principal watersheds near the western margin of the Chalk outcrop. Reason has been shown by the present author (1969a) for regarding this surface as of late Palaeogene (? Oligocene) age. To the south and east of the Vale of Wardour, near the margin of the Tertiary outcrop in the Hampshire Basin, the sub-Tertiary surface can again be distinguished (Wooldridge and Linton, 1955) but here the surface is largely of much earlier Reading Beds age. Thus, the Chalk appears to have been subject to erosion during much of the Palaeogene period. Moreover, the major part of the denudation of the Chalk outcrop appears to have taken place at that time. The product of this erosion was a polycyclic land surface. The reconstructed plane of this surface slopes from the highest summit levels on the western margin of the Chalk outcrop southwards and eastwards towards the Hampshire Basin, where it passes beneath the early Tertiary sediments.

In the early part of the Neogene period, a new land surface, the Higher Surface in the Vale of Wardour, was cut into the Chalk at the expense of the Palaeogene surface. This development appears to have occurred in response to changing base-level conditions, possibly the tilting of the earlier surface towards the Hampshire Basin. During the Neogene a further land surface, the Lower Surface in the Vale of Wardour, was developed and partially replaced the Higher Surface. The development of a second surface during the Neogene suggests a further change of base-level conditions, possibly indicating renewed tilting towards the south and east. The final stages in the evolution of the present landforms involved a slight uplift of the Lower Surface and its subsequent, progressive dissection during the Pleistocene period. The late Tertiary landscape may have been invaded by the Calabrian sea, but no evidence of such a transgression remains.

REFERENCES

ANDREWS, W. R. (1892) 'The origin and mode of formation of the Vale of Wardour', *Wilts. archaeol. nat. Hist. Mag.* 26, 258

BROWN, E. H. (1960) *The relief and drainage of Wales*

BURY, H. (1910) 'On the denudation of the western end of the Weald', *Q. Jl geol. Soc. Lond.* 66, 640–92

CHOLLEY, A. (1957) *Recherches morphologiques* (Paris)

DAVIS, W. M. (1895) 'On the origin of certain English rivers', *Geogrl J.* 5, 128–46

GIFFORD, J. (1957) 'The physique of Wiltshire' in R. B. PUGH (ed.), *The Victoria county history of Wiltshire*

GREEN, C. P. (1965) '*Some aspects of the denudation of the Chalk in the county of Wilts*', Unpubl. D.Phil. thesis, Univ. of Oxford

GREEN, C. P. (1969a) 'An early Tertiary surface in Wiltshire', *Trans. Inst. Br. Geogr.* 47, 61–72

GREEN, C. P. (1969b) 'The evolution of the Vale of Wardour', *Area*, 1 (2), 21–3

KIDSON, C. (1968) 'The role of the sea in the evolution of the British landscape' in H. CARTER, E. G. BOWEN and J. A. TAYLOR (eds), *Geography at Aberystwyth*

KING, L. C. (1962) *The morphology of the Earth*

LINTON, D. L. (1932) 'The origin of the Wessex rivers', *Scott. geogr. Mag.* 48, 149–66

LINTON, D. L. (1964) 'Tertiary landscape evolution', in J. W. WATSON and J. B. SISSONS (eds), *The British Isles*

LINTON, D. L. (1969) in discussion, GREEN, C. P. (1969b)

MACKINDER, H. J. (1902) *Britain and the British seas*

PINCHEMEL, P. (1954) *Les plaines de craie du nord-ouest du Bassin Parisien et du sud-est du Bassin de Londres et leur bordures. Etude de géomorphologie* (Paris)

PRESTWICH, J. (1891) 'On the age, formation and successive drift stages in the valley of the Darent', *Q. Jl geol. Soc. Lond.* 47, 126

PUGH, J. C. (1956) 'Fringing pediments and marginal depressions in the inselberg landscape of Nigeria', *Trans. Inst. Br. Geogr.* 22, 15–31

SMALL, R. J. (1961) 'The morphology of Chalk escarpments', *Trans. Inst. Br. Geogr.* 29, 71–90

SMALL, R. J. (1964) 'Geomorphology' in F. J. MONKHOUSE (ed.), *A survey of Southampton and its region*

WHITE, H. J. O. (1923) 'The geology of the country south and west of Shaftesbury', *Mem. geol. Surv. U.K.*

WOOLDRIDGE, S. W. (1927) 'The Pliocene period in the London Basin', *Proc. Geol. Ass.* 38, 49–132

WOOLDRIDGE, S. W. and D. L. LINTON (1938) 'The influence of the Pliocene transgression on the geomorphology of south-east England', *J. Geomorph.* 1, 40–54

WOOLDRIDGE, S. W. and D. L. LINTON (1939) 'Structure, surface and drainage in south-east England', *Inst. Br. Geogr.*, Publication No. 10 (revised and reprinted, 1955)

RÉSUMÉ. *Surface des sommets crayeux du Wessex*. Le dossier concernant l'emplacement envisagé de la pénéplaine mio-pliocène de Wooldridge et Linton (1955) est ré-ouvert. La géologie de surface des sommets crayeux, le relief et le ruissellement des eaux de la région sont décrits. En conséquence, la surface du sommet est divisée en deux éléments: le plus ancien de ces éléments est la conséquence d'un planage presque total qui n'a laissé que des résidus isolés, surmonté de vestiges des sédiments de l'ère paléogène, sur les principales lignes de partage des eaux. Le plus récent de ces éléments a été formé par un retrait des pentes des lignes d'écoulement analogues à celles d'aujourd'hui, mais ce phénomène a été moins étendu et a laissé des résidus importants appartenant à une surface antérieure. C'est ce qui forme les lignes de partage des eaux du paysage actuel. Les deux surfaces remontent à l'ère néogène. La surface plus récente s'est soulevée sur l'anticlinal, et l'actuel système de ruissellement des eaux s'est formé en réaction à ce soulèvement. L'importance de la transgression marine calabraise dans cette région est passée en revue, et il n'a pas été possible de trouver de preuves que les masses terrestres de la fin de l'ère tertiaire ont été modifiées d'une façon significative par cette transgression.

FIG. 1. Vallée de la Wardour, et collines crayeuses avoisinantes dans le sud du Wiltshire. Contours à 197, 227 et 257 mètres. Points foncés: vestiges de gravier contenant de la calcédoine

FIG. 2. Argile à silex sur la ligne de partage des eaux Nadder–Wylye

FIG. 3. Profile des collines crayeuses du nord et sud de la vallée de la Wardour. Le plan de la surface inférieure est indiqué

FIG. 4. Profils des crêtes des principales lignes de partage des eaux du Wiltshire sud. L'arrêt de la pente au pied de l'escarpement séparant les surfaces inférieures et supérieures est indiqué

FIG. 5. Ruissellement des eaux et schéma d'une vallée sèche à la frontière sud du Wiltshire. Les flèches indiquent les lignes de ruissellement d'origine de la surface inférieure du soulèvement. Terrain au dessus de 180 mètres hachuré. Points foncés comme pour la Fig. 1

ZUSAMMENFASSUNG. *Die Gipfelfläche der Wessex-Kreide.* Die vorgeschlagene Typlokalität für die Mio-pliozän Fastebene von Wooldridge und Linton (1955) wird wieder untersucht. Die Oberflächengeologie der Kreidegipfel sowie das Relief und die Entwässerung des Bereiches werden beschrieben. Dabei wird die Gipfelfläche in zwei Elemente untergeteilt. Das frühere ist das Ergebnis einer nahezu vollständigen Abhobelung, die nur stellenweise mit relikten Palaeo-Sedimenten bekappte Reste an den Hauptwasserscheiden zurückgelassen hat. Das spätere wurde durch Zurückfliehen des Hanges von ähnlichen Entwässerungslinien, wie sie heute bestehen, gebildet, war aber weniger ausgedehnt und hinterliess erhebliche Rest der früheren Fläche. Diese bilden die Hauptwasserscheiden in der gegenwärtigen Landschaft. Beide Flächen gehören der neogenen Epoche an. Die spätere Fläche ist über der Wardour Antiklinale gehoben, und die gegenwärtige Entwässerungs-struktur im Vale of Wardour war das Ergebnis dieser Hebung. Es wird der Status der calabrischen Transgression in dem Bereiche erörtert, und nichts deutet darauf hin, dass die späten terziären Landformen dadurch erheblich modifiziert wurden.

ABB. 1. Das Vale of Wardour und das benachbarte Kreidehochland in Süd-Wiltshire. Konturlinien in Höhen von 650, 750 und 850 Fuss über normal O. Fette Punkte: Hornsteinhaltige Schotterrelikte

ABB. 2. Ton mit Flintsteinen an der Nadder-Wylye Wasserscheide

ABB. 3. Profile des Kreidehochlandes nördlich und südlich des Vale of Wardour. Die Ebene der Unterfläche ist angedeutet

ABB. 4. Scheitellinienprofile der Hauptwasserscheiden in Süd-Wiltshire. Der Knick am Fusse der Steilwand, die die Ober- und Unterflächen trennt, ist angedeutet

ABB. 5. Entwässerungs- und Trockentalstruktur an der südlichen Grenze von Wiltshire. Pfeile zeigen die ursprünglichen Entwässerungslinien an der gehobenen Unterfläche an. Gelände oberhalb von 600 Fuss schattiert. Fette Punkte wie in Bild 1

The influence of the Calabrian transgression on the drainage evolution of south-east England

DAVID K. C. JONES

Lecturer in Geography, London School of Economics and Political Science

Revised MS received 5 July 1973

ABSTRACT. The discordant drainage pattern of south-east England has been explained in terms of superimposition from a high-level marine surface created by the early Pleistocene, or Calabrian, marine transgression. This paper argues that there are strong grounds for rejecting this hypothesis, especially as morphological and sedimentological evidence for the transgression is largely confined to the London basin while the discordant relationships are located in Sussex and Wessex. A number of major weaknesses inherent in the superimposition argument are examined and it is suggested that although the Calabrian incursion probably inundated considerable areas, it achieved insufficient erosion to remodel significantly the drainage pattern. The effects of the mid-Tertiary orogenic phase are then considered and the conclusion reached that there is little evidence to suggest that the buckling of weak strata led to drainage derangement. It is therefore argued that the major elements of the present drainage network are inherited from the concordant pattern created upon the withdrawal of the early Tertiary Sea. The discordant drainage relationships are thus explained in terms of antecedence, except those occurring near the Sussex and Wessex coasts which are the result of superimposition from either the Tertiary cover or erosion surfaces created by the sequence of post-Calabrian high sea levels which affected areas below 150 m.

THE subject of drainage evolution has caused considerable debate among geomorphologists in Britain. Much of this discussion has been focused on the south-east, where the extensive dissected chalklands are characterized by transverse and discordant drainage lines. Nearly a century ago W. Topley (1875) invoked drainage inception from a high-level marine surface to explain the transverse Wealden pattern. The existence of this surface, preserved on the higher summits of the area, was subsequently accepted by W. M. Davis (1895), although he considered it to have been formed sub-aerially. His conclusion, however, that the drainage is well adjusted and therefore the result of a long period of sub-aerial development failed to take sufficient note of the widespread discordant relationships which later caused H. Bury (1910) to reinstate the marine hypothesis. The controversy was to continue until the appearance in 1939 of Wooldridge and Linton's classic monograph *Structure, surface and drainage in south-east England* in which they presented the first comprehensive explanation of Tertiary landscape evolution. While accepting the sub-aerial origin of the high-level surface, which they called the 'Summit Surface' or 'Mio-Pliocene Peneplain', they concluded that the discordant drainage was the result of superimposition from a surface produced by later marine trimming. The postulated drainage evolution involves three distinct stages. A consequent drainage pattern was first established upon the Chalk cover exposed by the late Cretaceous uplift. Secondly, part of this network was deranged during the mid-Tertiary orogenic phase. Finally, the drainage pattern on the lower parts of the Mio-Pliocene peneplain was obliterated by the early Pleistocene, or Calabrian, transgression and the present discordant pattern created.

Their remarkable synthesis has since remained largely unmodified (K. M. Clayton, 1969). However, there is evidence to suggest that the hypothesis presented in *Structure, surface and drainage* should now be thoroughly reconsidered. Recent studies indicate that much of the denudation formerly attributed to the later Tertiary and Quaternary periods may, in fact, have been achieved during the early Tertiary. Similarly the capability of the sea to produce the broad erosion surfaces required by Wooldridge and Linton has been questioned (C. A. M. King, 1963), and thus raised serious doubts concerning the Calabrian hypothesis. Although the magnitude of the postulated erosional task may have resulted from a misplaced faith in the potency of marine erosion, it more probably resulted from an underestimation of the late Pliocene relative relief and therefore of the volume of rock removal required to produce a new drainage pattern. How a marine incurson could achieve such extensive drainage alterations and yet leave so little morphological evidence of its occurrence is perplexing, for adequate evidence has only been found within the London basin. It is, therefore, the intention of this paper to examine critically the accepted role of the Calabrian transgression in the geomorphological development of the south-east. It will be argued that the Calabrian incursion was an event of no great importance in the drainage evolution of the area and an alternative hypothesis involving antecedence will be proposed.

The general acceptance of the Calabrian hypothesis for over thirty years is due in large part to the many writings of David Linton, and as such is a monument to him. The discussion that follows is offered as a tribute to this great geomorphologist, for although contrary to his views it grew out of his work.

THE EXPLANATION OF DISCORDANT PATTERNS

Discordant drainage is usually explained in terms of antecedence, superimposition from a marine erosion surface or strata resting on an unconformity, glacial interference and, in some special cases, by river capture. While many of these processes have been proposed for the south-east, it is possible to indicate that some may only be of limited importance. Although it has been suggested that the South Downs and Wiltshire Chalk were glaciated (E. A. Martin, 1920; G. A. Kellaway, 1971), the generally held view is that glaciation was confined to north of the Thames and therefore cannot be used to explain the numerous discordant relationships elsewhere in the area. River capture was suggested for the Hampshire basin (C. Reid, 1902; H. J. O. White, 1910), but the idea was subsequently demolished by both Bury (1926) and Linton (1932) on the reasonable grounds that capture is a mechanism by which drainage becomes increasingly adjusted to structure. Further, to develop the existing pattern would require a great many captures and is therefore highly improbable.

To date the most convincing hypotheses have involved superimposition. That postulated by Wooldridge and Linton (1938, 1955) however, contains serious limitations, and for this reason antecedence can be considered as a possible mechanism. Antecedence has generally been avoided in the British literature, workers tending to be preoccupied with the panacea of marine trimming (C. Kidson, 1969). Linton (1932, 160), writing on the Wessex rivers, dismissed antecedence by stating that 'there remains the general objection that the Oligocene rivers were no better than sluggish streams flowing on a surface of but little gradient, and were in consequence quite incapable of maintaining themselves across the rising Miocene folds'. As will be shown, Linton's views on the nature of the land surface, the erosional capacity of the rivers and the implied rate of folding are all open to serious question. Wooldridge and Linton (1955, 64) took up a similar standpoint when they stated: 'The first hypothesis is, in fact, to suppose that the anomalous streams are antecedent. This view may be dismissed from present consideration as being unlikely and in such a case impossible to demonstrate.' This blinkered approach has been exposed by R. J. Small (1970, 251) who commented: 'The antipathy towards antecedence seems to go deeper, for geomorphologists are sometimes willing to propose theories of superimposition

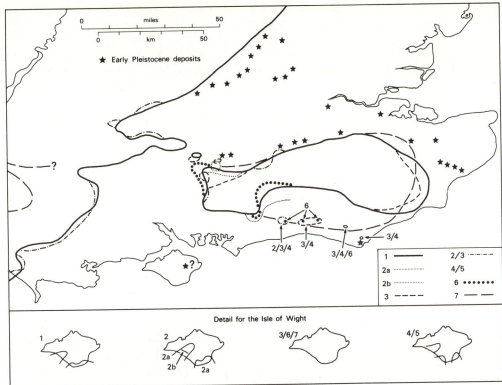

FIGURE 1. Suggested Calabrian coastlines. Note the particularly great variation in Sussex. Sources, 1, 2a, 4 and 5 Wooldridge and Linton (1955) Figs. 18, 19, 11 and 20; Wooldridge and Linton (1938) Figs. 4 and 7; 6 Linton (1956); 7 Brown (1960) and Kidson (1969)

on the very slenderest of evidence. Furthermore, one must be careful not to assume too readily that because a process cannot be easily demonstrated, it cannot actually occur.' It is therefore important that evidence for the suggested impact of the Calabrian be reviewed and, if found lacking, that serious consideration be given to the possibility that the discordant drainage pattern may, at least in part, be antecedent.

THE EXTENT AND EFFECTS OF THE CALABRIAN TRANSGRESSION

Problems of delimiting the Calabrian shoreline

The importance of the Calabrian stems largely from Wooldridge's early work in the London basin where he mapped the dissected bench and analysed the associated sediments, thereby establishing the validity of a late Tertiary 'marine trespass' (Wooldridge, 1927). Subsequent collaboration with Linton in the task of tracing the platform around the Hampshire basin led to the explanation of the discordant drainage in terms of superimposition from the Calabrian-trimmed surface (Wooldridge and Linton, 1938). This hypothesis represented the development of an earlier suggestion (Linton, 1932) involving superimposition from a thick series of aggradation gravels, and was extended to explain the discordant pattern of Sussex.

There are several major weaknesses in this important piece of work. Perhaps the most serious is that convincing morphological and sedimentological evidence for the transgression is confined to the London basin where the bench is extensively preserved and at several locations overlain by early Pleistocene marine sediments, although the Lenham Beds have been ascribed various ages back to the middle Miocene (B. C. Worssam, 1963). The shoreward margin of the platform is often marked by a low degraded bluff, above which the chalklands are mantled by a thick and variable regolith, known collectively as Plateau Drift and considered part of the Mio-Pliocene peneplain. Although Linton claimed to have traced the bench round the margin of the Hampshire basin to the Dorset coast (Linton, 1969), the feature is indistinct and doubts have been raised as to its existence. There are certainly no morphological features comparable to those preserved in the London basin and no proven marine deposits. Recent further examination of the morphology and superficial deposits on the north-western margin of Salisbury Plain has shown that the supposed marine features are indistinguishable from those of the so-called 'Mio-Pliocene Peneplain' (C. P. Green, 1969a). Along the south coast the situation is very similar. There are no clear morphological features and apart from a proven deposit near Beachy Head (F. H. Edmunds, 1927) and a dubious assemblage of battered cobbles on Mersley and Brading Downs, Isle of Wight, which correlate more satisfactorily with the 420 foot (130 m) marine level (C. E. Everard, 1954), Calabrian deposits are conspicuously absent (Fig. 1). This raises the important question of why evidence for a marine transgression should be relatively well preserved on the London basin chalklands and yet be virtually absent on the same rock type less than 100 km away.

For areas outside the London basin the discordant drainage pattern is often the main and sometimes the only line of evidence for postulating the past existence of the marine incursion that supposedly caused the discordant pattern. This unsatisfactory circular argument suffers from the further limitation that within the London basin the River Wey is the only discordant river ascribed to the effects of the Calabrian. Thus where the morphological evidence is sound, discordance is rare, and where it is unconvincing, discordance is widespread.

There are two possible major sources of error associated with the methods of shoreline delimitation used by Wooldridge and Linton in such areas. The first concerns the reality of the postulated 200 m datum which has been used throughout the south-east; the second relates to the use of suggested variations in drainage adjustment as a means of defining the shoreline.

The questions of sea-level and warping attain great importance outside the London basin because evidence for the Calabrian incursion is mainly of an indirect nature. During the 1930's it was generally considered that the south-east had been tectonically stable throughout the Quaternary. It therefore followed that erosional features of the same age were to be expected at similar elevations throughout the area. Thus it was assumed that the height of the Calabrian shoreline would be the same as found within the London basin and the marine limit could be marked accordingly on maps. The authors wrote: 'Throughout its extent in the London basin, and as far as the evidence goes in Wessex also, the upper or shoreward margin of the Pliocene marine plain is rarely much below, and practically never above 650 feet (200 m), and in no case is it known to rise to 700 feet (213 m). So generally is this the case that we may reasonably assume that the same is likely to be true for Sussex also, . . .' (Wooldridge and Linton, 1955, 80). The result of this unsatisfactory approach is the 'Gulf and Islands' coastline postulated for Sussex (Fig. 1), where the highest areas of the chalk cuesta are indicated as rising above the sea as a chain of islands. There is little evidence for such a coastline or such a sea-level.

A sea level of 200 m was suggested on the basis of morphological evidence in the London basin (Wooldridge and Linton, 1938) but reference has also been made to 690-foot (210 m) (A. J. Stephens, 1959) and 600-foot (183 m) sea levels (Wooldridge, 1960, 118). This large variation in altitude is partly due to the Calabrian bench being an old and degraded feature. Definition of the upper limit of the platform requires both accurate levelling and subsurface work, neither of which has been undertaken. It is difficult to estimate sea level from such an old feature and even more hazardous to measure it accurately enough to indicate whether or not

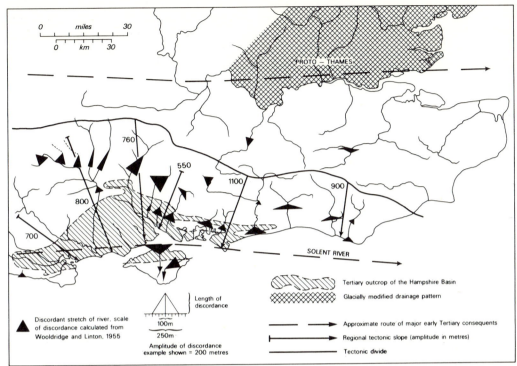

FIGURE 2. The location and magnitude of the major discordant relationships in south-east England

warping has occurred. Wooldridge, however, claimed that in the western and central portions of the London basin the bench showed no signs of warping, even though the sub-Pleistocene surface was known to decline to −75 m in East Anglia and to lie below −300 m in northern Holland. In an effort to explain these differences in elevation, Wooldridge proposed the existence of a north–south hingeline or flexure, known as the 'Braintree Line', separating the warped and unwarped parts of this surface (Wooldridge, 1928; Wooldridge and Henderson, 1955). There is little geophysical evidence for such a flexure; recent studies indicate that eastward down-warping associated with the subsidence of the North Sea basin has affected the whole of the south-east during the Pleistocene and most of the Holocene (D. M. Churchill, 1965; I. J. Smalley, 1967). It is therefore impossible to accept Wooldridge's statements on warping or the postulated coast-line configurations in Sussex and Wessex.

Both the delimitation of the Calabrian shoreline and the sequence of events put forward by Wooldridge and Linton hinge upon the presumed effects of mid-Tertiary folding on the drainage pattern. To illustrate the impact of the orogenic phase Linton constructed a map of the sub-Eocene surface that showed the effects of both regional and local deformation (Wooldridge and Linton, 1955; Linton, 1956). He suggested differential warping of this surface so that in the central Weald it stands some 1675 m above its lowest level in the Hampshire basin. Super-imposed on this macro-pattern are secondary folds showing greater development in the south and west. The authors suggested that the growth of these secondary folds disrupted the existing drainage pattern to produce a predominantly east–west, synclinal-oriented system (Fig. 4a). This concordant pattern, they argued, has persisted in modified form to the present day, except in those areas affected by the Calabrian transgression. This view must be challenged; for the contrasts in drainage adjustment are not as significant as postulated, discordant relationships being far from uncommon in the areas of supposedly adjusted drainage. The double crossing of the Penshurst anticline by the River Medway is a well-known example (Wooldridge and Linton, 1955), while discordant relationships are also displayed by the western Rother near Petersfield, the Wylye at Warminster, and several minor streams (Fig. 2). It is becoming increasingly clear that the level of adjustment postulated for the central Weald and other areas has been over-estimated, largely because it was based on the structural pattern indicated in the old 'reconnais-sance' series maps of the Geological Survey which, according to S. C. Holmes (1969), contains 'some major errors of geological structure and stratigraphy'. Therefore there are good grounds for arguing that the variations in drainage adjustment are insufficient to delimit the Calabrian transgression without the support of clear morphological and sedimentological evidence.

The morphology of the late Pliocene land surface

The Calabrian hypothesis depends upon the transgressive sea trimming a 'surface of low relief'. The greater the relief, the more difficult it would be for marine action to erase the late Pliocene drainage pattern and establish the discordant lines present today. It is therefore important to examine the argument that the late Tertiary sub-aerial phase led to the development of the Mio-Pliocene peneplain and a landscape characterized by low relative relief.

Although striking morphological evidence for such a sub-aerial surface exists in certain areas, for example on the Wiltshire Downs north of Mere, Wooldridge and Linton's statements appear contradictory as to its form and extent. They suggest that along the margins of the Wilt-shire chalk and the northern rim of the Vale of Pewsey the uplands display a polycyclic form (Wooldridge and Linton, 1955). The Mio-Pliocene surface is indicated as rising to approxi-mately 260 m O.D., above which stand residual eminences, or monadnocks, reaching a maximum elevation at Long Knoll of 288 m. In the Weald, however, it is suggested that virtually all upland areas rising above the Calabrian datum level represent remnants of the Mio-Pliocene peneplain: '. . . in the Western Weald Leith Hill (294 m), Hindhead (273 m) and Blackdown (280 m) indicate the elevations to which the peneplain rises over the Central Wealden axis' (Wooldridge and Linton, 1955, 43). This statement poses several questions, for these Lower Greensand hill masses have an elevation equal to that attained by the western monadocks and yet are ascribed to the later surface above which the monadocks rise. Similarly, it is difficult to envisage how the authors distinguished these hill masses from Butser Hill (271 m) which lies a few kilometres to the south and yet is 'probably a low residual rising slightly above the old plain' (Wooldridge and Linton, 1955, 43).

Recent work on the Chalk uplands indicates the presence of a much more complex pattern of surfaces than the sub-aerial peneplain with western residuals and exhumed sub-Eocene surface postulated by Wooldridge and Linton. Investigations in the Paris basin have revealed a series of early and mid-Tertiary surfaces, some overlain by proven Oligocene deposits (P. Pinchemel, 1954; A. Cholley, 1957). A similar pattern appears to be emerging in this country. A warped early Tertiary surface has been recognized in south-west Dorset (R. S. Waters, 1960), lying at approximately the same elevation as adjacent portions of the so-called 'Mio-Pliocene Peneplain', and has been tentatively dated as late Eocene (Bagshot). Similarly, the presence on the Berkshire Downs of surfaces originating during a composite Eocene cycle has been indicated (M. J. Clark et al., 1967). These latter workers came to the view that most of the erosional features hitherto claimed to have been produced during the late Tertiary cycle are in fact exhumed surfaces of early Tertiary origin. They concluded: 'The concept of post-Alpine (late Miocene and early Pliocene) peneplanation of the area cannot be fully discounted, but there are good grounds for seriously considering an alternative working hypothesis in which the main planation surfaces identifiable in the present landscape were developed at an earlier date.'

Further support for this view is to be found in Wiltshire where study of the north-western margins of Salisbury Plain led C. P. Green (1969b) to suggest the preservation of an early Tertiary (Oligocene) surface and the occurrence of both Miocene and Pliocene sub-aerial surfaces in the vicinity of the Vale of Wardour (Green, 1969a).

The higher parts of the chalklands therefore appear to represent a palimpsest of inclined and intersecting surfaces developed during the whole of the Tertiary. Thus the period of sub-aerial denudation that followed the Miocene orogenic phase does not appear to have achieved the degree of landscape modification and relief reduction attributed to it by Wooldridge and Linton. There is growing support for the view that the most potent phase of denudation occurred in the early Tertiary, especially as the presence of Palaeozoic and Mesozoic materials in the Eocene deposits of the London and Hampshire basins indicates that considerable denudation of the Chalk had already taken place.

If the conclusions of these studies are correct, the late Tertiary did not see a phase of extensive sub-aerial planation. The land surface over which the Calabrian Sea transgressed was therefore not only polycyclic in origin, but probably contained considerable relative relief. The Shooters' Hill, Cobham Wood and Rhode Common gravels of the London basin attain the lowest elevation for known Calabrian deposits within the area, being at about 120 m O.D., whereas the highest point in the south-east is Walbury Hill, 297 m. Thus even after excluding the effects of mass wasting on the elevation of the highest summits during the Quaternary, the *preserved* relief is still about 180 m. Small and Fisher (1970), however, recently stated in a discussion on the Washington gap that 'the Calabrian Sea affected a landscape already dissected to a height of 90–120 metres above present sea level'. A relief of at least 200 m may therefore be quite reasonably assumed for the late Pliocene land surface.

In these circumstances structural controls would have been important. The drainage and relief patterns would have been well established and would have profoundly influenced the shape of the coastline. The Calabrian transgression would have advanced into the synclinal lowlands of the London and Hampshire basins and fingered into the lower reaches of the river valleys. In Sussex a prominent chalk cuesta would have been succeeded in the north, if Small and Fisher are correct, by a Weald Clay vale at between 100 and 150 m. As extensive areas of downland at present rise above 200 m, reaching a maximum elevation of 271 m, the relief of this area could well have exceeded 180 m at the time of the transgression. The establishment of a new, discordant, drainage pattern upon such a landscape would represent an enormous feat of erosion. The lack of both morphological and sedimentological evidence for such an erosive phase in areas characterized by discordant drainage considerably weakens the Calabrian hypothesis.

Weaknesses in the marine trimming hypothesis

Although the geography of the shoreline will never be accurately known various alternatives have been suggested (Fig. 1). In several areas the proposed patterns are very different, including the debatable 'Gulf and Islands' coastline of Sussex, where the crest of the South Downs is indicated as rising above the sea as a chain of islands of variable number. It has been claimed that the former drainage pattern was obliterated, implying that marine action planed off extensive tracts of chalk and filled the Low Weald gulf with sediment. Upon the retreat of the sea the present trunk rivers extended southwards over the infill and across the platform developed between the islands (Wooldridge and Linton, 1955; Linton, 1956).

Although the 'Gulf and Islands' coastline has been generally accepted there has been little agreement on the precise geography, marked departures from the original concept being found in Brown (1960) and Kidson (1969) (Fig. 1). Linton (1956, 238) stated that 'possibly five small areas that rise above the 700 foot (213 m) contour remained unsubmerged', while the paleogeographic maps in *Structure, surface and drainage* indicate 0, 1, and 4 islands (Fig. 1) and it's figure 17 shows nine locations east of the River Arun which are claimed as remnants of the Mio-Pliocene peneplain. Inspection of the Ordnance Survey 1 inch maps reveals seven localities above 213 m and eleven more that rise to above 200 m (Fig. 5). As post-Calabrian mass wasting and scarp retreat have certainly destroyed further areas that stood at the same level at the close of the Pliocene, it would appear that the 'islands' must have been considerably more extensive than indicated by most authors (Fig. 5). Further, the very fact that the chalk cuesta stood up as a chain of islands indicates an undulating crestline, thereby implying the existence of north–south drainage lines within the South Downs prior to the transgression.

Assuming for the moment, however, that mid-Tertiary folding did fundamentally influence the later Tertiary drainage pattern, could the Calabrian have obliterated the existing pattern and superimposed a discordant one? Wooldridge and Linton argued that the two patterns were fundamentally dissimilar, thereby indicating that the influence of relief had to be negated. This could only be achieved, as King (1963) has shown, given non-resistant strata, a slow rate of transgression and adequate time. It has already been argued that the late Pliocene land surface was one of moderate relief, though the rocks would have yielded readily to marine erosion. The fossil evidence obtained from Lenham and Netley Heath indicates a transgressive sea of debatable duration (Worssam, 1963). Although it was described in *Structure, surface and drainage* as a 'brief Pliocene trespass', Linton wrote later that 'the incoming of the sea was doubtless slow' (1956, 240). The length of time that the sea has lingered in the landscape has tended to increase as it has been required to perform increasing amounts of erosion. However, it is doubtful if the sea could have achieved the suggested planation, even given that a slow transgression occurred over a reasonable length of time. The widest portion of platform preserved in the London basin is 13 km broad. Yet marine-trimmed surfaces of over 20 km and 65 km width are required respectively for Sussex and the Hampshire basin. Although planation surfaces of these widths lie within the erosive capacity of a slowly transgressive sea on 'soft' rock terrain, the requirement that the pre-existing drainage pattern be obliterated suggests that it was not trimming that was needed but the erosive removal of great quantities of strata. This is clearly far more difficult to accomplish, especially if Small and Fisher (1970) are correct and the late Pliocene land surface had been dissected to about 100 m O.D. However there is no evidence of such an intense phase of marine erosion in either Sussex or Wessex. There also remains the puzzle of how any islands survived at all if the sea was really capable of the scale of erosion postulated by Wooldridge and Linton. Further, if the relative relief at the time of inundation was about 180 m, the volume of sediment required to fill the Low Weald gulf would have been immense; yet there is no record of these materials having been found. It is also difficult to envisage where this sediment is supposed to have come from. All palaeogeographic maps show little drainage entering the gulf apart from the western Rother, and as the generally accepted Wealden coastline approaches closely to the structural divide (Fig. 2) it is difficult to postulate the existence of a more extensive southward-draining concordant network than exists at present.

These arguments indicate that even though the Calabrian Sea probably submerged much of Sussex and Wessex, it was largely impotent. The shoreline would have been controlled by structure and the existing drainage pattern. Although it probably achieved some erosional trimming, it is doubtful if it had any long-lasting effect on the drainage pattern, the rivers resuming their former courses during regression.

The late Tertiary drainage pattern

The last major limitation of the Calabrian hypothesis concerns the presumed nature of the Pliocene drainage pattern. Linton's reconstruction of the drainage pattern consequent upon the folds (Fig. 4a) shows well-developed, east–west oriented, segments. This pattern must be considered as representative of the early post-orogenic stage and he probably envisaged considerable drainage reorganization through river capture. If not, his statements concerning the western Rother are difficult to comprehend. He wrote, 'although Sussex can boast only half a dozen streams worth calling rivers, two of these, the two Rothers, are much older than the other four. They are accordant streams, well adjusted to the geological structures they traverse, and they acquired most of this adjustment in the long erosion cycle that was initiated by the mid-Tertiary folding' (Linton, 1956, 238). As no western Rother appears on Linton's Miocene drainage pattern, it must have developed during the Mio-Pliocene phase of sub-aerial denudation.

According to the established pattern of development, the pre-Calabrian drainage pattern in the southern Weald, of which the western Rother is a remnant, must have flowed either southwards across the chalklands or eastward along the Weald Clay vale. The first alternative suggests that the Rother flowed through one of the 'wind' or 'water' gaps that transect the South Downs, thereby indicating the pre-Calabrian existence of southward drainage. Further credence is given to this view by the recent work of Small and Fisher (1970) who indicate that the Washington windgap represents the line of a southward-flowing, pre-Calabrian consequent. The implications of southward drainage prior to the Calabrian are extremely damaging to currently held views, for although the upstanding areas would have been trimmed and the valleys suffered sedimentation, upon regression the rivers would almost certainly have resumed and re-excavated their former courses. It is most unreasonable to expect a completely new drainage system to develop which paralleled the original and yet was discordant.

If, however, it is assumed that the western Rother joined a trunk stream that flowed eastwards along the Weald Clay vale, serious problems arise. Strike-oriented drainage, although in adjustment with the small-scale folds, would not be concordant with the macro-structure and could only have developed through a series of river captures. Even if it had evolved it would have been beyond the capability of the Calabrian transgression to change it; its remodelling would have required the unlikely occurrence of another series of river captures (Thornes and Jones, 1969).

It is therefore suggested that southward-oriented drainage, including the present major rivers, existed prior to the Calabrian transgression. Moreover, as several of the discordant rivers in both Sussex and Wessex do not flow orthogonally to the postulated shoreline, yet further doubt is cast on the superimposition hypothesis.

These arguments demonstrate the inadequacy of the Calabrian hypothesis in Sussex and affect seriously its applicability in Wessex thus necessitating the provision of an alternative explanation of drainage evolution. Discordance is so widespread in southern Britain that a coherent hypothesis is required, and it is suggested that the most satisfactory alternative explanation involves antecedence. It should be noted that a similar hypothesis has been proposed for the discordant drainage of the south-west Peninsula (S. Simpson, 1964; 1969). However, the widespread occurrence of well-developed secondary structures in the south-east necessitates a careful examination of the applicability of the antecedent argument to this area.

THE EFFECT OF THE ALPINE FOLDING ON THE DRAINAGE PATTERN

Linton argued that the present discordant drainage pattern could not have been consequent upon the Alpine folds. Is it therefore possible that the pattern existed prior to the period of tectonic activity that produced the minor folds? A discussion of this question must draw on knowledge of the early Tertiary drainage pattern and the intensity and duration of fold activity. The late Cretaceous phase of differential uplift was accompanied by warping to produce the London basin, Hampshire basin and Wealden dome. Linton (1951) postulated the development on this surface of a system of eastward-draining consequents, of which the proto-Thames and Solent river are relevant to this paper. It is important to note that the remnants of these early consequents are still discernible, for the lower Thames and Kennett both lie close to the route of the envisaged proto-Thames while Clement Reid's Solent river appears to have been an important drainage line until post-glacial times (Everard, 1954). No worker, therefore, has suggested that the orogenic phase or the Calabrian had any profound effect on the major consequents of the region.

These drainage lines would have been the focus for a network of secondary, or lateral, consequents draining the margins of the basins. This early Tertiary consequent network would have been similar to the present pattern of trunk streams which are concordant with the macro-structure of the area but discordant with the micro-structures. Further, it should be noted that the concordant pattern of the northern Weald is similar to the discordant pattern of the south. It is difficult to envisage such similar patterns evolving in different ways; it is much simpler to suggest that the similarity indicates a common origin that predates both the Calabrian incursion and the orogenic phase.

Support for this view is to be found in the London basin where much of the drainage is implicitly accepted as being pre-Calabrian in origin. Neither Wooldridge nor Linton suggested that the rivers that once utilized the Chiltern 'wind' gaps, or the upper Lee, upper Thames, Mole, Darent, Medway or Kentish Stour, were consequent upon the Calabrian surface. Palaeo-geographical maps indicate pre-Calabrian consequents utilizing the lines of the Chiltern gaps (Wooldridge and Linton, 1955, 61), while evidence of inlets where the southern tributaries cross the North Downs suggests that these drainage lines already existed.

However, within the London basin and northern Weald the tectonic effects were generally slight and mainly confined to warping. The major question in the Hampshire basin and Sussex is whether or not the growth of the numerous periclinal folds was sufficiently rapid to disrupt the drainage pattern, as postulated by Wooldridge and Linton. If it was not, then the Calabrian hypothesis is unnecessary.

There are few quantitative data on the rate or duration of folding experienced by southern Britain during the early and mid-Tertiary. Wooldridge and Linton (1955) maintained that the late Cretaceous phase of uplift and broadscale warping was succeeded by a period of relative quiescence which lasted until the late Oligocene/early Miocene orogenic phase. More recent views, however, indicate that deformation may have occurred throughout the Tertiary and Quaternary (L. C. King, 1967; Smalley, 1967). Although the Upper Cretaceous certainly saw a period of considerable uplift and deformation, as is clearly indicated by the 'Eocene' overstep in the south-east, there is evidence to suggest that these movements continued during the early Tertiary, for the Palaeogene deposits of both the London and Hampshire basins indicate markedly oscillating sea levels. Similarly the inclusion of Mesozoic, and particularly Lower Cretaceous, materials in these deposits suggests the prior occurrence of considerable structural disturbance and erosion (Arkell, 1947). Finally, there is sound stratigraphic evidence for the early Tertiary development of certain fold structures affecting the chalk (Arkell, 1947; E. Williams-Mitchell, 1956), of which the development and partial breaching of the Purbeck anticline prior to the deposition of Upper Oligocene sediments is particularly impressive (Arkell, 1947).

There is also a growing body of evidence supporting the view that tectonic activity commenced at an earlier stage in the Mesozoic. A number of large faults and folds in Dorset, including the well-developed Marshwood Pericline, are known to have originated prior to the Upper Cretaceous (Wilson et al., 1958). Farther east there is mounting evidence for important intra-Cretaceous disturbances. Casey (1961) suggested that the Weald experienced slight folding before the deposition of the Lower Greensand, while Arkell (1947, 163) stated that during the interval between the deposition of the Wealden Beds and the Gault 'the Weymouth District was strongly folded and faulted'.

Thus it can be argued that the phase of tectonic disturbance began during the Lower Cretaceous and lasted through to the early Miocene (Wooldridge and Linton, 1955) or mid-Miocene (Clark *et al.*, 1967), with warping continuing up to the present. The macro-structural pattern, which may have originated during the Lower Cretaceous (A. P. Terris and W. Buller-well, 1965), was well established by the Eocene and continued to develop up to the Miocene. The superimposition of smaller-scale structures upon the broad-scale flexures probably began considerably earlier and lasted much longer than suggested by Wooldridge and Linton, although the major phase of tectonic activity could well have occurred during the late Oligocene and early Miocene. Therefore fold growth may be considered to have been a slow process which was associated with prolonged regional warping.

Wooldridge and Linton's arguments were based on the concept of a short phase of orogenic activity, the results of which were indicated on a map showing the deformation experienced by the sub-Eocene surface (Wooldridge and Linton, 1955; Linton, 1956). This map must be treated with suspicion, especially for those areas where few remnants of the surface and little early Tertiary material survive. From the fold pattern revealed, they postulated the development of a late Tertiary concordant network, assuming that *all* the drainage had suffered derangement. This hypothesis must be challenged for several reasons. First, the scale of the flexures is generally small and their development probably spanned several millions of years. Secondly, the pattern of regional warping would have rejuvenated the drainage which was itself consequent upon the surface produced during the early stages of warping (Fig. 2), thereby enhancing its ability to resist derangement. Thirdly, the strata are relatively unresistant to fluvial erosion. Fourthly, there is very little evidence to indicate that the early Tertiary drainage was ever actually disrupted by tectonic activity. Fifth, the concept of mid-Tertiary orogenic derangement raises the necessity of a later phase of drainage reorganization. As the Calabrian hypothesis has been shown to suffer great limitations, the task of explaining the discordant drainage becomes a severe problem.

THE HYPOTHESIS OF ANTECEDENCE

Wessex

Although the number of discordant relationships appears rather large, for Linton indicated 32 stretches, their scale is generally small. If the short discordant streams of the Weymouth low-lands are for the moment excluded from the discussion, then examination of Linton's map indicates that all but four result from the crossing of upfolds with amplitudes of less than 100 m (Fig. 2). The impact of these small and slowly growing flexures on the drainage pattern is a matter of conjecture, for there is little evidence as to the nature of the pre-Miocene land surfaces. However, it is reasonable to assume that they lay mostly above the present surface and that the Palaeogene accumulations had a more extensive outcrop. If the drainage pattern is inherited from the early Tertiary, many of the presently discordant rivers would have been flowing in valleys cut into Tertiary materials at the time when the folds were developing. In fact six of the discordant relationships still lie within the present Tertiary outcrop (Fig. 2). Maintenance of drainage lines across slowly rising folds composed of Tertiary materials would not have posed a great problem to even the smallest of Linton's 'sluggish Oligocene rivers', especially as warping would have rejuvenated them thereby increasing their capability to resist tectonic derangement (Fig. 2).

This would hold true for even the largest discordant relationships such as the crossing of the Winchester anticline by the River Itchen. The sub-Eocene surface lies at over 300 m O.D. on the crest of this fold, some 150 m above the syncline to the north. However, it is reasonable to

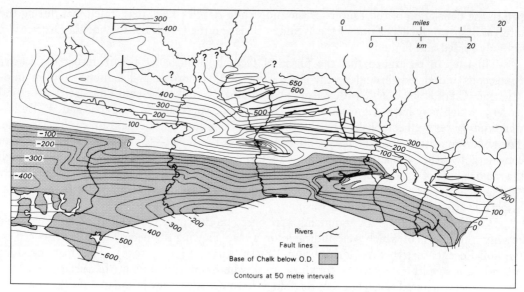

FIGURE 3. Contours on the base of the Chalk in Sussex. Note the strongly developed Pyecombe fold in the centre of the map and the complex Kingston–Beddingham structure to the south-east of it

assume that if the Itchen predates the fold then it originally flowed on Tertiary materials. With the growth of the flexure, incision would have constrained the river's ability to deviate from its course and the Chalk would eventually be exposed in the core of the growing fold.

Most geomorphologists baulk at the idea of antecedence because they doubt the ability of fluvial erosion to compete with uplift. Although several fine examples of antecedent drainage are well documented, such as the Himalayan Arun (L. R. Wager, 1937), Columbia river (P. B. King, 1959), Rhine (E. M. Yates, 1963) and Salzach (A. Coleman, 1958), the inhibitions remain. In the case of the Hampshire basin drainage the exposure of the Chalk in the cores of periclines need not have proved a severe obstacle to incision, as this limestone is relatively weak mechanically and only becomes a relief former through its permeability. When this permeability is inhibited, either through permafrost or saturation, surface flow has a great erosive capacity as is amply testified by the widespread distribution of deeply incised, chalkland dry valleys. The buried channels of the Arun, Adur, Ouse and Cuckmere which reach −30 m O.D. in places (D. K. C. Jones, 1971) further indicate how readily chalk can be eroded by fluvial action. Finally, the general acceptance of the Calabrian hypothesis includes the implicit understanding that the major discordant valleys cut into the Chalk have wholly evolved since the Calabrian. If the accomplishment of up to 180 m of incision in 2 million years is acceptable in association with the superimposition hypothesis, there is no reason why a similar rate of incision, i.e. 1 m per 11 000 years, should not be allowed in the antecedent argument. Most of the discordant rivers cut across folds with amplitudes of less than 180 m which, it has been suggested, took considerably longer than 2 million years to develop. Antecedence would therefore appear to be a reasonable hypothesis in the case of the Itchen discordance, more readily acceptable, in fact, than the removal of over a hundred metres of Chalk by a recent marine incursion that left few marks on the surrounding chalklands. Thus it can be suggested that the scale of discordance in the Hampshire basin poses no real problems to the antecedence argument.

Sussex

In order to ascertain the extent of discordance, a map of the basal contours of the Chalk (Fig. 3)

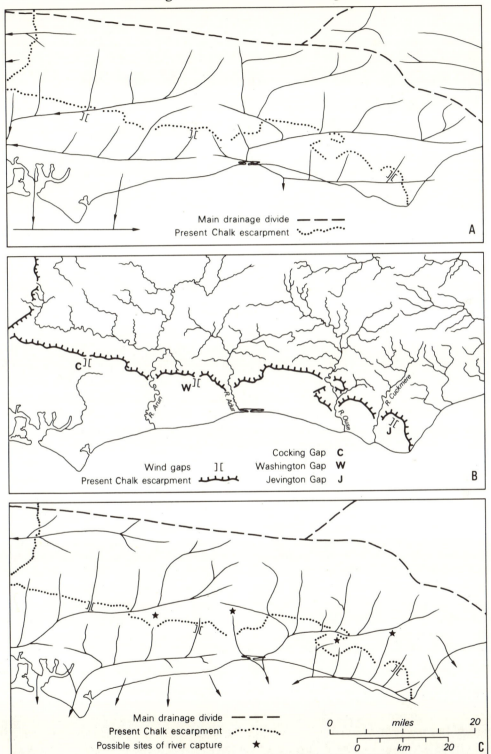

FIGURE 4. Variations in the Sussex drainage pattern (a) Linton's postulated Miocene concordant pattern, (b) the present drainage network and (c) the Late Tertiary concordant drainage pattern that would have been developed upon the pattern of folding depicted in Figure 3

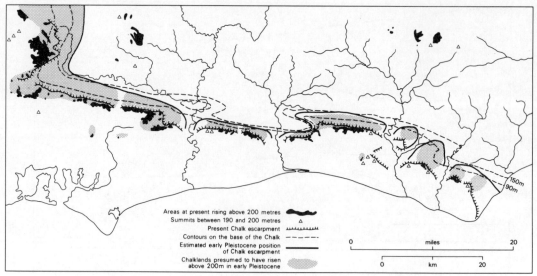

FIGURE 5. Reconstruction of the South Downs Chalk escarpment at the time of the Calabrian transgression, calculated on the basis of Figure 3 and Small and Fisher (1970). It is important to note that extensive areas of Chalk
would have remained unsubmerged by a 200 m sea level in this area

has been prepared from several sources (C. T. A. Gaster, 1937, 1939, 1944, 1951; D. W. Humphries, 1964; J. W. Reeves, 1948, 1958; Wooldridge, 1950). This sub-Cenomanian surface map
clearly shows that the southward-flowing rivers cross several well-developed periclinal structures
of variable amplitude. The Arun crosses the well-developed Littlehampton fold while the Adur
neatly transects the Henfield syncline and Pyecombe anticline, thereby crossing a fold limb with
a northward downthrow of 200 m. The Ouse cuts through the complex Kingston–Beddingham
periclinal structure which has an amplitude of about 150 m. Though the Cuckmere does not
display such an obvious disregard for structure, mapping of the Weald Clay by Reeves (1958)
revealed that it crosses one of a complex series of faulted folds between Alfriston (TQ 5203) and
Hellingly (TQ 5812).

　　The general arguments in favour of antecedence put forward in respect of Wessex are applicable to Sussex even though the sub-Eocene surface attains a greater elevation. Further support
for antecedence is to be found not only in the concordance of the main rivers to the macro-
structure, but also in the failure of the alternative hypotheses. This latter point is worth further
consideration.

　　If the rate of folding had been sufficiently rapid to derange all the drainage, as was suggested
by Linton (1956), the form of the sub-Cenomanian contours suggest that it would have produced
a predominantly southward-flowing network in Sussex (Fig. 4c). Although the pattern is
broadly similar to that indicated by Linton (Fig. 4a), it is markedly different in that the lines of
all the 'wind' and 'water' gaps are utilized by drainage lines. On cartographic evidence therefore,
it is possible to suggest the creation of the present trunk streams by invoking four river captures
(Fig. 4c), with two further captures to produce the Rother and Glynde. Although it is tempting
to suggest capture as an alternative to superimposition, for the process can produce discordance
under certain circumstances, any attempt to explain all the discordant relationships in Sussex and
Wessex by this mechanism would be to provoke the same major criticisms that Osborne White
received from Bury and Linton. Returning to Linton's reconstruction (Fig. 4a) it is difficult to
comprehend why the concordant pattern should have been obliterated and replaced by an almost

parallel discordant one. Although the new reconstruction (Fig. 4c) appears to offer more scope for the Calabrian hypothesis, the arguments considered earlier concerning the Pliocene relative relief weigh heavily against it. As neither superimposition nor capture appear reasonable explanations for the discordant pattern the simplest solution is to suggest that much of the drainage was not disturbed by fold growth and that neither of the concordant patterns shown in Figure 4 ever existed.

The suggested increase in adjustment towards the central Weald noted by Wooldridge and Linton can be explained in terms of increased subsequent stream development and superimposition from the Weald Clay without resorting to the existence of a hypothetical coastline. However, the answer may also lie in part with the selective tectonic derangement of drainage without invalidating the antecedent argument. The ability of drainage lines to maintain their courses in a basin suffering transverse fold development depends upon the rate of fold growth, the erosional resistance of the rocks and the erosional capabilities of the drainage lines. If the first two factors are approximately uniform throughout the basin, the most important variable will be the energy of the drainage lines; the generally small streams around the margins of the drainage basin being relatively easily diverted while the larger trunk stream and major tributaries are more likely to be capable of maintaining their courses. Thus the headstreams of the south coast drainage would be expected to show a greater degree of concordance near the structural divide (Fig. 2). It is worth noting in this context that the shoreline proposed by Wooldridge and Linton, on the grounds of drainage-structure relationships, parallels the structural divide fairly closely except in the south-western corner of the Weald where extensive areas of upland rise above 200 m.

The structural map (Fig. 3) furnishes further support for the antecedent hypothesis by indicating the possible occurrence of pressure-release warping. Changes in the slope of the surface were found along the margins of the chalk outcrop, especially in the vicinity of folds. Although some of these changes can be explained in terms of local variations in strata thickness and errors of outcrop delimitation, there exists the strong possibility that the larger denuded folds have experienced post-orogenic flexuring as a result of unloading. The removal of great thicknesses of competent strata from the crestal regions of periclines could lead to expansion and upward bulging of the underlying materials. Pressure-release features are known from other aspects of landform development and are probably more important than recognized at present. Bulging is known from mining activity while valley bulging has been reported with amplitudes of up to 30.5 m (S. E. Hollingsworth *et al.*, 1944). In an erosional context unloading could result in the migration of material from beneath a syncline to the axial regions of an adjacent anticline, thus leading to increased thicknesses of weaker material in the upfold, a steepening of limbs as a result of uparching and an apparent increase in the amplitude. Within the area under consideration the local removal of over 300 m of Chalk could well have caused the underlying Lower Cretaceous deposits to migrate. The following evidence can be put forward in support of this mechanism:

(1) Geological Survey cross-sections commonly show steeper dips in the vicinity of chalk escarpments. This feature has been particularly noted in Kent (H. J. W. Brown, 1925; H. G. Dines *et al.*, 1969) where the regional dip of 1–2° has been observed to increase to 8°, although the dip 'probably falls again in the Lower Chalk immediately south of the escarpment' (Dines *et al.*, 1969, 14). This marginal upturning could well be a consequence of erosional unloading due to scarp retreat.

(2) The map (Fig. 3) indicates that fold denudation appears to be associated with increased amplitudes and steeper dips. For example, the Dean Valley fold is a gentle flexure super-

imposed on the southward regional dip, while the denuded Pyecombe feature has the greatest amplitude of all folds shown, has steeply dipping limbs and rises abruptly with a marked change of dip at its margins.

(3) There is evidence of a marked local thickening of the Gault Clay beneath the Kingston–Beddingham periclinal structure. Whereas the local thickness of the Gault is given as 45 to 60 m (Gallois, 1965), well records in the core of the fold indicate greater thicknesses, the maximum being recorded at Courthouse Farm (TQ445,079) where no less than 94 m was encountered (White, 1926).

The possibility of post-orogenic flexuring has an important bearing on arguments about drainage evolution. Although it is suggested that much of the early Tertiary drainage pattern would have been able to maintain itself in the face of fold development, the growth of the larger structures still poses a problem: in the absence of field evidence it is difficult to prove that derangement would not have resulted. This difficulty is largely overcome if it is accepted that the denuded folds have been increasing in apparent amplitude since the end of the orogenic period, for the erosional performance required of the rivers during orogenic activity is greatly reduced.

The low-level discordant relationships of the south coast

Although it is proposed that the discordant drainage of Sussex and Wessex can be largely explained by antecedence, there are certain areas of low elevation, namely, the Weymouth lowlands, the Tertiary outcrop of the Hampshire basin and the western part of the Sussex coastal plain, where superimposition seems more likely. Although the rivers in these areas may have maintained their courses during fold development, the effects of superimposition from the Tertiary strata and the erosional surfaces produced by the numerous post-Calabrian marine incursions must also be considered.

The sea did not withdraw from much of the Hampshire basin Tertiary outcrop until the 112 m stage (Everard, 1954). Similarly, large areas of the Weymouth lowlands were submerged until after the 58 m stage (Sparks, 1952), while along the Sussex coast inundations continued throughout the Pleistocene. In these areas, therefore, the greater length of time for marine processes to act, the smaller widths of erosion surface required and the lower base levels involved, point to superimposition as the more satisfactory solution. This suggestion in no way invalidates the arguments set out in this paper, for the Calabrian was the first and highest of a series of marine inundations. Only where there were repeated inundations of non-resistant strata could the existing drainage pattern be obliterated. The erosion carried out by the brief submergence of extensive areas lying between the proposed 145 and 200 m shorelines would have been quite insufficient to modify the drainage pattern.

CONCLUSION

It is suggested that the explanation of discordance in southern England by superimposition from an early Pleistocene marine surface is neither necessary nor feasible. The accepted hypothesis depends upon the orogenic disturbance of drainage and the later trimming of a 'surface of low relief' by the Calabrian sea. It has been argued that the mid-Tertiary folding probably did not derange much of the existing drainage pattern because the folds are small and grew slowly in association with regional warping. It has also been argued that even if the drainage had been deranged by fold growth, the Calabrian sea did not have the capability to modify the pattern further. The lack of morphological and sedimentological evidence for the Calabrian suggests that it did nothing more than cause erosional embellishments on the lower portion of the early Pleistocene land surface.

Superimposition is not required to explain the discordant patterns if the principal of antecedence is accepted. The pattern of southerly-flowing discordant rivers which has been used as an argument for a marine plain can equally be used as an argument for antecedence, as the rivers are generally concordant with the macro-structure. For many reasons, therefore, antecedence commends itself as a simpler explanation of the discordant drainage pattern than does the invoking of a marine incursion of questionable potency.

Unfortunately there has been a general hostility towards the antecedence concept, well illustrated by the comment 'antecedence, as a hypothesis, should be the last resort of a geomorphologist seeking to explain an insequent pattern of drainage, as it is, except in ideal cases, undemonstrable' (Sparks, 1960). However, antecedence is known from many areas of the world (King, 1967). If rivers have been shown capable of maintaining their courses in the face of large-scale uplift, there seems little reason to doubt that the drainage lines of southern England, insignificant though they may appear, were not capable of doing the same in the face of slowly rising folds of small amplitude. Unfortunately there is no direct evidence for antecedence, there being no reports of warped terraces in the area. This is not really surprising considering that the main phase of flexuring ceased in the Miocene, leaving ample time for the evidence to be destroyed. In this respect the lack of evidence for an early Pleistocene transgression is very much more disturbing.

It is therefore suggested that the drainage pattern is largely derived from that of the early Tertiary. Areas in the vicinity of the major divides may have suffered drainage derangement during the early and mid-Tertiary earth movements, while drainage modifications through river capture may have occurred. Certain discordant relationships are the result of superimposition from a fairly homogeneous stratum, as was suggested for the Medway, while the low level discordances along the south coast are almost certainly the result of superimposition, either from later Pleistocene marine surfaces or the Tertiary cover. However, for the most part it is maintained that the drainage pattern reflects that produced on the regression of the Eocene sea, and that this pattern largely maintained its unity in the face of earth movements. Thus the majority of discordant rivers in southern England would appear to be best called anteconsequents (C. A. Cotton, 1948).

Although the Calabrian sea probably did transgress extensively over Sussex and Wessex, the coastline would have been controlled by the physiographic patterns of the time. Whereas the London and Hampshire basins would have been the sites of broad gulfs, the southern Weald was probably an area of rias, although the 'gulf and islands' coastline of Wooldridge and Linton may approximate to reality. Upon the regression of the Calabrian sea the rivers of southern Britain resumed their former courses.

ACKNOWLEDGEMENTS

The author is grateful to Professor M. J. Wise, Dr C. Embleton and Mrs J. A. Rees for reading and commenting on drafts of this paper. He also wishes to thank Mrs J. Baker for drawing the maps, Mrs J. Fox for typing the paper and Mr D. J. Sinclair for preparing the French translation.

REFERENCES

ARKELL, W. J. (1947) 'The geology of the country around Weymouth, Swanage, Corfe and Lulworth', *Mem. geol. Surv. U.K.*

BROWN, E. H. (1960) 'The building of southern Britain', *Z. Geomorph.* 4, 264–74

BROWN, H. J. W. (1925) 'Minor structures in the Lower Greensand of West Kent and East Surrey', *Geol. Mag.* 62, 439–51

BURY, H. (1910) 'On the denudation of the western end of the Weald', *Q. Jl geol. Soc. Lond.* 66, 640–92

BURY, H. (1922) 'Some high-level gravels of north-east Hampshire', *Proc. Geol. Ass.* 33, 81

BURY, H. (1926) 'The rivers of the Hampshire Basin', *Proceedings of the Hampshire Field Club* 10, 1–12

CASEY, R. (1961) 'The stratigraphical Palaeontology of the Lower Greensand', *Palaeontology* 3, 487–621

CHOLLEY, A. (1957) *Recherches morphologiques*, Paris, Colin

CHURCHILL, D. M. (1965) 'The displacement of deposits formed at sea level, 6,500 years ago in Southern Britain', *Quaternaria* 7, 239–49

CLARK, M. J., J. LEWIN and R. J. SMALL (1967) 'The sarsen stones of the Marlborough Downs and their geomorphological implications', *Southamp. Res. Ser. Geogr.* 4, 3–40

CLAYTON, K. M. (1969) 'Post-war research on the geomorphology of south-east England', *Area* 1 (2), 9–12

COLEMAN, A. (1958) 'The terraces and antecedence of a part of the river Salzach', *Trans. Inst. Br. Geogr.* 25, 119–34

COTTON, C. A. (1948) *Landscape*

DAVIS, W. M. (1895) 'On the origin of certain English rivers', *Geogrl J.* 5, 128–46

DINES, H. G., S. BUCHAN, S. C. A. HOLMES and C. R. BRISTOW (1969) 'The geology of the country around Sevenoaks and and Tonbridge', *Mem. geol. Surv. U.K.*

EDMUNDS, F. H. (1927) 'Pliocene deposits on the South Downs', *Geol. Mag.* 64, 287

EVERARD, C. E. (1954) 'The Solent river: a geomorphological study', *Trans. Inst. Br. Geogr.* 20, 41–58

GALLOIS, R. W. (1965) 'The Wealden District', *Mem. geol. Surv. U.K.*

GASTER, C. T. A. (1937) 'The stratigraphy of the chalk of Sussex, Part I', *Proc. Geol. Ass.* 48, 356–73

GASTER, C. T. A. (1939) 'The stratigraphy of the chalk of Sussex, Part II', *Proc. Geol. Ass.* 50, 510–26

GASTER, C. T. A. (1944) 'The stratigraphy of the chalk of Sussex, Part III', *Proc. Geol. Ass.* 55, 173–88

GASTER, C. T. A. (1951) 'The stratigraphy of the chalk of Sussex, Part IV', *Proc. Geol. Ass.* 62, 31–64

GREEN, C. P. (1969a) 'The evolution of the Vale of Wardour', *Area* 1 (2), 21–3

GREEN, C. P. (1969b) 'An early Tertiary surface in Wiltshire', *Trans. Inst. Br. Geogr.* 47, 61–72

HOLLINGWORTH, S. E., J. H. TAYLOR and G. A. KELLAWAY (1944) 'Large-scale superficial structures in the Northampton Ironstone Field', *Q. Jl geol. Soc. Lond.* 100, 1–35

HOLMES, S. C. (1969) 'The formative years in geomorphological research in south-east England', *Area* 1 (2), 11

HUMPHRIES, D. W. (1964) 'The stratigraphy of the Lower Greensand of the south-west Weald', *Proc. Geol. Ass.* 75, 39–59

JONES, D. K. C. (1971) 'The vale of the Brooks' in R. B. G. WILLIAMS (ed.), *Guide to Sussex Excursions*

KELLAWAY, G. A. (1971) 'Glaciation and the Stones of Stonehenge', *Nature, Lond.* 233, 30–5

KIDSON, C. (1968) 'The role of the sea in the evolution of the British landscape' in E. G. BOWEN, H. CARTER and J. A. TAYLOR (eds), *Geography at Aberystwyth*

KING, C. A. M. (1963) 'Some problems concerning marine planation and the formation of erosion surfaces', *Trans. Inst. Br. Geogr.* 33, 29–43

KING, L. C. (1967) *Morphology of the Earth* (2nd ed.)

KING, P. B. (1959) *The evolution of North America*

LINTON, D. L. (1932) 'The origin of the Wessex rivers', *Scott. geogr. Mag.* 47, 149–66

LINTON, D. L. (1951) 'Problems of Scottish scenery', *Scott. geogr. Mag.* 67, 65–85

LINTON, D. L. (1956) 'The Sussex rivers', *Geography* 41, 233–47

LINTON, D. L. (1969) 'The formative years in geomorphological research in south-east England', *Area* 1 (2), 1–8

MARTIN, E. A. (1920) 'Glaciation of the South Downs', *Trans. S. E. Union of Scientific Societies* 25, 13–30

PINCHEMEL, P. (1954) *Les plaines de craie du nord-ouest du Bassin Parisien et du sud-est du Bassin de Londres et leur bordures. Etude de géomorphologie*, Paris, Colin

REEVES, J. W. (1948) 'The surface problems in the search for oil in Sussex', *Proc. Geol. Ass.* 59, 234–69

REEVES, J. W. (1958) 'Subdivisions of the Weald clay in Sussex', *Proc. Geol. Ass.* 69, 1–16

REID, C. (1902) 'The geology of the country around Ringwood', *Mem. geol. Surv. U.K.*

SIMPSON, S. (1964) 'The supposed 690 foot marine platform in Devon', *Proc. of the Ussher Soc.* 1, 89–91

SIMPSON, S. (1969) 'Geology' in F. BARLOW (ed.), *Exeter and its Region*, 5–26

SMALL, R. J. (1970) *The study of landforms*

SMALL, R. J. and G. C. FISHER (1970) 'The origin of the secondary escarpment of the South Downs', *Trans. Inst. Br. Geogr.* 49, 97–107

SMALLEY, I. J. (1967) 'The subsidence of the North Sea basin and the geomorphology of Britain', *Mercian Geolst* 2, 267–78

SPARKS, B. W. (1949) 'The denudation chronology of the dip-slope of the South Downs', *Proc. Geol. Ass.* 60, 165–215

SPARKS, B. W. (1952) 'Stages in the physical evolution of the Weymouth lowland', *Trans. Inst. Br. Geogr.* 18, 17–29

SPARKS, B. W. (1960) *Geomorphology*

STEPHENS, A. J. (1959) 'Surfaces, soils and land use in north-east Hampshire', *Trans. Inst. Br. Geogr.* 26, 51–66

TERRIS, A. P. and W. BULLERWELL (1965) 'Investigations into the underground structure of southern England', *Advmt Sci.* 22, 232–52

THORNES, J. B. and D. K. C. JONES (1969) 'Regional and local components in the physiography of the Sussex Weald', *Area* 1 (2), 13–21

TOPLEY, W. (1875) *The geology of the Weald*

WAGER, L. R. (1937) 'The Arun river drainage pattern and the rise of the Himalayas', *Geogrl J.* 89, 239–49

WATERS, R. S. (1960) 'The bearing of superficial deposits on the age and origin of the upland plain of east Devon, west Dorset and south Somerset', *Trans. Inst. Br. Geogr.* 28, 89–95

WHITE, H. J. O. (1910) 'The geology of the country around Alresford', *Mem. geol. Surv. U.K.*

WHITE, H. J. O. (1926) 'The geology of the country near Lewes', *Mem. geol. Surv. U.K.*

WILLIAMS-MITCHELL, E. (1956) 'The stratigraphy and structure of the chalk of Dean Hill anticline, Wiltshire', *Proc. Geol. Ass.* 26, 221–7

WILSON, V., F. B. A. WELCH, J. A. ROBBIE and G. W. GREEN (1958) 'The geology of the country around Bridport and Yeovil', *Mem. geol. Surv. U.K.*

WOOLDRIDGE, S. W. (1927) 'The Pliocene period in the London Basin', *Proc. Geol. Ass.* 38, 49–132

WOOLDRIDGE, S. W. (1928) 'The 200-foot platform in the London Basin', *Proc. Geol. Ass.* 39, 1–26

WOOLDRIDGE, S. W. (1950) 'Some features in the structure and geomorphology of the country around Fernhurst, Sussex', *Proc. Geol. Ass.* 61, 165–90

WOOLDRIDGE, S. W. (1960) 'The Pleistocene succession in the London Basin', *Proc. Geol. Ass.* 71, 113–29

WOOLDRIDGE, S. W. and H. C. K. HENDERSON (1955) 'Some aspects of the physiography of the eastern part of the London Basin', *Trans. Inst. Br. Geogr.* 21, 19–31

WOOLDRIDGE, S. W. and D. L. LINTON (1938) 'Influence of the Pliocene transgression on the geomorphology of south-east England', *J. Geomorph.* 1, 40–54

WOOLDRIDGE, S. W. and D. L. LINTON (1955) *Structure, surface and drainage in south-east England*

WORSSAM, B. C. (1963) 'The geology of the country around Maidstone', *Mem. geol. Surv. U.K.*

YATES, E. M. (1963) 'The development of the Rhine', *Trans. Inst. Br. Geogr.* 32, 65–81

RÉSUMÉ. *L'influence de la transgression calabraise sur l'évolution hydrographique de Sud-est d'Angleterre.* L'explication du réseau fluvial discordant du Sud-est d'Angleterre se rattache à la transgression marine de la première époque pléistocène, autrement dite calabraise. Elle s'exprime en fonction d'un processus de superposition à partir d'une plaine d'abrasion marine de haut niveau. Cette communication fait l'exposé de cette hypothèse et en même temps elle présente des raisons majeures d'en nier l'authenticité. Notamment, les épreuves morphologiques et sédimentologiques touchant la transgression se limitent en grande partie au Bassin de Londres, pendant que les rapports discordants se trouvent dans le Sussex et en Wessex. Certaines faiblesses majeures sont propres a l'hypothèse de superposition; on les examine à fond et on propose que l'érosion résultante de l'incursion calabraise ne suffit pas pour un remodelage significatif du réseau fluvial, même qu'une bonne partie du pays était inondée. Ensuite, on prend en considération les conséquences de la phase orogénique du mi-tertiaire; quant au bouleversement hydrographique conséquent de la déformation des couches faibles, on conclut que les preuves font défaut. On propose, par conséquent, que les grands éléments du systéme fluvial contemporain viennent du dessin concordant effectué dès la retraite de la mer du Jeune Tertiaire. Donc, les rapports discordants du système fluvial s'expliquent en termes d'anté-cédence, excepté que les éléments voisins des côtes du Sussex et du Wessex résultent d'une superposition, soit du manteau tertiaire, soit des surfaces d'érosion faites par la séquence des hauts niveaux de mer post-calabraises. Les derniers atteignaient les terrains au-dessous de 150 m.

FIG. 1. Littoraux calabraises tentatifs. Notez la variation particulièrement grande dans le Sussex. Sources, 1, 2a, 4 et 5 Wooldridge et Linton (1955) Figs. 18, 19, 11 et 20, 2b et 3 Wooldridge et Linton (1938) Figs. 4 et 7; 6 Linton (1956); 7 Brown et Kidson (1969)

FIG. 2. Situation et magnitude des rapports discordants majeurs du Sud-est d'Angleterre

FIG. 3. Carte hypsométrique de la base de la Craie dans la Sussex. Au milieu de la carte, notez la flexure prononcée de Pye-combe; au sud-est se situe la structure complexe de Kingston–Beddingham

FIG. 4. Variations dans le réseau fluvial du Sussex (a) La configuration concordante Miocène de Linton, (b) Le système fluvial contemporain; (c) Le réseau fluvial concordant de l'ancien Tertiaire qui se serait, developpé sur le système de plisse-ment dépeint en Figure 3

FIG. 5. Reconstruction de l'escarpement de craie des South Downs a l'époque de la transgression calabraise, calculée sur la base de Figure 3 et de Small et Fisher (1970). Il est important de noter que des terrains très étendus resteraient au-dessus de mer dans cette region, donnè un niveau de mer de 200 m

ZUSAMMENFASSUNG. *Der Einfluß der Calabrischen Transgression auf die Entwicklung des Gewässernetzes im Südosten Englands.* Das verschiedenartige Gewässernetz im Südosten Englands ist erklärt worden in Form einer Überauflagerung einer hochliegenden Seeoberfläche, erzeugt bei der frühen Pleistozän oder Calabrian Seeüberschreitung. Diese Abhand-lung erörtert, dass es starke Gründe für eine Ablehnung dieser Hypothese gibt, besonders weil morphologische und sedimentologische Bewiese für eine Überschreitung sich grösstenteils auf das Londoner Becken beschränken, während die verschiedenartigen Beziehungen sich in Sussex und Wessex befinden. Eingige der anhaltenden Hauptschwächen in diesem Bewiesgrund der Überauflagerung werden geprüft und es wird vorgeschlagen, dass, obwohl der Calabrian Einfall möglicher-weise beträchtliche Gebiete überschwemmte, erzielte es ungenügende Erosion, um das Gewässernetz genügend umzubilden. Die Effekte der Mittels-Tertiären orogenischen Phase werden dann betrachtet und der Schluss gefasst, dass es wenig Beweis dafür gibt vorzuschlagen, dass das Verbiegen von schwachen Schichten zu Gewässerzerrüttung führt. Es wird deshalb vorgebracht, dass die Hauptelemente des heutigen Gewässernetzes vom übereinstimmenden Muster ererbt sind, welches durch das Zurrückziehen der frühen Tertiären See erzeugt wurde. Die Beziehungen der verschiedenartigen Gewässerformen werden dadurch in Form eines Rücklaufes erklärt, mit Ausnahme jener, die in Nähe der Sussex und Wessex Küsten vor-kommen und welche das Ergebnis einer Überauflage sind, entweder von der Terziären Decke oder Erosionsoberflächen, die durch die Aufeinanderfolge von Post-Calabrian hohen Meeresspiegeln hervorgerufen wurden, welche Gebiete unter 150 m befielen.

ABB. 1. Vorgeschlagene Calabrian Küstenlinien. Bitte beachten Sie die besonders grossen Variationen in Sussex. Quel-
lenangaben, 1, 2a, 4 und 5 Wooldridge und Linton (1955) Abb. 18, 19, 11 und 20; 2b und 3 Wooldrige und Linton (1938)
Abb. 4 und 7; 6 Linton (1956); 7 Brown (1960) und Kidson (1969)

ABB. 2. Die Lage und Crösse der Hauptpunkte der verschiedenartigen Beziehungen in Südost England.

ABB. 3. Höhenlinien auf der Grundfläche der Kreide in Sussex. Zu beachten—die stark entwickelte 'Pyecombe' Falte im
Zentrum der Karte und der Komplex der Kingston-Beddingham-struktur im Südosten davon.

ABB. 4. Variationen des Gewässernetzes in Sussex. (a) angenommenes Miozän übereinstimmendes Muster von Linton, (b) das
gegenwärtige Gewässernetz und (c) das spät-Tertiäre übereinstimmende Gewässernetz, das sich entwickelt hätte auf dem
Faltmuster, beschrieben in Abb. 3.

ABB. 5. Rekonstruktion des South Downs Kreide Steilhang zur Zeit der Calabrian Überschreitung, kakluliert auf der Basis
von Abb. 3 und Small und Fischer (1970). Es ist wichtig zu beachten, dass ausgedehnte Kreidegebiete nicht überschwemmt
worden wären bei einem 200 m Meeresspiegel in diesem Gebiet.

IV. GLACIAL GEOMORPHOLOGY

Introduction

IT WAS in Scotland where he went to live and work in 1929 that David Linton began his field observations and systematic mapping of glacially-eroded forms, initially in order to elucidate the role of ice in shaping the Scottish landscapes. Subsequently he extended his investigations to other parts of Highland Britain, Norway, New Zealand and Antarctica. His published work in glacial geomorphology reflects alike this wide, first-hand knowledge of the forms of glacial erosion and the ability of a field scientist who 'combined artistic and scientific perception in his eye for landscape' (Obituary, 1972).

Field-work in the Scottish Highlands in 1939, leading to the recognition of glacially-breached watersheds in the Cairngorms and Grampians, formed the basis of a notable research paper which revived interest in the morphological consequences of glacial diffluence and transfluence, phenomena 'of widespread occurrence—probably as widespread as that of glaciated mountains—and of fundamental importance' (Linton, 1951). After postulating the essential condition leading to either phenomenon as 'inadequacy of the available means of discharging all the accumulated névé and resulting glacier ice', he defined diffluence 'as the crossing of a major or a minor watershed by a distributary ice-tongue . . . Where inadequacy of discharge, however, is so pronounced as to lead to a rising névé line in the source region escape will ultimately be made over relatively high-level cols, often involving the crossing of major watersheds. New ice tongues arise following new routes: there is no question here of bifurcation of an existing valley glacier, and it is therefore appropriate to use the second term—transfluence. But the controlling factors . . . remain the same and the new outlet will continue to grow . . . until equilibrium is achieved. But . . . a situation that is remedied by transfluence is often a desperate one and by the time equilibrium has been reached the pre-glacial col may have been spectacularly lowered and a major breach torn in an otherwise continuous divide' (Linton, 1951, p. 5). The argument is supported by lucid descriptions and perceptive interpretations of selected instances of both phenomena and by his own highly distinctive, and illuminating, illustrations.

This effective introduction and refinement of the long-neglected concept of glacial diffluence/transfluence to explain watershed breaches in Scotland was soon followed by G. H. Dury's detailed studies of similar features in other parts of Britain and Ireland (G. H. Dury, 1953, 1955, 1957). However, Linton's next major research publication in glacial geomorphology was concerned to relate nature and amount of landscape modification by ice to degree of glacierization and to offer a hypothesis for the origin of radiating valleys in glaciated lands (Linton, 1957). 'The ice of the smaller sorts of ice-bodies, corrie glaciers and glaciers of alpine type, moves downhill under the influence of gravity along paths which are almost wholly determined by the pre-existing topography, and the erosive work they do can modify only the style and not the ground plan of the landscape. On the other hand the ice of an ice-body large enough to cover the whole of the land mass which nourished it and to reach the open sea will also descend under gravity but in a simple radial outward movement *en masse*, unimpeded by the topography submerged beneath it . . .

Between these two extremes are ice-bodies whose central parts are ice-domes wholly submerging the uplands . . . but whose marginal portions are constrained to some degree by the pre-existing topography. The tendency for free radial outflow exists powerfully but it is canalized

159

by the relief of the land along favourably disposed pre-glacial valleys, or along new troughs produced by the integration of pre-existing valley elements by breaching of watersheds. Such ice-domes may thus do much to impart a radiate pattern of the valley system of a glaciated region . . .' (Linton, 1957, p. 310). The smaller ice-domes of the Lake District and the south-west Highlands of Scotland coincided with the areas of greatest present as well as Pleistocene precipitation but the larger examples, as in southern Norway, 'spread far to leeward of the original areas of maximum snowfall. The smaller domes may be safely inferred from the coincidence of radiate troughs and an area of excessive modern precipitation. The larger reveal their former existence by the coincidence of a radiate trough system with the evidence of ice dispersal provided by the travel of erratic stones' (Linton, 1957, p. 311).

David Linton's contribution to the volume of essays in memory of A. G. Ogilvie consists of an assessment of the degree of glacial modification of the landscapes of Scotland (Linton, 1959). It brings together evidence afforded by the distribution of the three classes of landform, corries, major troughs and ice-moulded forms, which he 'regarded as having been mainly produced under distinctively different degrees of glacierization of the Scottish landmass'. The corries, a western phenomenon, 'must have characterized early and very late stages . . .: the great radiating troughs effected the discharge, not of groups of corries, but of areas of Highland Ice covering areas of the order of a thousand square miles or more, or even of considerable ice-domes: the ice-moulded lowlands and low plateaux testify to the massive advance of the ice on a broad front virtually co-extensive with the whole western seaboard from an ice-divide extending from Sutherland to Galloway. At all three stages the areas of maximum ice accumulation lay in the west . . . At all times the west was the area of glacial activity; the east, even when ice-clad, was relatively passive' (Linton, 1959, p. 42). Confirmation of this view was provided by his recognition of the little-modified pre-glacial forms, tors and relics of deeply weathered regoliths in the eastern Highlands. 'So I came to the conviction that ice erosion in Scotland has been markedly differential, being everywhere intense in the west but over wide upland areas in the east so slight as to be fairly called negligible' (Linton, 1959, p. 45).

For his Presidential Address to the Institute of British Geographers (Linton, 1963) David Linton drew on his extensive field experience to present a major research contribution on the forms of glacial erosion. The paper treats successively ice-moulded forms, troughs, corries, and a mode of glacial structure not previously discussed—'the truncation and progressive elimination of not just arêtes and mountain spurs but of what were once the actual divides between pre-glacial valleys' (Linton, 1963, p. 22; 1968). It is noteworthy as much for its demonstration of his method of approach and reasoning as for its contribution to our knowledge of glacially-eroded forms which are 'the products and the witnesses of an association of processes whose nature and mode of operation we understand imperfectly enough, but whose efficacy is such that even in the short span of time represented by an ice age they can remove mountains. These processes are in all cases the work of ice in motion. . . . Yet away from the great avenues of movement, or the slopes where the descent of even small masses is rapid, the ice lies sluggish and essentially passive. Even in the Antarctic ice may be seen to be both powerfully erosive and effectively protective' (Linton, 1963, p. 26)—the conclusion he had arrived at years previously in respect of the Pleistocene ice in Scotland.

REFERENCES

DURY, G. H. (1953) 'A glacial breach in the northwestern Highlands', *Scott. geogr. Mag.* 69, 106–17
DURY, G. H. (1955) 'Diversion of drainage by ice', *Sci. News, Harmondsworth* 38, 48–71
DURY, G. H. (1957) 'A glacially breached watershed in Donegal', *Ir. Geogr.* 3, 171–80
LINTON, D. L. (1951) 'Watershed breaching by ice in Scotland', *Trans. Inst. Br. Geogr.* 17, 1–16
LINTON, D. L. (1957) 'Radiating valleys in glaciated Lands', *Tidschr. K. ned. aardrijsk. Genoot.* 74, 297–312

LINTON, D. L. (1959) 'Morphological contrasts of eastern and western Scotland' in R. MILLER and J. W. WATSON (eds) *Geographical essays in memory of Alan G. Ogilvie*, 16–45

LINTON, D. L. (1963) 'The forms of glacial erosion', *Trans. Inst. Br. Geogr.* 33, 1–28

LINTON, D. L. (1968) 'Divide elimination by glacial erosion' in H. E. WRIGHT JR and W. H. OSBURN (eds) *Arctic and alpine environments*, Proc. 7th Congr. INQUA, 241–8

(1972) 'Obituary: David Leslie Linton', *Trans. Inst. Br. Geogr.* 55, 171–8

Zones of glacial erosion

KEITH M. CLAYTON

Professor of Environmental Sciences, University of East Anglia

Revised MS received 10 August 1973

ABSTRACT. Investigations of areas which have been glaciated suggests that they have been altered to a varying extent by glacial erosion. A qualitative scale of five zones of modification of the landscape by erosion is proposed. The scheme is described in relation to parts of New York and Pennsylvania, and also to the northern part of Great Britain. It is emphasised that the scheme is descriptive of the landforms left by the melting ice, and is not likely to be related in any simple way to the internal dynamics of continental ice sheets.

In August 1964, during the 20th International Geographical Congress, David Linton and I contributed a paper (K. M. Clayton and D. L. Linton, 1964) to the geomorphology section with the title 'A qualitative scale of intensity of glacial erosion'. The paper as read consisted of two separate, although closely related, contributions. First I described the landforms of the Finger Lakes area, and presented a case for regarding them as a sequence of landscapes increasingly altered by glacial erosion from south to north. This arrangement was summarized in a map of 'Zones of Glacial Erosion' which showed the boundaries of the five zones (one of which was divided) described in the published abstract. David Linton followed me, discussing the survival of pre-glacial forms such as tors and boulder fields in relation to the glacial limit and the least modified parts of the glaciated landscape. Most of his maps referred to Scotland, and he, too, concluded with a map of zones comparable with the one I had made for the Finger Lakes.

The ideas we put forward in 1964 had arisen quite independently. I had developed them in working in upstate New York in 1961 and put them forward in a paper I read to the Association of American Geographers at Miami Beach in 1962. David Linton saw the paper early in 1963, and he referred to some of its conclusions in his Presidential Address to the Institute of British Geographers at Swansea in January 1963 (Linton, 1963). Arising from his own recognition of degrees of alteration by glacial erosion, David Linton suggested we present the joint paper I have already described. On the day we read the paper, he had seen my maps and argument, but I had not seen his. It says much for the convergence of our ideas that these two contributions formed a coherent paper. It had always been our intention to publish a revised version of the paper as read and by July 1965 we had agreed a revised zonal scheme and I had written my part of the text. Other writing commitments and the overall pressure on his time prevented David Linton from completing his part of the text, although I found some notes on Scotland (which I have not attempted to use) in his file after his death. The paper which follows is necessarily an incomplete version of the paper we had aimed to write together in that, while I include his maps of Scotland, I do not feel able to discuss them in the way he would have done. The model is presented in full, but one of the examples is not worked out in complete detail. Sadly, it is not the only contribution our subject has lost through the untimely death of this most able geomorphologist.

THE FORMS OF GLACIAL EROSION

It has long been a commonplace of geomorphological textbooks to stress that, while few individual forms produced by glacial erosion are unique to that process, the assemblage of forms is convincing demonstration of their origin. Part of the problem in attributing individual forms to glacial erosion is the way in which they vary. Time since deglaciation, the varying influence of structure, and the degree to which pre-existing non-glacial forms have been modified all affect the resulting landform. This is true of individual landforms and also of the landform assemblage of a glacially-modified landscape. As in so much of geomorphology, the range of possible combinations of even these three major variables is so great that it is not always easy to establish the extent to which the landforms of an area are the result of its pre-glacial, glacial or post-glacial evolution. Identification of the part played by the very different processes in glacial and non-glacial conditions is necessary if geomorphological understanding is to get very far.

One device used by geomorphologists is the classification of landforms on a genetic basis: thus such features as cirques, roche moutonnées, diffluence troughs, truncated spurs and so on may be identified in the landscape, and conclusions drawn about the work of ice in shaping the present land surface. In rather homogeneous rocks and where relative relief is appreciable, features such as these may be rather easily and securely identified and we find little disagreement about the role of ice in fashioning the landscape. The cirques, troughs and lake basins of the Lake District or Snowdonia are cases in point. There are many upland areas, however, which it would seem must have been ice covered, but where there is no agreement about the geomorphological work done by that ice cover. The Welsh uplands outside Snowdonia are a good example: here, in the absence of many readily-identifiable cirques and with thick bedded screes on what may or may not be trough walls, opinions range from the view that this is wholly a periglacial landscape, to the other extreme of attributing all the landforms (other than a periglacial decoration) to erosion by ice. Arguments based on the recognition of specific landforms, attributed to one particular process, fail to converge because there is doubt about the interpretation of individual landforms. A 'cirque' may be the result of nivation or areal glaciation; periglacial slope deposits and tills may possess common characteristics. A frequent cause of confusion is that many landscapes include both 'glacial' and 'periglacial' forms, and individual research workers may unwittingly select the evidence they require to sustain their own hypothesis. Confusion has also arisen through the assumption that adjustment to structure is a criterion for non-glacial forms: in fact ice is as capable of exploiting structure (in the broad, Davisian sense) as any other erosional process. We must acknowledge that in glaciated areas, glacial landforms exist alongside non-glacial landforms and are not readily separated from them.

One other example of this uncertainty is worth elaborating because it occurs in the area selected here to illustrate the concept of 'zones' of glacial erosion. Writing a regional geomorphology of the United States W. D. Thornbury (1965) faced the problem of a conflicting literature on the glaciated upland landscapes either side of the Mohawk valley, northern New York State. The position of the glacial limit implies that both the Adirondacks and the Catskills were covered by ice, and this is supported by the widespread occurrence of till and fluvioglacial deposits (including moraine and kame forms) in the valleys of these two upland areas. Further, several authors had suggested that at least for part of the time they were glaciated, these two uplands had acted as centres of dispersal for ice derived from local snowfall. This implies (and it was said was supported by) centrally-located cirques, although for lithological or other reasons (perhaps a brief period of radial flow after the main Laurentian ice sheet retreated?) these cirques were not very well developed. Against this clear evidence of former ice cover, if rather poorly supported by associated forms of glacial erosion, some authors had argued that in the Adirondacks at least a pre-glacial landscape had survived, decorated by glacial and fluvioglacial drifts and erratic boulders, and cut into by a very few poorly-developed cirques, but fundamentally with the overall pre-glacial relief preserved. Thornbury, faced by the necessity of disposing of these contrasting views in the sentence or two his text allowed, wrote: 'Glacial cirques are present

TABLE I

Zones of glacial erosion

Zone	Lowlands	Uplands
O	No erosion Head on weathered rocks and slopes Outwash in concavities Rare occurrences of till on weathered rock	No erosion Outwash on valley floors Solifluxion deposits on slopes Boulder fields and tors on divides
I	Ice erosion confined to detailed or subordinate modifications Concavities drift mantled but convexities may show some ice moulding Occasional roches moutonnées Ice-scoured bluffs in favourable locations	Ice erosion confined to detailed or subordinate modifications Suitable valley slopes ice steepened Entrenched meanders and spurs converted to rock knobs Interfluves still commonly *Zone* O
II	Extensive excavation along main flow-lines so that concavities may be drift free or floored by outwash or post-glacial deposits. Isolated obstacles may be given ovoid or cutwater forms if of soft rock, or crag-and-tail with associated scour troughs if hard Margins of larger masses converted to ice-scoured bluffs or planar slopes	Conversion of pre-glacial valleys to troughs common, but usually confined to those of direction concordant with ice flow Some diffluence; transfluence rare Interfluves may be *Zone* I or even *Zone* O, and separated from troughs by well-marked shoulders
III	Pre-glacial forms no longer recognizable but replaced by tapered or bridge interfluves with planar slopes on soft rocks, and by rock drumlins and knock-and-lochan topography on hard	Transformation of valleys to troughs comprehensive giving compartmented relief with isolated plateau or mountain blocks. *Zone* O may still persist on interfluves at sufficiently great heights
IV	Complete domination of streamlined flow forms even over structural influences	Ice moulding extends to high summits Upland surfaces given knock-and-lochan topography (sometimes of great amplitude at lower levels) Lower divides extensively pared or streamlined

on the faces of many of the Adirondack peaks, but in general they are not so strongly developed as those in the White and Catskill Mountains' (Thornbury, 1965, 182). My own view happens to be that the Catskills have no true cirques, so the 'not so strongly developed' may seem to raise major problems for the view that the Adirondacks have been intensely glaciated. However, presence or absence of cirques is not the sole criterion for a glacially-eroded landscape, and I shall return to the Adirondack/Catskill problem later in this paper.

A ZONAL SCHEME

Areas which have been glaciated commonly show a strong contrast between those parts of the landscape affected by ice action, and those facets that survive comparatively unmodified from the landscape overrun by the ice. The contrast can be marked, whether the areas affected by glacial erosion have been attacked by abrasion or plucking. The recognition of such differences is part of the field technique involved in the understanding of glaciated landscapes, and it is probably true to say that the stronger such contrasts are, the easier it is to appreciate the contribution of ice erosion to the formation of the present landscape. Where ice action is weak it is obvious that interpretation is difficult and will depend on subtle evidence that must be most carefully assessed. It has not, perhaps, been understood that a landscape that has been completely transformed by the passage of ice is also difficult to identify. My own experience suggests that landscapes which owe their entire gross form to ice action have often not been recognized as such at all, or have at least been subject to appreciable misinterpretation. It may be helpful to report that David Linton took this view too.

Without the contrast of unaltered slopes, of glacial forms alongside, and so readily differentiated from non-glacial forms, the contribution of glacial erosion is easily misjudged. Fortunately it appears that such landscapes form the end member of a sequence of landscapes showing progressive modification by ice, and further that these landscapes occur frequently in a simple geographical sequence. If these zones can be recognized and mapped in a systematic way, we have a new tool to aid us in understanding landscapes which have been glaciated in the past.

The sequence of zones proposed is set out in Table I. As is to be expected, there is a transition from one zone to the next, although on the basis of the definitions we give here, zonal boundaries can be drawn as line boundaries on maps of the scale 1:250 000. It would also be possible to subdivide these zones in many areas, as a comparison of the map of the Finger Lakes area shown here (Fig. 1) with the map drawn before the present zonal scheme was devised (Clayton, 1964, inset Fig. 4) will show. However, it seems likely that such subdivisions are only of local application, and they may well reflect local relief or structure as much as the intensity of glacial erosion. They are not part of any general scheme and are not discussed here.

THE GLACIATED AREA OF NEW YORK AND ADJACENT PARTS OF PENNSYLVANIA

Although the map covers a much wider area, it will be best to begin with the glaciated Allegheny plateau, including the Finger Lakes themselves. This area has been described in greater detail in a paper published nearly ten years ago (Clayton, 1965), to which the reader is referred for a more detailed description of the field evidence for a sequence of glacial landscapes. It is of particular interest because the zones are arranged in a very simple way, and the relationship between them is easily seen (Fig. 6).

Ice moving south from Canada spread out south of the Lake Ontario basin (as is well shown by the drumlins of the lowland south of the lake) and moved southward to invade the higher ground of the Allegheny Plateau. The lake shore is at 75 m; the northern edge of the upland is marked by the 150 m contour in the east near Syracuse, by the 200 m contour in the west, south-west of Rochester. The plateau itself rises above 500 m almost everywhere, reaches 600 m in several areas, and attains 750 m in Potter County, Pennsylvania. Today there is a great embayment in the escarpment occupied by the Finger Lakes. It has been shown that this embayment, with its associated lake troughs, was cut by the ice, and in pre-glacial times the existing north-facing escarpment east and west of the Finger Lakes must have continued across from Syracuse to west of Canandaigua Lake. The easy route down the Mohawk valley led to far less penetration by active ice in the area east of Syracuse, and the escarpment in the area south of Utica approaches the form of the unmodified Allegheny scarp. Similarly, the very well-developed dendritic river pattern of the plateau between Utica and Binghamton probably continued westwards into the area now occupied by the Finger Lakes and the very broken plateau to the south.

The ice streamed over this plateau, extending southwards into Pennsylvania. The glacial margin lies 200 km south of the original escarpment at Williamsport, Pa., and 300 km from it in New Jersey. Immediately south of the Finger Lakes the ice failed to reach the southern boundary of the Allegheny Plateau, and the direction of flow towards the New Jersey boundary suggests that it was ice from the Hudson valley that was responsible for the more southerly extension of the glacial margin in New Jersey. The margin of the glaciated area shows evidence of multiple advances, and three major drift sheets have been mapped in Pennsylvania; the older Kansan and Illinoian advances extended 65 km farther south down the Susquehanna valley than did the Wisconsin advance. The glacial margins here have been mapped on the evidence of the drift sheets and their associated soil development: drift landforms are poorly developed and features

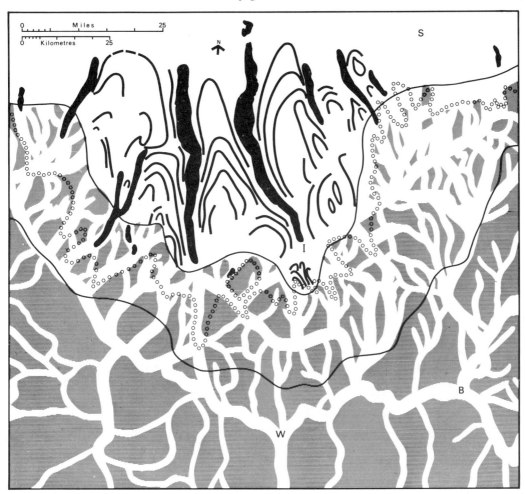

FIGURE 1. The Finger Lakes and part of the glaciated Allegheny Plateau. Darker shading—lakes; heavy lines—contours on streamlined interfluves between lakes (every 61 m); line of open circles—Valley Head moraine; lighter shading—remnants of pre-glacial plateau separated by glacial troughs which are left white. The two continuous lines divide the area into three zones of glacial erosion; IV to the north, III in the middle, II to the south. S = Syracuse; I = Ithaca; W = Waverly; B = Binghamton

due to ice erosion are entirely absent near the limit of the ice advance. It is interesting to note that fine tors and boulder fields are developed in part of this marginal zone.

The area studied in detail at this stage comprises most of Zone II, and all of Zones III and IV. Zone IV extends northwards towards (and beyond?) Lake Ontario, and no northern boundary (if one exists) has been mapped. This point is taken up in a later comment. The most striking feature of the Zone Map is the parallelism of the boundaries of these Zones. Zone III has a range of width between 11 and 27 km, with most of the values lying close to the mean of 20 km. This parallelism is a direct reflection of the symmetry of ice flow and the overall structural, lithogical and topographical uniformity of the Allegheny Plateau. The simple, radiating flow pattern that developed was unaffected by any broadscale variations in the resistance of the plateau to the passage of the ice, or to the processes of glacial erosion.

The contrasts between Zones II, III and IV are considerable (Fig. 1). In Zone II the ice moved southwards most easily wherever existing river valleys led in a southward direction. This is shown by the modification to a trough form (complete with good examples of truncated spurs) of tributaries joining the Susquehanna or the Cohocton-Chemung from the north. But it is marked equally by the valleys that lead across the water-parting south of the Susquehanna into Pennsylvania. Here the trough form is seen not only in the river valleys themselves, but also in deep cols across the water-parting. Thus the Appalachin Creek leads south to the Wyalusing Creek and is followed by State Route 858 from Apalachin (west of Binghamton) to Rushville, Pa. The col is at 375 m, over 120 m below the local plateau level. A similar col to the west, followed by State Route 187, is also 120 m below the local plateau level, as is that on Route 267 to the east. Although these cols may well have carried some meltwater, their form strongly suggests erosion by ice.

These north–south through troughs (glacial breaches) are about 8 to 16 km apart. They are virtually the only major recognizable results of erosion in Zone II. No doubt other valleys roughly aligned with the regional direction of ice movement have been modified to some smaller extent, but where this does not include a col that breaks the general height of the water parting, it is impossible to attribute such features to glacial erosion with any certainty. Similarly, some minor plateau landforms may be the work of ice, but most of the landscape closely resembles the unglaciated plateau to the south, and it seems likely that, despite a continuous ice cover, the plateau was essentially untouched by glacial erosion.

Such an interpretation receives support in the landscapes of Zone III. Here there is hardly a valley that has not been modified by the passage of ice, whatever its direction, however, sharply it may bend. Chauncy Holmes assembled impressive evidence of the rapid changes of direction of glacial striae as the ice conformed to some of these valley bends (C. D. Holmes, 1937). He also discredited views that successive ice advances from different directions had picked out and modified valleys running in different directions; substituting the hypothesis that the ice sheet filled and moved through all this trough complex at the same time. The flow divided and then reunited as the ice made its way across this intensely dissected upland. Where the trough junctions have not been concealed by subsequent deposition, the lack of conformity of trough-floor levels is striking evidence that this is an ice-cut landscape, conforming to the law of cross-sections, not the law of accordant junctions. Discordant junctions are prominent landscape features, particularly where the flow divided, as south-east of Watkins Glen (where the fine trough of Havana Glen is 200 m above the present drift-filled floor of the Watkins Glen valley) or where the Nanticoke Creek trough heads 100 m above the main Tioughnioga valley at Whitney Point, north of Binghamton.

This maze of intersecting troughs, often marked by very steep and often very straight side walls, contrasts with the plateau surface which seems almost unmodified by the passage of the ice sheet. Flow was apparently concentrated in the valleys where the depth of ice was greatest, a conclusion reinforced by the contrasting lithology of valley and plateau drifts. The asymmetry of minor relief features on the upland surface is characteristically related to the dominantly southerly dip; the steeper slopes face north, not to the south as would be expected were these hummocks shaped by glacial erosion. The increased frequency of modified troughs (and the tendency for the troughs to be wider farther north) leads to a progressive decrease in the area of the intervening remnants of the original plateau. Along the northern edge of Zone II, these trough-bounded segments of the plateau surface reach 500 km^2 and are rarely less than 250 km^2. Along the southern margin of Zone III their sizes range from 75–180 km^2, while in the dissected area some 25 km south of Syracuse in Zone IV, there are two dozen fragments that average less than 5 km^2 apiece.

This progressive northward decrease in the area of the plateau remnants continues to the point where they cease to exist at all. This is the boundary between Zones III and IV. In terms of

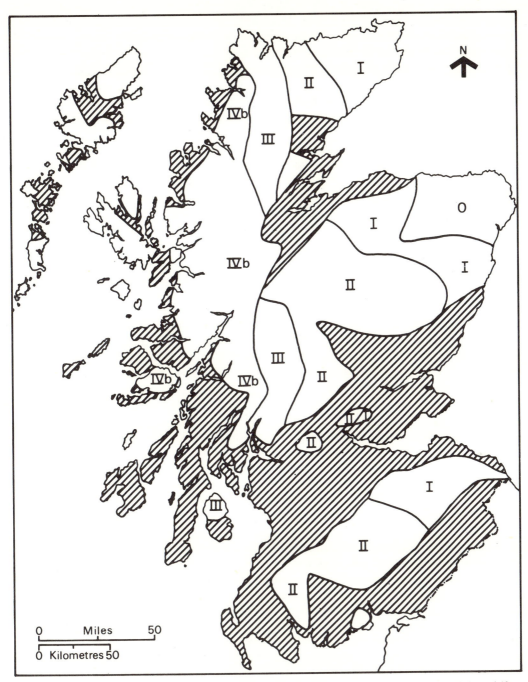

FIGURE 2. Scotland: zones of glacial erosion (D. L. Linton, 1964). The shaded areas are ice-moulded or drift covered to summits

ice movement, the elimination of these plateau fragments shows that active, erosional ice flow had extended over the whole land surface. It was still more intense in the deeper troughs than on the interfluves, but in Zone IV the whole landscape suffered recognizable glacial erosion. The extension of flow to the whole landscape meant that the pattern of flow could become simplified, the obstruction offered by the plateau remnants had been overcome. Once overcome at all, they could be reduced from above, not just attacked from the sides as seems to have been the case in Zone III, where the reduction could only come from trough widening. The lowering which this overall ice movement caused is considerable. East and west of the central Finger Lakes survive patches of the former upland, the rise of the scarp-forming rocks northward (due to the southerly dip) giving an overall cuesta form. South of Syracuse the highest point is now 585 m, south-east of Lake Skaneateles. The corresponding point west of the Finger Lakes is Canadice Hill, 685 m south of Honeoye Lake. On the same latitude, the highest point of the central Cayuga–Seneca interfluve is at only 245 m, and even 16 km to the south it only rises to 460 m.

Even more striking than the amount of lowering is the overall transformation of the land surface. The process of modification of the surface under the ice was no doubt far from complete when the Wisconsin ice-sheet melted away. But appreciable progress had been made in the conversion of the trough-cut plateau into a streamlined surface that provided a more efficient shape for maximum ice discharge. This can be seen from the symmetrical and simple shapes of the troughs and divides, and from the radiating arrangement of the major landforms. The symmetry of the radial pattern is far from perfect, but it is definite enough to be one of the most significant aspects of the Finger Lakes. The size and depth of the lake decreases from the central interfluve outwards; the distance between lakes (or the alluviated troughs that have replaced them) decreases in the same directions. The central interfluve between Lakes Cayuga and Seneca is more completely streamlined and lowered than those to both the east and west. Some of these relationships are expressed diagramatically in Figure 1.

SCOTLAND

This detailed discussion of the Finger Lakes area demonstrates the field relationship of the five zones. It will be noted that the Zones succeed each other in sequence, and that they are related in a simple way to the pattern and intensity of ice movement. The maps of Scotland prepared by David Linton in 1964 but not previously published (Figs 2 and 3), show that the same orderly sequence occurs there, although the pattern is more complex than around the Finger Lakes. Fundamentally the sequence is a north–south banding, the most modified landscapes in the west, the least modified in the east. The east-west contrast in the development (and elevation) of glacial landforms in Scotland was described by David Linton in his contribution to the Ogilvie Memorial Volume (Linton, 1959). The zonal map adds precision to the points developed in that paper by establishing the east–west sequence and locating the boundaries on the map.

A striking feature of the map is the orderly arrangement of the zones, and the way in which the zonal boundaries within separate upland areas form part of a regional, Scotland-wide pattern. The interviewing lowlands are largely drift covered, and cannot so readily be placed within the zonal pattern. At the time of the 1964 paper, Linton was inclined to take these lowlands out of the scheme into a division he then called IVa. Later he extended the zonal scheme into the arrangement set out in Table I, and it is possible to sketch in the zonal boundaries within the lowland areas using the criteria set out there. Figure 5 shows Scotland and northern England with the zonal boundaries extended across the lowlands. In order to establish the fundamental upland/ lowland contrast the boundaries are broken in lowland areas.

In several of his papers, in particular one published in 1957, David Linton emphasized the close relationship between contemporary and Pleistocene precipitation in the oceanic marginal

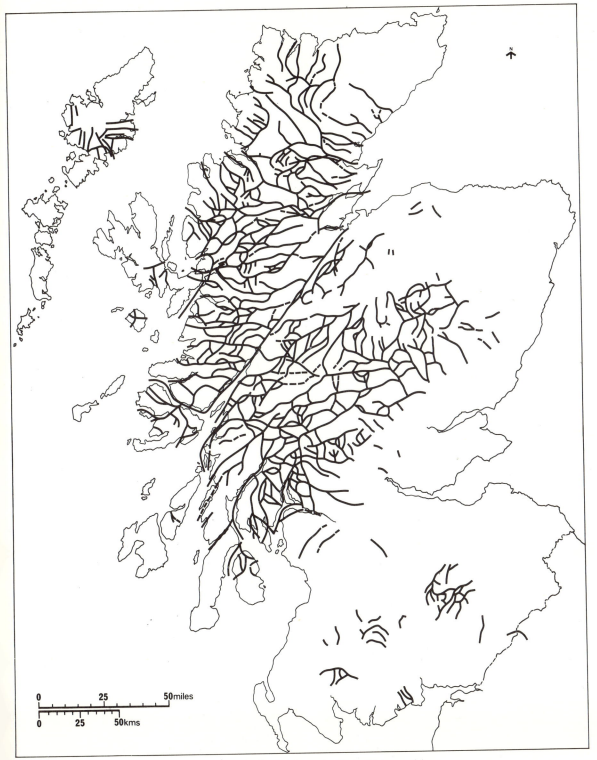

FIGURE 3. Scotland: glacial troughs (D. L. Linton, 1964)

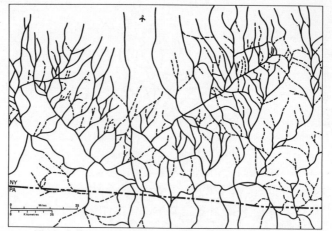

FIGURE 4. Finger Lakes and adjacent areas of upstate New York: glacial troughs

FIGURE 5. Scotland and northern England: zones of glacial erosion (as defined in Table I)

uplands of north-west Europe, and the control this had on the pattern of ice movement. His main areas of radial dispersal from ice domes sustained by very heavy precipitation are necessarily areas classified as Zone IV, since the radial movement of ice created its own system of radiating troughs whatever the pre-existing pattern. This is clearly the fundamental influence on the zonal pattern, the dispersal of ice away from centres of maximum precipitation resulting in less vigorous movement so that the land surface is modified less and less. Transformation was less complete in Scotland than in the Finger Lakes area, so that, for example, the map of glacial troughs does not show any great contrast between Zones III and IV. To take one example, the great line of Loch Lomond was established by the integration of transfluent troughs (D. L. Linton and H. A. Moisley, 1960), but the lowering of divides west and east of the loch did not proceed so far that the original west–east troughs were eliminated. A stage beyond that of the through-valley network in the Finger Lakes area has been reached, but the simplification of the trough network achieved in the Finger Lakes area itself does not occur in Scotland (Figs 3 and 4).

UPSTATE NEW YORK

Figure 6 extends the zonal boundaries of the Finger Lakes area to a much larger part of New York State and on the same scale as Figure 5. While the map of the Finger Lakes area is the result of two years' field study, this wider area has been covered by traverses carried out in 1965, 1969 and 1972, and the reliability of the zonal boundaries is much lower than for the Finger Lakes. However, the pattern is unlikely to be altered however much individual boundaries are shifted. It will be seen that there is a northward shift in the boundaries east of the Finger Lakes until the longitude of the Champlain–Hudson valley is reached. This shift is related to three features, the falling off of ice movement away from the vigorous dispersion from the Lake Ontario basin, the obstruction offered by the Adirondack Mountains, and the diversion of regional ice flow west and more particularly east of the higher ground that culminates in the Catskills. The major difficulty in constructing this map is the status of the north–south lowland of the Hudson valley. If this broad vale was in existence in much its present form before the glacial period, then the features within it represent relatively minor modification by the passage of ice and it lies, perhaps, in Zone I. At the other extreme the Hudson valley may be the product of glacial erosion, an Icelandic trough, to use Linton's (1963) terminology, of formidable proportions. On that basis it should lie in Zone III, or at least in Zone II. The existence of roches moutonnées in Central Park, New York, may suggest that the main part of the Hudson valley farther north is properly classified as Zones III and IV, and that is the view taken here.

In an early section of this paper, attention was drawn to the very different views taken of the extent of glacial modification in the Adirondack Mountains. That those views remained unresolved despite the field evidence available suggested that simply gathering more information on landforms of glacial or non-glacial genesis would not resolve the problems on a convincing way. The regional pattern of zones mapped in Figure 6 depends on evidence gathered both within and outside the Adirondacks. It shows that the great amount of glacial modification they have undergone contrasts strongly with the Catskills. The absence of well-developed cirques, which Thornbury (1965) happens to have concentrated on, is not due to little or to ineffective glaciation. It is the result of the style of glaciation: the Adirondacks were overrun most vigorously by the Laurentide ice sheet, and are dominated by the effect of a regionally transfluent ice sheet of considerable depth. Only the very highest summits are little modified by the passage of this ice, not because they were not covered, but because the ice was able to flow round them. Summits that rise to intermediate levels are completely ice moulded: compare for example the rather castellated ridges of Whiteface Mountain (1485 m) with the smooth, gently rounded, glacial

174 K.M.CLAYTON

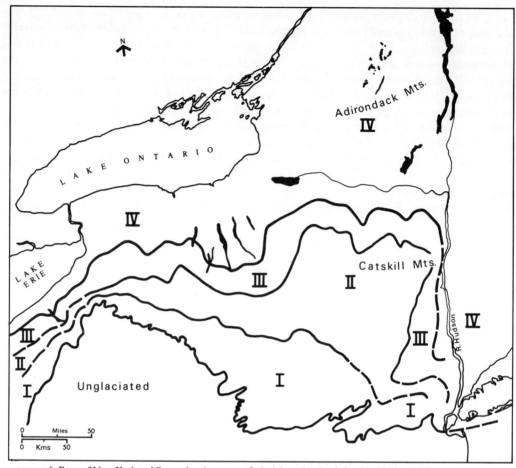

FIGURE 6. Parts of New York and Pennsylvania: zones of glacial erosion (as defined in Table I). The outermost line marks the limit of glaciation as shown on the glacial map of North America east of the Rockies, 1959. The boundary between Zones I and O has not been located and is therefore omitted from the map

pavement that forms the summit of Blue Mountain (1146 or 600 m above Blue Mountain Lake). At lower levels huge transfluent troughs break through between even the highest mountains, perhaps the overhanging walls of Avalanche Lake (715 m) below the extraordinary rock slope to the west of Colden Mountain (1410 m) are the best example of this. On the west side of this trough the summit of Mount Algonquin (1558 m) is heavily glacierized.

CONCLUDING COMMENTS

This paper has presented a qualitative classification of landscapes affected by glacial erosion. The criteria are geomorphological and thus the system of zones is an assemblage of a particular specialized type of landform region. It is important not to lose sight of this point: this is not a regionalization of ice-sheet dynamics, of the period or frequency of glaciation or of rock lithology in relation to the processes of glacial erosion. No doubt the landforms are related to these and other factors, but only locally will the relationship be a simple one. In time it should be useful to compare these patterns with the models of J. T. Andrews (1972) or G. S. Boulton (1972), although the fact that the forms have been created by the accumulative effect of successive ice sheets of different dimensions (and perhaps with differing regimes) will complicate interpretation.

Recently W. A. White (1972) has speculated that the greatest amount of erosion occurred in the central areas of the great continental ice sheets. This is an interesting suggestion (whether true or false), and some of the evidence he produces (e.g. areas of Precambrian outcrops) comes from the central areas of these great ice sheets, and so may be relevant to the problem. Unfortunately much of his evidence comes from the peripheral areas, and he uses the Finger Lakes in particular. Whatever the virtues of the zonal sequence I have championed in this paper, it is applied on the ground on the basis of landform assemblages; it is not a scheme resting on such secure theoretical foundations that it can be extended across two or three thousand kilometres of the Canadian Shield. A northward increase in intensity of glacial modification of the landscape across a 150 km belt south of Lake Ontario is no evidence of the relationships farther north. An hypothesis that erosion reached a peak in the Lake Ontario basin (which would seem to be a sizeable glacial excavation) and declined to the north as well as to the south, is just as consistent with the Finger Lakes evidence as the hypothesis White so arbitrarily adopts. It would be unwise to attempt to extend the area surveyed other than by the same process as that used to produce Figure 6.

REFERENCES

ANDREWS, J. T. (1972) 'Glacial power, mass balances, velocities and erosion potential', *Z. Geomorph. Suppl.* 13, 1–17
BOULTON, G. S. (1972) 'The role of thermal régime in glacial sedimentation', *Spec. Publ. Inst. Br. Geogr.* 4, 1–20
CLAYTON, K. M. (1965) 'Glacial erosion in the Finger Lakes Region (New York State, U.S.A.)', *Z. Geomorph.* 9, 50–62
CLAYTON, K. M. and D. L. LINTON (1964) 'A qualitative scale of intensity of glacial erosion', *Abstract of papers supplement*, *20th International Geographical Congress United Kingdom*, 18–19.
GEOLOGICAL SOCIETY OF AMERICA (1959) *Glacial map of the United States east of the Rocky Mountains, Scale 1 : 1 750 000*
HOLMES, C. D. (1937) 'Glacial erosion in a dissected plateau', *Am. J. Sci.* 33, 217–32
LINTON, D. L. (1957) 'Radiating valleys in glaciated lands', *Tijdschr. K. ned. aardrijksk Genoot.* 74, 297–312
LINTON, D. L. (1959) 'Morphological contrasts of eastern and western Scotland' in R. MILLER and J. WATSON (eds) *Geographical essays in memory of Alan G. Ogilvie*, 16–45
LINTON, D. L. (1963) 'The forms of glacial erosion', *Trans. Inst. Br. Geogr.* 33, 1–28
LINTON, D. L. and H. A. MOISLEY (1960) 'The origin of Loch Lomond', *Scott. geogr. Mag.* 76, 26–37
THORNBURY, W. D. (1965) *Regional geomorphology of the United States*
WHITE, W. A. (1972) 'Deep erosion by continental ice sheets', *Bull. geol. Soc. Am.* 83, 1037–56

RÉSUMÉ. *Zones de l'érosion glaciarie.* Des recherches dans les régions qui ont été soumises à la glaciation suggèrent qu'elles ont été changées jusqu'aux points divers par l'érosion glaciare. On propose d'établir une échelle qualitative de cinq zones où le paysage a été modifié par l'érosion. Le projet est décrit relativement à quelques parties de New York et de la Pennsylvanie, et aussi par rapport au nord de la Grande Bretagne. On souligne le fait que le projet est descriptif des fondations de terrain laissées par la glace fondante et qu'il est peu probable qu'il se rapporte en aucune façon simple à la dynamique interne des couches de glace continentales.

FIG. 1. Les « Finger » lacs et une partie du Plateau glacé d'Allegheny. Hachures foncées—lacs; lignes épaisses—courbes de niveau sur les formes aérodynamiques entre les lacs (tous les 61 m); ligne de cercles ouverts—la moraine de Valley Head; hachures moins foncées—le restant du plateau d'avant la glaciation séparé par des auges glaciaries qui sont montrées en blanc sur la carte. Les deux lignes continuelles divisent la région en trois zones de l'érosion glaciarie; IV au Nord, III au milieu, II au sud. S = Syracuse; I = Ithaca; W = Waverly; B = Binghamton
FIG. 2. L'Ecosse: zones de l'érosion glaciarie (D. L. Linton, 1964). Les parties hachurées sont lisse par le glaciarie ou ils ont moraines de fond aux leurs sommets.
FIG. 3. L'Ecosse: auges glaciaries (D. L. Linton, 1964)
FIG. 4. Les « Finger » lacs et régions adjacentes de la partie au nord du l'état de New York: auges glaciaries
FIG. 5. L'Ecosse et le Nord de l'Angleterre: zones de l'érosion glaciarie (comme definies dans Table I)
FIG. 6. Des parties de New York et de la Pennsylvanie: zones de l'érosion glaciaire (comme definies dans Table I). La ligne, la plus à l'extérieur marque la limite de la glaciation montrée sur la carte glaciaire de l'Amérique du Nord à l'est des montagnes Rocheuses 1959. La ligne de démarcation entre les Zones I et O n'a pas été située et donc n'est pas indiquée sur la carte

---

ZUSAMMENFASSUNG. *Gebiete der gletschererosion.* Untersuchungen von vergletschert gewesenen Gebieten legen den Schluss nahe, dass sie von der Gletschererosion in unterschiedlichem Ausmass verändert worden sind. Es wird also eine qualitative Skala von fünf durch Erosion bewirkte Modifikationszonen der Landschaft vorgeschlagen. Das Scheme wird erläutert im Bezug auf Teile New Yorks und Pennsylvanias sowie auch auf den nördlichen Teil Grossbritanniens. Es wird jedoch betont, dass das Schema Landschaftsformen bescrheibt, wie sie von schmelzendem Eis hinterlassen wurden und dass es sich kaum auf irgend eine einfache Weise auf die interne Dynamik kontinentaler Eisdecken wird anwenden lassen.

ABB. 1. Die Finger-Seen und ein Teil des vergletscherten Allegheny Plateaus. Dunkel schattiert—Seen; starke Linien—Umrisse der stromlinien-förmigen Riedel zwischen den Seen (alle 61 m); Linie aus offenen Kreisen—die Valley Head Moräne; heller schattiert—Überreste des voreiszeitlichen Plateaus abgeteilt durch Gletscherrinnen, die weiss ausgespart sind. Die zwei durchgängigen Linien unterteilen das Gebiet in drei Abschnitte der Gletschererosion; IV nördlicher Abschnitt, III Mittelabschnitt, II südlicher Abschnitt, S = Syracuse, I = Ithaca; W = Waverly; B = Binghamton

ABB. 2. Schottland: Gebiete der Gletschererosion (D. L. Linton 1964). Die schattierten Gebiete sind glatt oder mit Grundmoränen bedeckt

ABB. 3. Schottland; Trogtäler (D. L. Linton, 1964)

ABB. 4. Finger-Seen und die anliegenden Gebiete des nördlichen Teils von New York State: Trogtäler

ABB. 5. Schottland und Nordengland: Gebiete der Gletschererosion (definiert wie in Tafel I)

ABB. 6. Teile der Staaten New York und Pennsylvania: Gebiete der Gletschererosion (definiert wie in Tafel I). Die äusserste Linie bezeichnet die Grenze der Vergletscherung nach der Gletscherkarte Nordamerikas östlich der Rocky Mountains (1959). Die Grenze zwischen den Gebieten I und O ist noch nicht festgestellt und wurde daher auf der Karte nicht eingetragen

Landscapes of glacial erosion in Greenland and their relationship to ice, topographic and bedrock conditions

D. E. SUGDEN

Lecturer in Geography, University of Aberdeen

Revised MS received 1 August 1973

ABSTRACT. A classification of landscapes of glacial erosion is given for Greenland (Table I), together with a distribution map (Fig. 9). Well-developed *mountain valley glacier* landscapes are thought to have escaped inundation by the full Pleistocene Greenland ice sheets. Of the landscapes modified by ice sheets, *areal scouring* is thought to reflect the former widespread existence of basal ice at the pressure melting point, *selective linear erosion* the former existence of basal ice with only restricted zones at the pressure melting point, and landscapes with *little or no sign of erosion* the former existence of basal ice below the pressure melting point. The distribution of the various types of landscape is analysed at two scales. At a Greenland scale, areal scouring occurs mainly in the south-west and selective linear erosion mainly in the north and east (Fig. 10). This is thought to reflect the behaviour of the former Pleistocene ice sheet with greater activity and turnover of ice in the maritime south-west than in the continental north and east. The broad variation in bedrock altitude between west and east Greenland may also play a role at this scale through its influence on ice thickness. At more local scales, topographic form, bedrock altitude and rock permeability are variables which have caused local variations from the general trend.

AN explanation of landscapes of glacial erosion created by ice sheets is an elusive prospect. Much of the field evidence lies in remote latitudes or worse still beneath the opaque ice sheets of Antarctica and Greenland. Partly as a result, not only is a simple morphometric description of forms rare, but the nature of the processes operating at the base of a glacier is recognized to be one of the most important unsolved problems in glaciology at present. Not surprisingly, understanding is at a less sophisticated level than in some other branches of geomorphology. D. L. Linton (1963a) has imaginatively and effectively applied a Davisian model of 'structure, process and stage' to the landscapes in west Antarctica and has stressed in particular the importance of evolutionary development. Elsewhere in the literature it is common to find landforms of glacial erosion grouped under such headings as 'upland', 'low relief', 'coastal', etc. Such terms lean heavily on only one of the many variables affecting glacial erosion, usually topography, and ignore others such as ice conditions and rock type. The use of such terms illustrates the lack of firm theory on which to hang a more meaningful classification. This paper approaches the problem from an initial classification of the landforms of Greenland based on their form. The forms are then viewed as elements of glacial systems and are related at the outset to two main types of glacier system: ice sheets and mountain valley glaciers.

Assuming adequate snow with conditions suitable for its accumulation, ice will build up over a land area in the form of a broad dome-shaped ice cap or ice sheet whose consistent shape reflects the basic flow properties of glacier ice. The crucial characteristics of such ice masses so far as erosion is concerned are that they submerge the underlying topography and that any erosional effects are superimposed onto the underlying land surface. Clearly for an understanding of the erosional effects of such an ice-sheet system it is important to examine the behaviour of the system as a whole. Thus in Greenland it is logical to begin a study of landforms eroded by ice

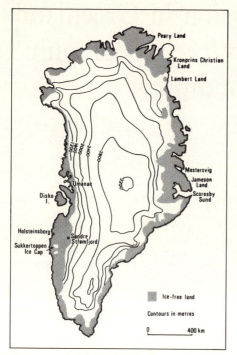

FIGURE 1. Location map of Greenland showing the ring of ice-free land and approximate ice-sheet surface contours

sheets at a sub-continental scale. Mountain valley glaciation occurs wherever for some reason an ice sheet or ice cap cannot build up. Normally the reasons will reflect inadequate snow, insufficient time or more commonly topography whose overall gradient is so steep that it cannot accommodate the gentle slopes of an ice sheet or ice cap. Viewed at a Greenland scale such relief is relevant only in so far as it may represent those parts not submerged by an ice sheet.

The landforms considered in this paper lie in the ring of land between the edge of the present ice sheet and the coast (Fig. 1). Virtually the whole of the area was submerged by ice during the Pleistocene glacial maxima (B. Fristrup, 1966). A particular advantage of a study in this area is the nearness of the relatively well-known, contemporary ice sheet which may be used as a model for comparison with the ice sheets of the glacial maxima. A disadvantage of course is that little can be said about erosional landforms beneath the ice cover.

THE LANDFORMS

Before any sort of analysis is possible it is necessary to classify the landform types. The following classification is based on examination of the fine collection of Greenland air photographs held by the Geodetic Institute in København and on field observations in Greenland and Europe. Table I shows that initial subdivision is based on the distinction between landform associations developed beneath an ice sheet and those associated with mountain valley glaciers.

Ice sheets

Areal scouring describes topography everywhere affected by glacial erosion by an ice sheet (Fig. 2). This type of scenery has been well described by Linton (1963b) who called it 'knock-and-lochan' topography. It is characterized by the way in which structural lines in the bedrock have been exploited to form linear depressions with often complex and angular junctions. Often these depressions form rock basins and are studded with irregular lakes. The eminences between the depressions are moulded by ice action and often consist of bare, striated, rock surfaces. Some-

TABLE I

Classification of landform types

System	Type of erosional landscape
Ice sheet	areal scouring
	selective linear erosion
	little or no sign of erosion
Mountain valley glaciers	integrated trough networks
	individual cirque (corrie) basins

FIGURE 2. Landscape of areal scouring in west Greenland. Reproduced with the permission (A. 481/73) of the
Geodetic Institute, Denmark

times the eminences are smooth on all sides but commonly they are shaped like roches moutonnées with a steep, plucked lee side; the detail of the plucked face is strongly related to the joint
patterns in the rock (K. G. Pike, personal communication, 1973). The relief amplitude of such
scenery is generally limited. For example, if one excludes large-scale undulations of several
kilometres across, the area east of the Sukkertoppen ice cap in west Greenland has an amplitude
of the order of 100 m. It is difficult to know how much scouring has been accomplished by ice in
such areas. There are indications that frequently the ice has only etched a pre-existing surface
and may have removed less than *c*. 100 m of rock debris. Over thousands of square kilometres in
the area east of the Sukkertoppen ice cap, the main rivers maintain integrated drainage patterns
reminiscent of non-glacial humid environments. These patterns occur regardless of the abundant
signs of scouring and indeed the only exceptions are in the vicinity of distinct troughs. Since
integrated patterns on such a scale are unlikely to have formed on a newly-created glacial surface
without greater modification of the glacial forms, one is led to conclude that the ice was unable
to change the land surface sufficiently to derange a pre-existing pattern. This view is confirmed
by evidence from the summits between the rivers. Adjacent summits tend to be conformable in
altitude and a generalized contour map of their elevations produces a surface which is conformable
with the direction of river flow. Both lines of evidence imply that a pre-existing surface has been
modified by ice without major transformation. Similar conclusions have been reached elsewhere,
for example by J. B. Bird in Arctic Canada (1967).

Landscapes with evidence of selective linear erosion are characterized by upper rolling slopes
or plateau surfaces dissected by deep troughs. There are magnificent examples in east Greenland
(Fig. 3) and a map of the trough pattern has been published elsewhere (D. E. Sugden, 1968).
There are two types of trough in east Greenland. Those aligned approximately east–west (in the
direction of ice flow) are long, slightly sinuous, narrow and deep. S. Funder (1972) notes that the
depth of these troughs is of the order of 2500–3500 m while their width is about 5 km. Those
aligned north–south (apparently at right angles to the overall direction of ice flow) follow lines of
structural weakness and are straighter, shallower and wider. The upland plateau remnants be-

FIGURE 3. Landscape of selective linear erosion, Nathorsts Land, east Greenland. Reproduced with the permission (A. 481/73) of the Geodetic Institute, Denmark

tween the troughs generally show little or no sign of glacial erosion and it is common to find the surface covered with regolith. In similar scenery in the Cairngorm Mountains in Scotland fragile pre-glacial remnants such as rotted rock and tors survive on the upland surfaces (Sugden, 1968). Characteristically the break of slope between the plateau and the trough is remarkably fresh and abrupt. This is particularly well brought out in Figure 3. It is relevant to note that such troughs are thought to have been cut beneath ice sheets. The case has been argued on morphological grounds in east Greenland (J. H. Bretz, 1935; H. W:son Ahlmann, 1941), in Scotland (Sugden, 1968) and in part of North America (K. M. Clayton, 1965). Often recent echo-sounding of modern ice sheets is revealing troughs beneath the ice (D. J. Drewry, 1972).

Landscapes characterized by little or no sign of glacial erosion are devoid of areal scouring forms or troughs. Such areas commonly comprise coherent fluvial landscapes with smooth slopes covered in regolith. The main problem involved with discussion of this type of landscape is the difficulty of demonstrating whether or not such areas were covered by ice. In Greenland it is generally accepted that the whole continent was covered by ice during the Pleistocene maxima, with the possible exception of northern Peary Land (Fristrup, 1966).

Mountain valley glaciers

Perhaps the most spectacular erosional landscapes of all are those associated with erosion by mountain valley glaciers. The main features are brought out in Figure 4, which shows a network of glaciers overlooked by frost-etched mountain peaks. Characteristically the valley glaciers begin in an arcuate collecting ground and eventually several combine to form a major trunk glacier. Such glacier systems are restricted to local massifs and individually are rarely more than about a few tens of kilometres in length. Most valley glacier landscapes in Greenland support active glaciers today. Some, however, are free of ice and display many of the features well known in the Alps.

The final landscape category includes glacial landforms cut by isolated cirque glaciers. In

FIGURE 4. Landscape of mountain valley glaciation, Borggraven, east Greenland. Reproduced with the permission (A. 481/73) of the Geodetic Institute, Denmark

such cases well-developed troughs are rare and the main feature is the cliffed cirque basin etched into a massif. As would be expected in a case where the cirque glaciers are isolated, there are commonly remnants of the initial surface of the massif between the cirques; this surface may form a plateau or gently undulating upland.

The categories mentioned above are intended to be no more than idealized models. In reality there is much overlap between the various groups and at times assignment into one category is difficult. For example, troughs may occur in areas of extensive areal scouring and in this paper such landscapes have been included in that category.

THEORETICAL CONSIDERATIONS

Since landforms of glacial erosion are the result of processes operating at the interface between the sole of a glacier and the bedrock, it is logical to assume that the nature of the processes and the results of their operation will vary as glacier conditions and bedrock conditions vary.

Basal ice conditions

It is likely that basal slip is a necessary prerequisite for effective glacial erosion beneath an ice sheet. Further, whether or not basal slip occurs depends on a critical threshold related to the effective temperature of the basal ice. If the ice is at the pressure melting point basal slip may occur; if it is below the pressure melting point basal slip is unlikely (J. Weertman, 1957, 1964; G. S. Boulton, 1972). When basal ice is below its pressure melting point movement within the ice takes place within the lower layers of the ice mass rather than between ice and rock. When basal ice is at the pressure melting point movement may also involve basal slip between ice and rock. A film or layer of water is formed at the base and its relative abundance is thought to play a crucial role in reducing friction and influencing the rate of basal slip. The greater the quantity of basal water the greater the role of basal slip. This conclusion has been reached in a number of theoretical studies by L. Lliboutry (1965, 1968a) and Weertman (1957, 1964, 1966).

With regard to the landforming processes occurring beneath a glacier it is reasonable to suppose that the more favourable conditions are for basal melting the greater will be the opportunity for the transport of basal debris. There are two reasons for this. First, if one imagines a point on the ground beneath an ice sheet, the more readily ice slips over its base the greater the total amount of ice that will pass that point during a glaciation. Secondly, from an areal point of view favourable basal ice conditions are likely to mean that widespread areas of bedrock are affected by basal slipping. It is important to stress that this does not mean that the full potential of the ice for transport of material is necessarily used. Thus there are no grounds for expecting conditions most favourable for basal slipping to be associated with the greatest amount of glacial erosion, measured as the volume of debris removed. As recent work suggests, the process by which debris is entrained into the ice may be linked to rather sensitive thermal conditions (Boulton, 1972). However, assuming that there is debris in the sole of the ice it is fair to expect that areas subjected to widespread basal slip will bear extensive evidence of the ordeal.

Before attempting a geomorphological assessment of ice sheet erosion, it is clearly vitally important to review the conditions affecting temperatures at the base of ice sheets. Such basal temperatures are a response to heat obtained from three sources: (1) the base, (2) the surface, and (3) internal deformation.

(1) Geothermal heat is derived from the bedrock base and although it varies from place to place on the earth's surface is generally regarded as a constant providing c. 38 cal/cm^2/year. In ice below the pressure melting point this gives a vertical gradient of 1°C per 44 m at the base of an ice sheet (G. de Q. Robin, 1955). If the ice is at its pressure melting point this heat cannot be absorbed into the glacier and will melt a layer of ice with an average thickness of 6 mm/year at the bottom of the glacier (W. S. B. Paterson, 1969).

(2) Sources of heat from the surface are affected by the depth of the ice, the initial temperature of the firn and the rate of accumulation (Robin, 1955). Since temperatures in ice sheets below the pressure melting point rise with increasing depth, the thicker the ice sheet the higher the basal temperatures will be (assuming similar surface temperatures). The temperature of the firn beneath the thin surface zone of seasonal variations is influenced by the seasonal climatic conditions at the surface. In areas of no summer melting the firn is close to the mean annual temperature. As the climate becomes more maritime with more snow and more summer melting the firn temperature rises rapidly until it reaches 0°C (Paterson, 1969). The rate of accumulation is important because it controls the vertical velocity of ice and thus the rate at which cold surface firn or ice is carried down into the glacier mass. High rates of accumulation tend to reduce the vertical temperature gradient within the ice.

(3) The third source of heat is derived from internal movement within the ice (W. F. Budd, 1969). In ice below its pressure melting point most heat is generated close to the bed in the zone of maximum deformation. For most purposes it can be regarded as an addition to the geothermal heat flux, with an ice velocity of 20 m/year approximately equivalent to the geo-thermal heat flux. In general the heat produced is likely to increase from the centre to a maximum near the equilibrium line of the ice sheet where ice velocities will tend to be greatest. Once slip over the bedrock has commenced, heat produced by that slip is equal to $S\tau/k$, where S = speed of slipping, τ = average shear stress at the glacier bed, and k = a conversion factor, energy to heat units (Boulton, 1972). This latter source of

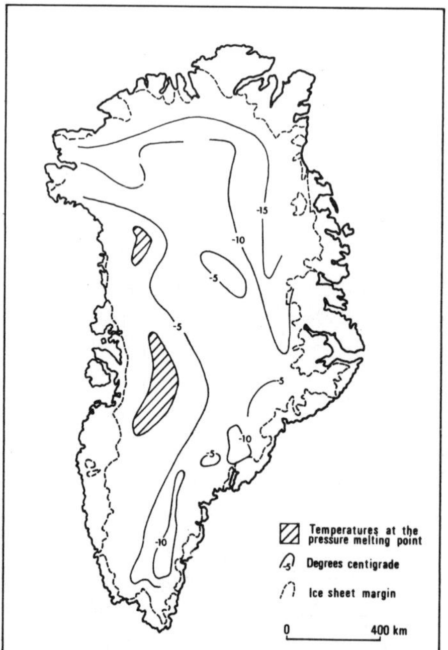

FIGURE 5. Calculated basal ice temperatures beneath the Greenland ice sheet assuming steady-state conditions. Reproduced by courtesy of D. Jenssen

heat is of considerable importance for conditions at the ice/rock interface. Slipping will not occur till the ice has been raised to its pressure melting point. But then the very process of slipping releases more heat which ensures its continuation.

The basal ice temperatures of the contemporary Greenland ice sheet have been calculated from mathematical simulation models and give some guide to the main trends. Figure 5 shows the result of one such dynamic model where it is assumed that the ice sheet is in steady-state (D. Jenssen, personal communication, 1972). Details of the models and other alternatives are given by Budd, Jenssen and U. Radok (1971), where they are applied to the Antarctic ice sheet. In Greenland there are detailed variations in the pattern of basal ice temperatures depending on the model used and the exact values fed into the input data. However, the main pattern is consistent from model to model. Figure 5 shows that temperatures are at, or close to, the pressure melting point in a zone parallel to the west coast and under the central northern part of the ice sheet. Basal temperatures decrease towards the north-east. Other models show a similar pattern but with the zone of pressure melting more extensive and closer to the western edge of the ice sheet.

The pattern is consistent with what is known about the climate. Figure 6a is an estimate of precipitation totals over Greenland except the Thule area (S. J. Mock, 1967). With the exception of southern Greenland south of lat. 67°, precipitation is high near the west coast and falls off towards the north-east. Figure 6b shows the pattern of 10 m snow depth temperatures and is a reasonable approximation to mean annual temperatures. The pattern is largely a reflection of surface altitude, but latitude is also important with a tendency for temperatures to fall off towards the north (Mock and W. Weeks, 1966). Taken together these maps reflect a maritime climate in the south and west and a progressively more continental climate towards the north and east. The weather dynamics responsible for this trend are considered in detail by P. Putnins (1970).

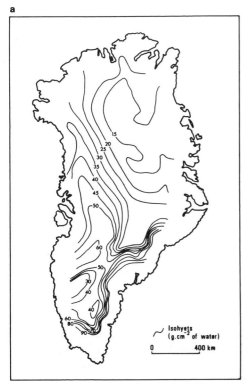 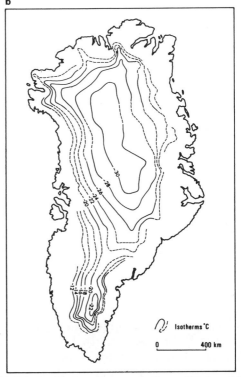

FIGURE 6. Some climatic data for the Greenland ice sheet. (a) Calculated isohyets in g. cm^{-2} of water (Mock, 1967)
(b) Calculated isotherms (Mock and Weeks, 1966)

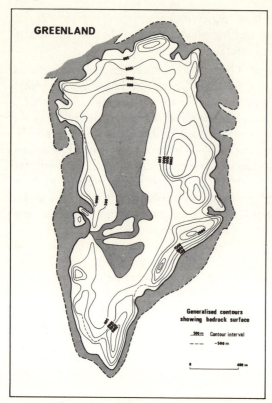

FIGURE 7. Approximate bedrock topography of Greenland, showing low altitude of ice sheet base in central and north Greenland

The maritime climate favours relatively high surface temperatures and a high turnover of ice with consequently a great deal of heat generated by internal deformation. Also, in view of the high velocities of many west Greenland glaciers, heat is also generated by basal slip. High basal ice temperatures are likely under such conditions. Towards the north-east where ice turnover is less and surface temperatures lower, basal ice temperatures tend to be lower.

The relatively high basal temperatures under the central northern ice dome where precipitation is low need further explanation. Here the highest parts of the ice dome lie over areas of low-elevation bedrock and ice thicknesses are high (Fig. 7). In this case ice thickness is the critical factor, as it is in areas of low basal temperatures in south-east Greenland which reflect thin ice.

Although these models apply to the present Greenland ice sheet and not to those of the Pleistocene maxima, there is no reason to suppose that the atmospheric circulation was very different in this area during the Pleistocene. So long as most depressions approached from the south and west and there was pack ice off the north and east coasts there would be a major contrast between the maritime west and south and the continental north and north-east. The main difference is likely to have been in the south-west where the modern ice sheet is partially protected from a full maritime climate by coastal mountains. The contrast is well illustrated by comparing for example the high precipitation total at the coastal town of Holsteinsborg (355 mm) and the low total at Søndre Strømfjord at the edge of the contemporary ice sheet some 130 km inland (152 mm). During the Pleistocene maxima the ice sheet extended beyond the present coast and would have experienced a full maritime climate. Under these circumstances basal ice temperatures beneath the south-western periphery of the ice sheet are likely to have been higher than at present.

Bedrock variables

The shape and nature of the rock bed is an important variable affecting processes at the ice-rock interface. As Lliboutry (1968b) has pointed out, undulations in the bed affect ice thicknesses and thus lead to local variations in basal ice temperatures. It is possible to envisage a situation where an ice sheet covers an irregular land surface and the basal ice reaches the pressure melting point only over the sites of low-lying depressions and valleys. Clearly such a situation can be expected to favour irregular flow within the ice mass with a tendency towards highest velocities over the sites of valleys and depressions.

Rock permeability can play an important role in influencing the relative abundance of water existing at the ice/rock interface, assuming the ice is at the pressure melting point. If the rock is sufficiently permeable for water formed at the base to be evacuated through the rock then basal slipping may be greatly reduced (Boulton, 1972). The hydraulic conductivity of such beds is a function of rock permeability, thickness of the permeable beds and the hydraulic gradient beneath a glacier (Weertman, 1966). On a larger scale it will also be influenced by the attitude of the beds and in particular the ease with which water drainage may be evacuated in a horizontal direction.

Relationship of the landform associations to theory

If the theoretical considerations are accepted it is possible to suggest conditions under which the different landform associations occur. Areal scouring reflects slipping over the whole land surface. This is most likely to occur when conditions are such that the basal ice is at the pressure melting point almost everywhere, a situation that is to be expected in maritime climatic environments and where the ice is thick. Elsewhere, on a smaller scale, it is likely to occur where local factors favour local zones of ice at the pressure melting point, such as in local deeps in ice over depressions or in the vicinity of outlet glaciers where frictional heating occurs. Probably slipping is also favoured by abundant water at the base and thus is more likely to occur where rocks are impermeable rather than permeable (Boulton, 1972).

Areas with no sign of erosion are more likely to occur where conditions are the opposite. At a macro-scale these areas will tend to occur where climatic conditions are continental while more locally they will reflect patches of thin or largely stagnant ice and areas of highly permeable rocks.

Areas of selective linear erosion may be envisaged as an intermediate category where conditions are so marginal that basal slipping can only occur in places. Following R. Haefeli (1968) Figure 8 shows a hypothetical ice sheet with the ice just attaining pressure melting point over the sites of depressions. Assuming a depression is favourably orientated for evacuation of ice, basal slipping will occur in the position marked. This would generate sufficient heat to maintain the basal ice at the pressure melting point. Further slipping and selective erosion would increase the thickness of the ice at these points and accentuate the contrast further. It is quite possible to envisage such ice streams providing sufficient drainage for the ice sheet in these areas and thus there is no reason for the ice over the intervening plateaux to thicken and its basal layers to reach the pressure melting point. Under such circumstances the intervening plateaux can be regarded as local examples of landscapes with no sign of glacial erosion. Such selective linear landscapes would be favoured by climatic conditions midway between maritime and continental and more locally by relatively thin ice and/or irregular topography. In terms of rock type, local variations in permeability might be expected to lead towards a similar result.

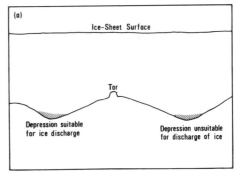

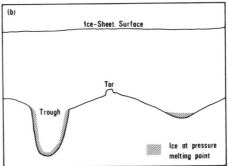

FIGURE 8. Diagram to illustrate the development of a landscape of selective linear erosion. (a) Ice sheet builds up and basal ice reaches the pressure melting point over the sites of depressions; (b) The depression on the left is suitable for ice flow and basal slipping occurs and erodes a trough. Ice will not slip over the adjacent upland areas unless the ice sheet thickens further (following Haefeli, 1968)

ANALYSIS OF THE LANDFORM PATTERNS IN GREENLAND

If the gist of the theoretical discussion is correct then it should be possible to test the conclusion against the evidence in Greenland, for one would

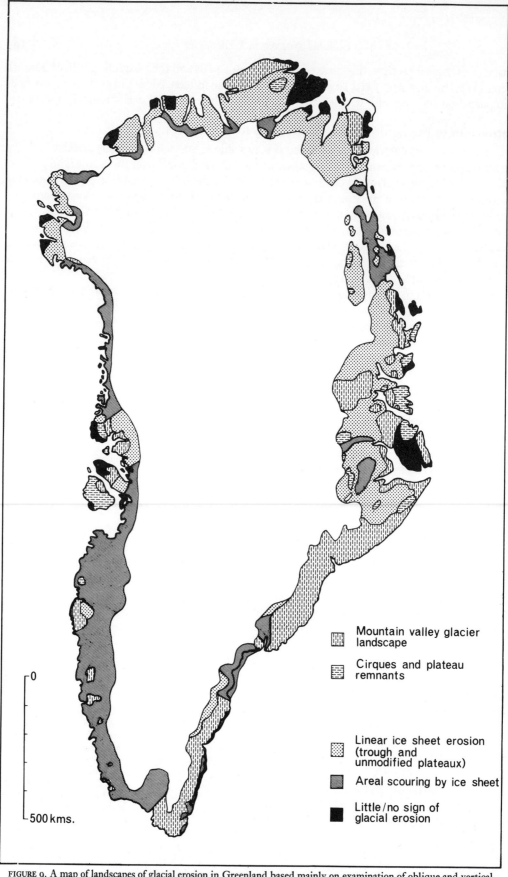

Mountain valley glacier landscape

Cirques and plateau remnants

Linear ice sheet erosion (trough and unmodified plateaux)

Areal scouring by ice sheet

Little/no sign of glacial erosion

0

500 kms.

FIGURE 9. A map of landscapes of glacial erosion in Greenland based mainly on examination of oblique and vertical air photographs. See text for description of landscape categories

expect certain types of landform association to occur in certain areas. Figure 9 is a map of the various landform types in Greenland. The map relies on the interpretation of several thousand vertical and oblique aerial photographs, together with field examination of part of east (Scoresbysund-Mestersvig) and west Greenland (Sukkertoppen and Umanak). Clearly the mapping of such a large area will contain mistakes especially in the less accessible and less known northern Greenland. Therefore for these areas the map relies also on interpretation and photographs in papers by workers who have visited them (e.g. W. E. Davies and D. G. Krinsley, 1961; R. L. Nichols, 1954, 1969; R. B. Cotton and C. D. Holmes, 1954; P. J. Adams and J. W. Cowie, 1953). Although there may be local areas where the classification can be disputed, it is hoped that the main pattern is substantially correct.

To facilitate analysis it is helpful to look at the problem at two scales. The macro-scale pattern is concerned with Greenland as a whole while the meso-scale pattern concerns local deviations from the Greenland pattern and is dealing with areas whose size is of the order of ten to a few hundred kilometres. Smaller areas are not considered in this paper.

The macro-scale pattern

Viewing Greenland as a whole there are several clear trends. Excluding for the moment areas of mountain valley glaciation and concentrating on landforms eroded by an ice sheet, the percentage of the land area classified into one or other ice-sheet categories can be calculated for unit areas. Such calculations were made for 105 squares representing land areas of 90 × 90 km and the results are plotted as linear trend surfaces for the two main categories of landform type. Higher order trend surfaces were avoided largely because of excessive distortion introduced by edge effects and the peripheral distribution of the data (J. C. Davis, 1973). Such problems are minimized by limiting interpretation to linear trend surfaces (D. J. Unwin, personal communication, 1973).

Figure 10a shows the linear trend surface for the percentage of the land area affected by areal scouring. The calculations involve only those landscapes affected by ice sheets. The fit of the surface is good: 55 per cent *RSS*, *Significance* 99·9 <F (Unwin, 1970). The surface brings out the broad trend and clearly shows how most of the available land area in the west and south has been subjected to areal scouring while the relative importance of this type of erosion falls off dramatically to the north-east, where values of less than 10 per cent are indicated. Figure 10b shows the linear trend surface for the percentage of available land area classified as landscapes of selective linear erosion. Again areas of mountain valley glaciation are excluded from the calculations. Although the fit of the surface is poorer it is still good: 31 per cent *RSS*, *Significance* 99·9 <F. It is worth pointing out that the main trend is complementary to that of areal scouring and that it varies in the opposite direction. Here values are highest in the north-east and lowest in the south-west.

Interpretation of such trends can only be tentative but the pattern is just what one would expect if the landscape categories are related to the former presence or absence of pressure melting at the base of the Pleistocene ice sheets. On a Greenland scale there are two variables which vary broadly in sympathy with the main trend of

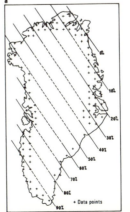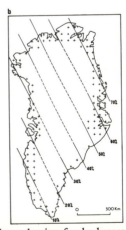

FIGURE 10. Linear trend surfaces showing for landscapes modified by ice sheets: (a) the percentage land area classified as areal scouring; (b) the percentage land area classified as selective linear. Areas of mountain glaciation have been excluded from these calculations

landform variation and which could account for this variation. One is the ice-sheet climate and the other is bedrock altitude. The variation of basal ice characteristics from south and west to north and east is shown on Figure 5 and is broadly similar to the variation in the amount of areal scouring (Fig. 10a). As suggested on page 184, during the Pleistocene maxima the similarity is likely to have been still closer, for the ice sheet in the south-west is likely to have had basal temperatures higher than those of today. The similarity of two maps is no proof of a causal link between the two. However at the macro-scale the pattern in Greenland is consistent with the hypothesis that areal scouring tends to occur beneath maritime areas of ice sheets while selective linear erosion or no apparent erosion tends to occur beneath ice which is nourished under a more continental regime. In addition, bedrock altitude happens to vary broadly from west to east in Greenland (Fig. 7). Although this trend does not mirror the landform trends exactly, it is likely to contribute to the overall landform pattern. A Kolmogorov-Smirnoff test was carried out to compare the average summit altitudes in thirty sample areas of both selective linear erosion and areal scouring. The test suggested that there was a significant difference between the two sets of altitudes (99 per cent level) and that there was a tendency for altitudes associated with selective linear erosion to be higher than those associated with areal scouring. This association is to be expected for, given a constant ice-sheet surface altitude, basal ice is more likely to be at the pressure melting point over low bedrock areas than over high altitude bedrock areas (Boulton, 1972). In the case of Greenland both the behaviour of the ice sheet and bedrock altitude contribute to the macro-scale pattern. The relative importance of the two could best be ascertained by examination of other areas where the two trends are known not to be approximately complementary.

Other variables can be regarded as relatively unimportant contributors to the macro-scale pattern unless they too can be shown to vary in a consistent and similar manner over Greenland as a whole. Thus structure and rock type are likely to be unimportant at this scale for they do not change consistently from west and south to north and east (K. Ellitsgaard-Rasmussen, 1970).

The distribution of areas of mountain valley glaciation forms a reasonably coherent pattern at the macro-scale. Well-developed landscapes in this category occupy areas which are unlikely to have been covered by any Greenland ice sheet. Figure 11 attempts to show the relationship between the altitude of such landscapes and the surface of the reconstructed Pleistocene Greenland ice sheet(s). The ice-sheet profiles, one theoretical, assuming a horizontal base (J. F. Nye, 1952), and one a composite profile based on actual Greenland profiles, are drawn assuming that the seaward edge of the ice sheet was delimited by what is now the − 200 m submarine contour. In most cases this contour is close to the top of the edge of the continental slope and is thus likely to be a reasonable approximation. In the north the limited knowledge of submarine conditions makes any reconstruction doubtful but in these regions it is probably as good an estimate as any (A. Weidick, personal communication, 1971). The altitudes of the mountain valley glacier landscapes were reached by averaging the altitudes of the five highest summits in 10 × 10 km squares chosen from a stratified

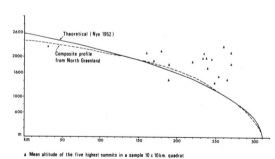

▲ Mean altitude of the five highest summits in a sample 10 x 10 km. quadrat

FIGURE 11. Surface profiles constructed for the maximum Pleistocene Greenland ice sheet(s) and their relationship to the altitude of areas of well-developed mountain valley glaciation. The profiles are drawn assuming the ice sheet extended over the continental shelf to a depth now marked by the −200 m submarine contour. The mean altitude of the five highest summits in each sample area is represented by one triangular symbol. Most of the summit areas are clearly above the profiles and are thus likely to have escaped inundation by the full ice sheet(s)

random sample of areas of this landscape type. The altitudes are entered at the appropriate distance inland from the −200 m submarine contour.

It is notable that the summit altitudes of most areas of mountain valley glaciation rise decisively above both ice-sheet profiles. The main exception is in Kronprins Christian Land, which is a little-known area with few accurate altitudes or submarine contours. It is reasonable to suggest therefore that these areas represent land areas of Greenland which have not been effectively submerged by any Greenland ice sheet. As a result the upstanding massifs have been shaped by local valley glaciers throughout the glacial age.

It is important to note that this is a generalization applying only to areas of well-developed mountain valley glaciation with integrated valley glacier systems. There are many other areas where local valley glaciation has operated during 'inter-glacial' periods such as the present. Generally these areas are characterized by cirque forms and rather simple valley patterns. The latter have been excluded from this discussion. In addition, the generalization ignores many factors which should be considered in a detailed examination of mountain glaciation. For example it is likely that the surface profile of an ice sheet is shallower and less convex in the vicinity of outlet glaciers (J. T. Buckley, 1969). Also glacio-isostasy will influence absolute summit heights.

The meso-scale pattern

The main residuals from the two trend surfaces highlight some of the areas where local factors override the general trend (Fig. 12). The residuals fall into three main groups:

(1) The Umanak area of the west coast which consists of a landscape of selective linear erosion (and of little sign of erosion) surrounded by a zone of areal scouring.
(2) The east coast of Greenland in approximately lat. 66° where there is more selective linear erosion and less areal scouring than expected.
(3) Areas in northern Greenland where there is more areal scouring and less selective linear erosion than would be expected. The most important anomaly here is the Lambert Land area.

In addition to these residual areas which tend to measure some hundreds of kilometres across there are much smaller-scale anomalies which can be picked out by visual comparison of the trend surfaces and the map of the original data (Fig. 9). These include examples of landscapes with no sign of glacial erosion, small areas of areal scouring surrounded by selective linear landscapes and *vice versa*.

Without far more detailed knowledge of Greenland it is impossible to explain individual anomalies at this stage. However, it is possible to highlight the various situations and to discuss the factors which could explain them.

Local contrasts in areal scouring and selective linear erosion

There are many instances where a contrast between a landscape of areal scouring and one of linear erosion is associated with a change in bed-rock altitude. In northern and eastern Greenland the association commonly takes the form of a low-lying area of areal scouring surrounded by

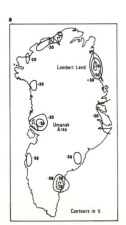
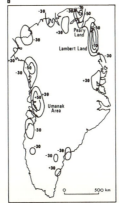

FIGURE 12. Positive and negative residuals from the linear trend surfaces of landscapes of (a) areal scouring and (b) selective linear erosion

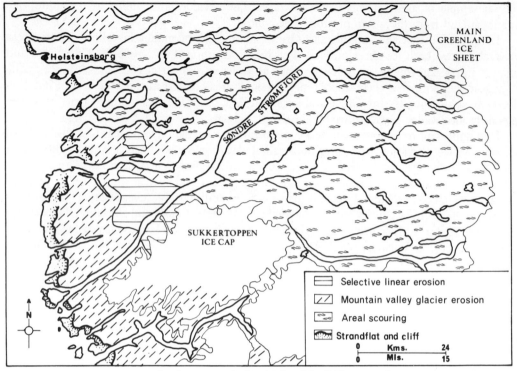

FIGURE 13. Landscape types in the vicinity of the Sukkertoppen ice cap

higher ground characterized by linear erosion, while in the west an upland marked by linear erosion may be surrounded by a zone of low-lying areal scouring. An example of the latter occurs in the Sukkertoppen ice-cap area. Here the surface of a plateau falls north-eastwards from an altitude of *c.* 1700 m in the west to an altitude of 500–600 m in the east. The whole area was enveloped by a more extensive ice sheet flowing towards the west (Weidick, 1968; Sugden, 1971). The distinct contrast in landscape types is shown in Figure 13. Most of the lower plateau area is extensively scoured regardless of differences in rock type and structure. The western part of the plateau was sufficiently high and sufficiently close to the ice-sheet edge to escape inundation by the main ice sheet. As a result it now forms spectacular mountain valley glacier scenery. Between the mountain glacier scenery and the areal scouring is an area of high ground (now partly occupied by the Sukkertoppen ice cap) characteristic of selective linear erosion. The altitude and position of this bit of upland and the fact that it was relatively close to the ice-sheet edge suggest it was only thinly covered by ice. It is reasonable to postulate that basal ice over the upland was cooler than basal ice over lower ground further inland and may not have reached its pressure melting point over the uneroded plateau areas. Such an interpretation is strengthened by the fact that the changes in landscape types indicated occur regardless of the slight changes in bed-rock type shown on the Tectonic Geological map of Greenland (Ellitsgaard-Rasmussen, 1970). It is tempting to suggest that the other main occurrence of selective linear erosion on the west coast, the Umanak area, is also related to the fact that this area is much higher than average, for the area contains some of the highest summits in north-west Greenland. However, it is also underlain by distinctive rocks.

In eastern and northern Greenland where selective linear erosion is dominant, areal scouring is confined to depressions, usually in the vicinity of troughs and outlet glaciers (Fig. 9). When the whole area was covered by the full Pleistocene ice sheets the basal ice in the vicinity of the depressions would have been warmer than basal ice over the adjacent upland. This would be due partly to the greater ice thickness and partly to the internal heat generated by ice flowage concentrated in the valleys at the expense of the upland areas.

Local areas of little or no glacial erosion

In Greenland all occurrences of landscapes with little or no sign of erosion are local in distribution and thus are likely to reflect critical local factors. If such areas simply reflect the lack of basal slipping then local factors will need to be sufficiently important to prevent such slipping.

Rock type is likely to be an important local factor for the reasons stated on p. 184. Although it is impossible to examine its role in any depth without detailed field and laboratory work, there are one or two suggestive relationships which are worth noting. In particular there is evidence which conforms to the expectation that outcrops of porous rocks will locally reduce glacial erosion. In east Greenland for example, the bulk of the areas with no sign of glacial erosion lie along a zone of relatively young marine sediments and volcanics, extending through time from Upper Permian to Tertiary (J. Haller, 1971). It is reasonable to assume that these relatively undisturbed rocks are more permeable than the older folded metamorphics immediately to the west. This is certainly true of parts of Jameson Land where the younger rocks include limestones and sandstones (E. Kempter, 1961). The association is even more striking when one realizes that, with the exception of a few fjord outlets, virtually all the remaining parts of the zone are devoid of ice-sheet forms. Indeed if it were not for minor etching by local cirque glaciers virtually the whole area would have been mapped in the category of little or no erosion (Fig. 14). It may also be significant that the areas with no sign of glacial erosion in the vicinity of Disko Island lie on Tertiary volcanics while in Peary Land they occupy zones of young rocks (undifferentiated) and limestones. It is important to state, however, that some of these parts of Peary Land may never have been glaciated (Fristrup, 1966). Topography is an important local variable which is likely to have some effect on the distribution of landscapes of little or no glacial erosion. The position and altitude of a massif may be such that it is covered only thinly by ice, and basal ice temperatures may never rise to the pressure melting point. Special cases of this are the plateau areas between the troughs in landscapes of linear erosion. Also, topography may affect basal ice conditions through its role in causing ice to diverge or converge (Linton, 1963b). Such a role has been hinted at by reference to the channelling effect of depressions and valleys which drain more ice than surrounding areas and thus allow more internal heat to be generated. A narrow promontory or peninsula will fulfil the opposite role and thus tend, other variables being equal, to be less subject to erosion. In this context it may be significant that the sixteen main instances of little or no glacial erosion occur mainly on peninsulas, i.e. on sites where ice might be expected to diverge.

At this stage it is impossible to assess the relative role of different local factors in preventing basal slip. None the less it is interesting to consider their role in relation to the macro-scale pattern of basal slipping in Greenland. It is likely that variations in local conditions need to be less dramatic in the north and east to prevent basal sliding, than

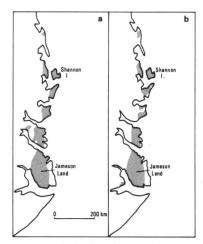

FIGURE 14. A visual correlation of (a) a zone of young (permeable?) rocks and (b) areas with little or no sign of ice-sheet erosion in East Greenland. (b) is derived from Figure 9, but areas of cirque erosion have been ignored

for example in the west and south. Thus it is reasonable to suppose that the existence of areas with no signs of glacial erosion in the vicinity of Umanak and Disko Island must represent a powerful combination of local factors. In this context it is interesting to note that the relevant areas lie on young basaltic (porous?) rocks, on peninsulas, and in general are high. It may be necessary for all these local factors to work together in order to override successfully the macro-scale tendency for basal slipping in the west.

CONCLUSION

This paper began with an initial classification of landscapes of glacial erosion based on morphological characteristics. On theoretical grounds it was then suggested that these variations could be attributed to different conditions in the basal layers of the ice. It was proposed that

(1) areal scouring is related to basal ice at its pressure melting point capable of slipping over a film of water at the ice/rock interface;

(2) landscapes of selective, linear erosion are related to situations where only zones of the basal ice are at the pressure melting point, and

(3) landscapes with no sign of glacial erosion are related to basal ice which is below the pressure melting point and thus does not slide at the ice/rock interface.

The distribution of the various landscape types was then analysed at two scales to test this hypothesis, apparently successfully. As a cautionary note it is important to stress that such a correlation on its own cannot prove a causal link between basal ice conditions and the resultant landforms. However, it can be stated that the hypothesis has survived a test in a significant part of the world's glaciated landscapes.

ACKNOWLEDGEMENTS

I am grateful to the Geodetic Institute in København for permission and space to work in their air photographs library and indebted to Dr A. Weidick for clarifying specific problems and providing some little-known literature on northern Greenland. Field-work in east and west Greenland was accomplished during the course of the Oxford University East Greenland Expedition, 1962, and the Aberdeen University West Greenland Expedition, 1968. It is a pleasure to acknowledge the help of all members of the expeditions and of all bodies whose support made them possible. I would like to thank all those who have kindly given valuable criticisms of early drafts of the paper and advice on techniques, in particular A. Dawson, A. Gemmell, J. Gordon, P. Hamilton, D. Jenssen, J. von Weymarn and D. Unwin. Lastly, I thank the Carnegie Trust for the Universities of Scotland who partially financed a visit to København and contributed towards the cost of illustrations.

REFERENCES

ADAMS, P. J. and J. W. COWIE (1953) 'A geological reconnaissance of the region around the inner part of Danmarks Fjord, Northeast Greenland', *Meddr Grønland* 111 (7)

AHLMANN, H. W:son (1941) 'Studies in North-East Greenland, 1939–40', *Geogr. Annlr* 23, 145–209

BIRD, J. B. (1967) *The physiography of Arctic Canada*

BOULTON, G. S. (1972) 'The role of thermal régime in glacial sedimentation', *Spec. Publ. Inst. Br. Geogr.* 4, 1–19

BRETZ, J. H. (1935) 'Physiographic studies in East Greenland' in L. A. BOYD (ed.) *The fiord region of East Greenland, Spec. Publ. Am. geogr. Soc.* 18, 161–266

BUCKLEY, J. T. (1969) *Gradients of past and present outlet glaciers*, Geol. Survey of Canada, Paper 69-29

BUDD, W. F. (1969) 'The dynamics of ice masses', *ANARE* Scientific Reports Ser. A. (4) Glaciology Publ. 108

BUDD, W. F., D. JENSSEN and U. RADOK (1971) 'Derived physical characteristics of the Antarctic ice sheet (Mark 1)', University of Melbourne Met. Dept. 18

CLAYTON, K. M. (1965) 'Glacial erosion in the Finger Lakes region (New York State, U.S.A.)', *Z. Geomorph.* 9, 50–62

COTTON, R. B. and C. D. HOLMES (1954) 'Geomorphology of the Nunatarssuak area' in *Final Report Operation Ice Cap 1953*, Program B. Dept. Army Project 9-98-07-002. Stanford Research Inst. 27–52

DAVIES, W. E. and D. G. KRINSLEY (1961) 'Evaluation of arctic ice-free land sites, Kronprins Christian Land and Peary Land, North Greenland, 1960', *Air Force Surv. Geophys.* 135

DAVIS, J. C. (1973) *Statistics and data analysis in geology*

DREWRY, D. J. (1972) 'The contribution of radio echo sounding to the investigation of Cenozoic tectonics and glaciation in Antarctica', *Spec. Publ. Inst. Br. Geogr.* 4, 43–57

ELLITSGAARD-RASMUSSEN, K. (1970) *Tectonic/geological map of Greenland*, Copenhagen, Geological Survey of Greenland

FRISTRUP, B. (1966) *The Greenland ice cap*, Copenhagen, Rhodos

FUNDER, S. (1972) 'Deglaciation of the Scoresby Sund fjord region, north-east Greenland', *Spec. Publ. Inst. Br. Geogr.* 4, 33–42

HAEFELI, R. (1968) 'Gedanken zur Problem der glazialen Erosion', *Felsmechanik u. Ingenieurgeol.* 4, 31–51

HALLER, J. (1971) *Geology of the East Greenland Caledonides*

KEMPTER, E. (1961) 'Die Jungpalaozoischen Sedimente von Süd Scoresby Land', *Meddr Grønland* 164 (1)

LINTON, D. L. (1963a) 'Some contrasts in landscapes in British Antarctic Territory', *Geogrl J.* 129, 274–82

LINTON, D. L. (1963b) 'The forms of glacial erosion', *Trans. Inst. Br. Geogr.* 33, 1–28

LLIBOUTRY, L. (1965) *Traité de Glaciologie*, 2, Paris, Masson

LLIBOUTRY, L. (1968a) 'General theory of subglacial cavitation and sliding of temperate glaciers', *J. Glaciol.* 7, 21–58

LLIBOUTRY, L. (1968b) 'Steady-state temperatures at the bottom of ice sheets and computation of the bottom ice flow law from the surface profile', *J. Glaciol.* 7, 363–76

MOCK, S. J. (1967) 'Calculated patterns of accumulation on the Greenland ice sheet', *J. Glaciol.* 6 (48), 795–803

MOCK, S. J. and W. WEEKS (1966) 'The distribution of 10 meter snow temperatures on the Greenland ice sheet', *J. Glaciol.* 6 23–41

NICHOLS, R. L. (1954) 'Geomorphology of south-west Inglefield Land' in *Final Report Operation Ice Cap 1953*, Program B. Dept. Army Project no. 9-98-07-002. Stanford Research Inst. 151–208

NICHOLS, R. L. (1969) 'Geomorphology of Inglefield Land, North Greenland', *Meddr Grønland* 188 (1)

NYE, J. F. (1952) 'A method of calculating the thickness of the ice-sheets', *Nature, Lond.* 169, 529–30

PATERSON, W. S. B. (1969) *The physics of glaciers*

PUTNINS, P. (1970) 'The climate of Greenland' in S. ORVIG (ed.), *World survey of climatology* 14, 3–128

ROBIN, G. DE Q. (1955) 'Ice movement and temperature distribution in glaciers and ice sheets', *J. Glaciol.* 2, 523–32

SUGDEN, D. E. (1968) 'The selectivity of glacial erosion in the Cairngorm Mountains, Scotland', *Trans. Inst. Br. Geogr.* 45, 79–92

SUGDEN, D. E. (1971) 'Deglaciation and isostasy in the Sukkertoppen ice cap area, West Greenland', *Arctic Alpine Res.* 4, 97–117

UNWIN, D. J. (1970) 'Percentage RSS in trend surface analysis', *Area* 2, 25–8

WEERTMAN, J. (1957) 'On the sliding of glaciers', *J. Glaciol.* 3, 33–8

WEERTMAN, J. (1964) 'The theory of glacier sliding', *J. Glaciol.* 5, 287–303

WEERTMAN, J. (1966) 'Effect of a basal water layer on the dimensions of ice sheets', *J. Glaciol.* 6, 191–207

WEIDICK, A. (1968) 'Observations on some Holocene glacial fluctuations in West Greenland', *Meddr Grønland* 165 (6)

RÉSUMÉ. *Paysages d'érosion glaciaire au Groenland et leurs rapports avec les conditions de la glace, la topographie et la roche de fond.* On donne une classification des paysages d'érosion glaciaire pour le Groenland (Table 1), avec une carte de distribution (Fig. 9). On estime que les paysages de glaciation alpine n'ont pas été affectés par l'inondation de l'indlandsis de l'ère pléistocène au Groenland. De tous les paysages modifiés par l'inlandsis, on considère que l'érosion aréolaire prouve qu'autre-fois il existait partout dans la région de la glace basale qui était au point de fusion; l'érosion linéaire sélective prouve l'existence autrefois de la glace basale dont quelques parties seulement étainet au point de fusion; et les paysages qui montrent très peu ou aucune érosion prouvent l'existence autrefois de la glace basale dont aucune partie n'était au point de fusion. La distribu-tion des divers genres de paysages est étudiée à deux échelles. Selon l'échelle de Groenland l'érosion aréolaire a lieu surtout dans le sud-ouest et l'érosion linéaire sélective surtout dans le nord et dans l'est (Fig. 10). On estime que cela reflète l'action de l'indlandsis pléistocène d'autrefois qui produisait une plus grande activité et une plus grande quantité de glace dans le sud-ouest maritime que dans le nord et dans l'est continentaux. La grande différence de la profondeur des couches de fond entre l'est et l'ouest de Groenland a peut-être un rôle à jouer à cette échelle par son influence sur l'épaisseur de la glace. Aux échelles plus locales, les formes topographiques, la profondeur des couches de fond et la perméabilité des roches sont autant de variables qui out causé les variations locales qui diffèrent de la tendance générale.

FIG. 1. Carte de Groenland indiquant les régions libres de glace et les contours approximatifs de la surface de l'inlandsis

FIG. 2. Paysage typique d'érosion aréolaire dans l'ouest de Groenland. Copyright Geodetic Institute, København

FIG. 3. Paysage typique d'érosion linéaire sélective dans l'est de Groenland. Copyright Geodetic Institute, København

FIG. 4. Paysage typique de glaciation alpine. Copyright Geodetic Institute, København

FIG. 5. Températures calculées de la glace basale sous l'inlandsis de Groenland dans des conditions supposées stables. Citées avec la permission de D. Jenssen

FIG. 6. Données climatiques pour l'inlandsis de Groenland:
(a) Isohyètes calculées en g. cm^{-2} de l'eau (Mock, 1967)
(b) Isothermes calculées (Mock and Weeks, 1966)

FIG. 7. Topographie des couches de fond approximatives de Groeland indiquant le peu d'altitude du fond de l'indlandsis dans le centre et dans le nord de Groenland

FIG. 8. Schema illustrant l'évolution d'un paysage d'érosion linéaire sélective:
(a) L'inlandsis s'accumule et la glace basale atteint le point de fusion sur les régions des dépressions.
(b) La dépression à gauche favorise le mouvement de la glace et le glissement de la glace basale a lieu qui produit un creuset. La glace ne peut glisser sur les régions élevées des alentours à moins que la couche de glace ne s'épaississe davantage

FIG. 9. Carte de paysages d'érosion glaciaire au Groenland basée principalement sur une étude de photographies aériennes obliques et verticales
FIG. 10. Surfaces des tendances linéaires indiquant les caractéristiques suivantes des paysages modifiés par l'inlandsis:
(a) Le pourcentage de la superficie classé dans la catégorie d'érosion aréolaire.
(b) Le pourcentage de la superficie classé dans la catégorie d'érosion linéaire sélective.
On a exclu les régions de glaciation alpine de ces calculs
FIG. 11. Profile de surface dessinés pour l'inlandsis Pléistocène au Groenland à son maximum et leurs rapports avec l'altitude des régions bien développées de glaciation alpine. En dessinant les profils on a supposé que l'inlandsis s'étendait sur le plateforme continentale à une profondeur aujourd'hui indiquée par le contour sous-marin −200 m. L'altitude moyenne des 5 sommets les plus hauts de chaque région étudiée est marquée par un symbole triangulaire. La plupart des régions autour des sommets sont évidemment au-dessus des profils et ont donc en toute probabilité échappé à l'inondation totale de l'inlandsis
FIG. 12. Résidus positifs et négatifs des surfaces de tendances linéaires des paysages:
(a) d'érosion aréolaire, et
(b) d'érosion linéaire sélective
FIG. 13. Types de paysage aux environs de la calotte glaciaire de Sukkertoppen
FIG. 14. Correlation visuelle:
(a) d'une zone de roches jeunes (et perméables)
(b) des régions avec très peu ou aucune évidence d'érosion par l'inlandsis dans l'est du Groenland. (b) est tiré de la figure 9, mais on n'a pas considéré les régions d'érosion des cirques

---

ZUSAMMENFASSUNG. *Landschaften der Glazialerosion in Grönland, und deren Verhältnis zu Eis-, Gesteins- und topographischen Bedingungen.* Landschaften der Glazialerosion werden für Grönland klassifiziert (Tabelle 1), und eine Karte deren Verbreitung wird dargestellt (Abb. 9). Gut entwickelte Landschaften der Talvergletscherung sind mutmasslich der Überflutung durch die vollen pleistozänen Grönlandeisdecken entronnen. Unter den Landschaften, die durch Eisdecken überformt worden sind, reflektieren nach Meinung des Authors die *glaziale Flächenerosion* das ehemals weitverbreitete Bestehen von basalem Eis am Druckschmelzpunkt, die *selektive Linearerosion* das frühere Vorkommem in nur beschränkten Gebieten von basalem Eis am Druckschmelzpunkt und die Landschaften mit *wenigen oder gar fehlenden Anzeichen von Glazialerosion* das ehemalige Bestehen von Eis unter dem Druckschmelzpunkt. Die Verbreitung der verschiedenen Landschaftstypen wird in zweierlei Massstäben analysiert: Im Massstab ganz Grönlands tritt glaziale Flächenerosion hauptsächlich im Südwesten auf, während selektive Linearerosion überwiegend im Norden und Osten erscheint (Abb. 10). Dieses spiegelt vermutlich das Verhalten der früheren pleistozänen Eisdecke wieder, die eine stärkere Aktivität und einen grösseren Umsatz von Eis im maritimen Südwesten als im 'kontinentalen' Norden und Osten aufwies. Die breite Variation in der Höhe des Gesteins zwischen West- und Ostgrönland mag ebenfalls durch ihren Einfluss auf die Mächtigkeit des Eises eine Rolle in diesem Massstab spielen. Auf mehr örtlicher Ebene stellen topographische Form, Gestein und Gesteinsdurchlässigkeit Grössen dar, die lokale Abweichungen vom allgemeinen Trend verursacht haben.

ABB. 1. Lokalisierungskarte von Grönland, die den Ring des eisfreien Gebiets und die ungefähren Höhenlinien auf der Eisdecke zeigt
ABB. 2. Typische Landschaft der glazialen Flächenerosion in Westgrönland. Copyright Geodetic Institute, København
ABB. 3. Typische Landschaft der selektiven Linearerosion in Ostgrönland. Copyright Geodetic Institute, København
ABB. 4. Typische Landschaft der Gebirgstalvergletscherung in Grönland. Copyright Geodetic Institute, Købehavn
ABB. 5. Errechnete Temperaturen basalen Eises unter der Grönlandeisdecke, in der Annahme eines unveränderlichen Zustandes. Reproduziert mit der freundlichen Genehmigung von D. Jenssen
ABB. 6. Einige Klimawerte der Grönland-Eisdecke
(a) Errechnete Isohyeten in g. cm^{-2} des Wassers (Mock, 1967)
(b) Errechnete Isothermen (Mock and Weeks, 1966)
ABB. 7. Ungefähre Gesteinstopographie Grönlands, die die niedrige Höhe der Basis der Eisdecke in Zentral- und Nordgrönland zeigt
ABB. 8. Das Diagramm illustriert die Entwicklung einer Landschaft der selektiven Linearerosion
(a) Die Eisdecke baut sich auf, und basales Eis erreicht den Druckschmelzpunkt in Depressionsgebieten
(b) Die Depression zur Linken erlaubt Eisfliessen, basales Gleiten kann auftreten und einen Trog herausarbeiten. Wenn die Eisdecke nicht an Mächtigkeit gewinnt, wird das Eis nicht über die benachbarten Hochlandgebiete gleiten
ABB. 9. Eine Karte der Landschaften glazialer Erosion in Grönland, die hauptsächlich auf der Untersuchung schräg und vertikal aufgenommener Luftbilder basiert. Für die Beschreibung der Landschaftskategorien siehe Text
ABB. 10. Lineare Trendebenen, die für die Landschaften, die durch Eisdecken modifiziert worden sind, folgendes zeigen:
(a) Das prozentuale Gebiet, klassifiziert als glaziale Flächenerosien
(b) Das prozentuale Gebiet, klassifiziert als selektive Linearerosion
Die Gebiete der Gebirgstalvergletscherung sind von diesen Berechnungen ausgeschlossen
ABB. 11. Oberflächenprofile, konstruiert für die maximale(n) pleistozäne(n) Grönlandeisdecke(n), und deren Verhältnis zu der Höhe von Gebieten mit gut entwickelter Gebirgstalvergletscherung. Die Profile sind in der Annahme gezeichnet, dass die Eisdecke sich über den Kontinentalschelf bis zu einer Tiefe erstreckte, die jetzt als die −200 m—Tiefenlinie gilt. Die mittlere Höhe der fünf höchsten Gipfel in jedem einzelnen Untersuchungsgebiet ist durch ein Dreiecksymbol repräsen-

tiert. Die meisten Gipfelgebiete liegen eindeutig über den Profilen und sind damit vermutlich der Überflutung durch die volle(n) Eisdecke(n) entronnen

ABB. 12. Positive und negative Restwerte, errechnet von den linearen Trendebenen für die Landschaften der:

(a) glazialen Flächenerosion

(b) selektive Linearerosion

ABB. 13. Landschaftstypen in der Nachbarschaft der Sukkertoppen-Eiskappe

ABB. 14. Eine visuelle Korrelation der:

(a) Zone junger (durchlässiger?) Gesteine und

(b) Gebiete mit wenigen oder gar fehlenden Anzeichen von Erosion durch die Eisdecke(n) in Ostgrönland. (b) ist von Abb. 9 abgeleitet, jedoch wurden Gebiete mit Karerosion ausgelassen

The glacial history of the Shetland Islands

GUNNAR HOPPE

Professor of Physical Geography, University of Stockholm

Revised MS received 28 June 1973

ABSTRACT. The glacial history of the Shetland Islands was investigated mainly by means of roches moutonnées and glacial striae. The studies have been based on experience of similar research in Scandinavia and other glaciated areas, particularly of the results obtained in areas near the highest shoreline. Older sets of striae, especially those on Bressay, indicate an ice flow from the east which probably means from Scandinavia. Most striae, however, radiate from the central parts of the island group proving the existence of a local ice cap at the end of the glaciation. This local ice cap was maintained over the islands because wastage was controlled mainly by melting, whereas intense calving over deeper waters was responsible for the fast deglaciation of the surrounding seas. The Shetland ice cap probably disappeared 12 000–13 000 years ago.

THE extent of the ice sheets of northern and arctic Europe has long been a major problem of natural history, both for geological-geomorphological and biological reasons (see for instance A. and D. Löve, 1963). The Department of Physical Geography, Stockholm University, has been involved with this problem for several decades, particularly in respect of Scandinavia, Svalbard and Iceland (G. Hoppe, 1971).

As regards Scandinavia the discussion has mainly concerned the existence or non-existence of *refugia* for plants and animals in western and northern Norway during the last glaciation (Würm, Wisconsin). As the evidence which was collected in the supposed *refugia* was interpreted in different ways, the solution of the problem was thought to lie beyond the Norwegian coast. This was part of the background to a small expedition with six participants to the Shetland Islands in 1964. Shetland lies only about 300 km from Norway, and the depth of the intervening sea only exceeds 200 m in the narrow Norwegian Channel. This depth was reduced to about 100 m during the glacial maxima. Well-preserved glacial striae have been observed by the author on Lihesten at the mouth of Sognefjord at an elevation of 700 m, an indication that the ice sheet stretched far away into the North Sea. Furthermore the possible overriding of the Shetland Islands by a Scandinavian ice sheet had been discussed in the literature. For the problem under consideration, then, Shetland seemed to be a key area.

The problem of glacierization of different areas was tackled in many different ways. We sought direct evidence of glaciation, such as glacial and glaciofluvial features of erosion and accumulation, and utilized indirect evidence, especially isostatic movements as demonstrated by the presence of raised marine beaches and sediment sequences. Dating of the various features was also attempted.

There are no raised beaches on Shetland so the possibilities for studies of isostatic movements are limited, though some information can be obtained from the sea bottom (G. Hoppe, 1965). Moreover, a thick cover of peat obscures much of the landform. Under such circumstances the search for glacial evidence was confined to roches moutonnées with their striae and other indicators of ice-movement directions. Using experience gained in other areas our intention was to obtain a general idea of glacial history rather than a full coverage of all details of ice-flow development. The restricted time at the disposal of the expedition also made it impossible to study transport directions of till and erratics, particularly because trajectories can be extremely complicated and can be derived from more than one glaciation as well as from other kinds of transport.

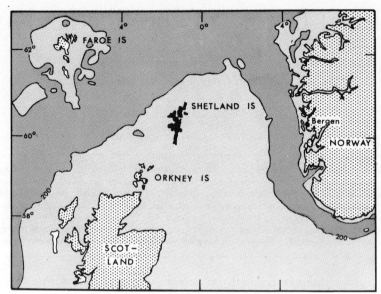

FIGURE 1. The Shetland Islands are situated approximately 300 km west of Norway, and only in the Norwegian
Channel does the depth exceed 200 m

The field studies covered all parts of the Mainland, and parts of Bressay, Whalsay, Out Skerries
and Unst.

EARLIER INVESTIGATIONS

Shortly after the glacial theory was proposed in continental Europe, similar reasoning was applied
to explain various features on the Shetland Islands. C. W. Peach (1865) mentioned the ice-worn
aspect of the rocks, the striae and the existence of boulder clay in certain localities. J. Croll (1870
and elsewhere) tried to demonstrate that both the Orkney and Shetland Islands were overridden
by coalescent ice sheets from Scandinavia and Scotland. The basic paper, however, is that by
B. N. Peach and J. Horne (1879) entitled 'The glaciation of the Shetland Isles'. On the basis of
observations of glacial striae, lithological content of boulder clay, and erratics in all parts of
Shetland, from Unst in the north to southernmost Mainland in the south, from Foula in the west
to Out Skerries in the east, they concluded that during a 'primary glaciation' Shetland was
overridden by ice from Scandinavia. This glaciation was represented by a movement to the
south-west over eastern Shetland, but after having passed the higher parts the direction of ice
flow turned to the north-west. After this ice sheet 'had ceased to be confluent with the local
glaciers of Shetland, the latter lingered on for a time, filling all the main valleys and flowing off
the land in all directions' (1879, 810). The study of Peach and Horne is in many respects admir-
able, or even visionary, especially as the main results seem to be correct in spite of the fact that
many of the observations cannot be duplicated and were probably erroneous for reasons which
are not easily explained.

Another early study which deals with the glaciation of Shetland is that of A. Helland (1879)
who supported the concept of Peach and Horne of a Scandinavian ice sheet which extended to
this island group. T. M. Finlay (1926a) observed glacial striae, erratics and what he describes as a
corrie basin on Foula. He interpreted the absence of erratics on the highest parts of that island as
evidence that the 'main ice sheet did reach the island without completely over-riding it' (p. 564).
Furthermore, in a short note (1926b) Finlay also mentioned his observation of a tönsbergite

boulder of two tons at Dalsetter, southern Mainland, at a height of 100 feet above present sea level. The late Professor T. Barth, Oslo, after examination of a hand specimen of the boulder, has confirmed Finlay's identification (letter to the author, 20 October 1964).

The Geological Survey of Great Britain (Scotland) has been mapping Shetland since the late 1920s. Recently the first two sheets of the drift edition appeared, covering 'Northern Shetland' (1968) and 'Western Shetland' (1971). The latter sheet especially, has a rich documentation of glacial striae which has been of importance for the following discussion.

In connection with his comprehensive geological work on Shetland D. Flinn has also dealt with its glacial history. In his 1964 paper two alternatives are given: either that the Scandinavian ice sheet flowed across Shetland from east to west at an early stage of a 'glacial' maximum previous to the last glaciation and 'was replaced in Shetland by a local ice cap which flowed eastward on the east side' or that 'the Scandinavian ice sheet crossed Shetland in the last glacial maximum' (1964, 338). In Flinn (1967), which deals mainly with an 'arcuate belt of deeps' south of Shetland, a map showing ice flow 'in the northern North Sea during the last glacial maximum' is given. 'In the Shetland area, Norwegian ice was deflected to the north and south by a local ice cap but possibly overwhelmed the extremities in the early stages.' In his 1970 paper he extends his ideas to include a 'subsidiary centre of dispersion in the Shetland ice cap in the general area of Foula' (1970, 275).

Inspired by Flinn, R. Chapelhow studied 'The glaciation in North Roe, Shetland' (1965) by observing glacial striae and boulder clay. The most important result of her study was the discovery at Fugla Ness of a peat bed between two glacial tills; the contents of this bed have since been studied by different scientists. H. J. B. Birks and M. E. Ransom (1969) referred it to the Hoxnian Interglacial on paleobotanical evidence, whereas N. R. Page (1972) has obtained two C^{14} datings on wood and peat respectively of $34\,800^{+900}_{-800}$ (T − 1092) and $37\,000^{+1200}_{-1000}$ (T − 1093) years BP.

Finally it may be mentioned that exploration groups from the Brathay Field Study Centre, Ambleside, Westmorland, have undertaken study projects, including glacial geology, on Foula during the last few years (information from R. J. Metcalfe, C. Dingwall and H. Proudman). A great deal of information about glacial striae and erratics has thereby been collected.

GLACIAL STRIAE: OBSERVATIONS AND CONCLUSIONS

The significance of glacial striae has been investigated in Sweden since the last century (see for instance G. De Geer, 1897; E. Ljungner, 1949; G. Hoppe, 1948, 1967; B. Strömberg, 1971). Such studies have been facilitated by unique possibilities to relate striae and other features indicating directions of ice movement to landforms and deposits indicating successive positions of the ice front i.e. recessional moraines (especially of the De Geer type), glacial varves, eskers, drumlins, drainage channels, etc.

There are important differences between glacial striae formed below and above the highest shoreline. When an ice front ends in water a vertical cliff is formed by calving and melting. Thereby an unstable situation develops which results in rapid flow normal to the cliff. The ice in this situation has considerable erosional capacity, which means that the formation of roches moutonnées and glacial striae continues until the ice disappears. The consequence is that the youngest striae below the highest shoreline (naturally, older sets of striae may be preserved also) were formed not very far inside the ice cliff—they are more or less submarginal. Furthermore, they relate to the time of ice recession and, most important, their direction is strictly controlled by the orientation of the ice cliff. So-called calving bays, formed where the sea depth is greater than in the surroundings, are normal features at recent tidal glaciers under recession, such as around Glacier Bay, Alaska. Calving bays also developed during the retreat of the Pleistocene ice sheets where depth conditions were favourable. Changes of orientation of ice front must thus be regarded as quite normal. For the same reason more than one set of striae, all of them normally rather young, are very common in areas below the highest shoreline.

When an ice front leaves the water, part of the very characteristic pattern of subaquatic recession is completely changed, whereas the ice-flow pattern does not change abruptly but adjusts gradually to the new environment (examples in Hoppe, 1967, 1968). Quite soon the ice front changes from concave, in the depressions (calving bays), to convex (tongues). Even above the highest shoreline the ice movement near the front is directed normal to the edge, but as the marginal zone thins the velocity decreases, even to the point of immobility. Naturally the ice then becomes incapable of scratching; the same incompetence will result if the ice changes from being temperate, which is the probable situation inside an ice cliff standing in water, to being cold. Another characteristic of the thinning of an ice sheet/cap is a successively stronger influence of topography (example in Hoppe, 1967).

In places like Shetland, where the ice may have moved in one direction or in the opposite direction, the determination of the *sense* of the movement (cf. Flint, 1971, 90) is of fundamental importance. This determination of the direction in which the ice flowed has been made by observation of the plucking at joints; studies of crescentic marks, especially gouges, have been of some importance too.

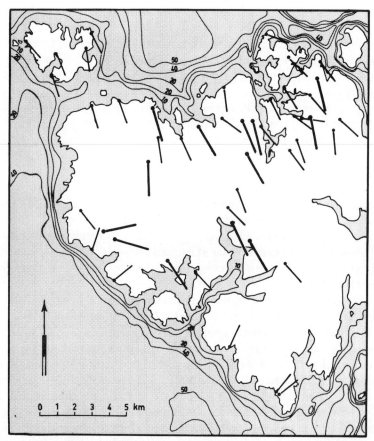

FIGURE 2. Glacial striae on west-central Mainland. Heavy symbols show observations by G. Hoppe and thin symbols observations by the Geol. Survey. Dots represent points of observation. Submarine contours in fathoms

Case studies

West-central Mainland (Fig. 2)   The glacial striae of Figure 2 originate from our own investigations in 1964 and from the drift map of western Shetland published by the Institute of Geological Sciences (1971). Through the kindness of W. Mykura the observations of the Survey were made available to us at the time of our own field-work.

The depths of the sea around west-central Mainland normally reach 30 to 40 fathoms or even more within 1 to 3 km of the coastline. This means that even if we do not know the exact position of the shoreline immediately after the ice receded, it can be assumed that the margin of the ice stood in water in the vicinity of the present shoreline; in accordance with experience gained in other areas this must also have strongly influenced the early recession on land.

In the coastal areas of west-central Mainland all striae are directed more or less normal to the present shoreline as well as to the submarine contours outside it. This pattern confirms the correctness of the hypothesis outlined above. And since the striae are believed to have been scratched not far inside and normal to the ice front its orientation can be roughly reconstructed.

Striae directed to the north-west also appear in the southern parts of the area, not far from the south-west coast. This must be a result of earlier deglaciation in the north-west than in the south-west and is in accordance with the general depth conditions west of Shetland; the 200 m contour and the edge of the continental shelf are, for instance, oriented north-east–south-west (Fig. 1).   No older sets of striae have been observed in the area.

Esha Ness (Fig. 3)   All the striae on this peninsula point to the north-west or north-north-west. This is about normal to the present shoreline to the north and to the depth contours outside it. Under the general retreat from north-west to south-east there was not sufficient time to establish an east–west-oriented ice front parallel to the south coast of the peninsula; nor are there any striae directed towards the south.

Gulber Wick area (Fig. 5) Our observations of striae demonstrate an ice movement from north to south in the northern part of the area around Gulber Wick while the striae in the western and southern parts show a movement from the north-west. On the southern side of Gulber Wick are numerous bedrock outcrops with two sets of striae: the younger one from the north-west as mentioned, and an older one from N 20–35°E. This older movement will be discussed under 'Lerwick–Bressay area'.

If the younger striae are used for reconstruction of the ice front, they demonstrate a concavity in the Gulber Wick area. When such concavities, instead of a convex lobe, appear in valley areas they prove an ice recession in water and the existence of so-called calving bays. The depth conditions in the Wick and outside seem to demonstrate that the bay may have developed not earlier than during the recession over Gulber Wick. As the present depth of the Wick does not exceed 15 fathoms (about 27 m) the conclusion may be that the shoreline in the area at the time of ice recession was not much lower, and in any case not more than 25–30 m lower, than it is at present. It must be admitted that this reasoning may be more speculative than elsewhere in this paper. However, it demonstrates a way to get information about the level of the shoreline at the time of deglaciation of Shetland; it should be added that the whole situation gives a strong impression of a shoreline not much below that of today.

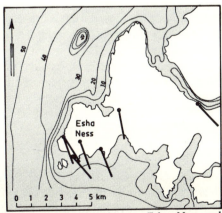

FIGURE 3. Glacial striae on Esha Ness and surroundings. Submarine contours in fathoms

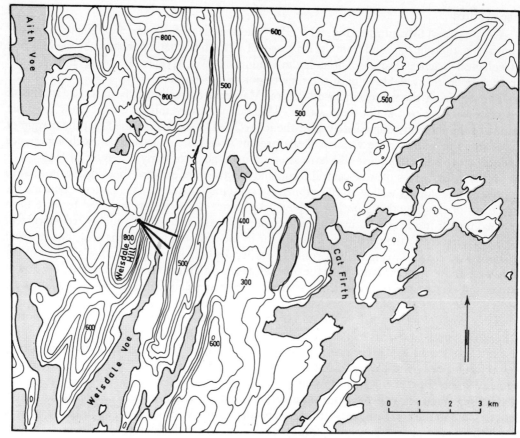

FIGURE 4. Glacial striae on Weisdale Hill. Contours at 50 ft intervals

Weisdale Hill (Fig. 4)   This locality is mentioned by Peach and Horne: 'Near the gap in the ridge overlooking the head of Weesdale Voe, the polished surfaces and striations are as fresh as if the ice had but recently passed away' (1879, 793). This is an adequate description. Representative striae point N 35°W in full accordance with the statement of these authors, but there are also striae, probably influenced by local topography, pointing N50–70°W. In other words the ice over the area has moved from south-east to north-west. The sense is absolutely clear both because of stoss-and-lee topography in macro- as well as micro-scale and by the plucking at cracks in the bedrock surfaces.

Glacial striae and grooves were observed to a height of at least 230 m (750 feet) above sea level, which means that this ridge of hills, in places exceeding 250 m in elevation, must have been totally covered by the ice sheet. Still more important is the statement that the nearest topographic 'high' on the ice sheet must have been situated east of Weisdale Hill, and that there is only lower land on Shetland east of the Hill.

Bressay–Lerwick area (Fig. 5) On the eastern side of Bressay the youngest striae show a movement towards the east or south-east, i.e. normal to the coastline and the nearest submarine curves, again demonstrating the pattern of ice recession in water. The influence of the small-scale local topography in the neighbourhood of the shoreline is indicated also by striae pointing to the north-east at the head of the Voe of Cullingsburgh.

In the shallow Bressay Sound area the youngest movement was still oriented in easterly

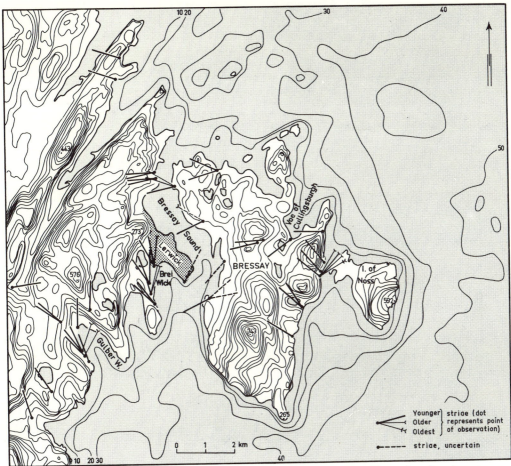

FIGURE 5. Glacial striae in the Bressay–Lerwick–Gulber Wick area. Contours at 50 ft intervals; submarine contours in fathoms

directions. However, on the western side of Brei Wick the youngest striae were formed by a movement from north-north-west which means that an ice cliff receded to the north-west in the sea south of the area under consideration.

In the Bressay–Lerwick area older sets of striae also appear. Both north and south of the ferry-crossing to the Isle of Noss there are several places with a young system from N40–50°W, but on well-developed stoss sides facing north-east there are also extremely well-preserved striae from the north-east. They include at least two sets; an older one, mainly represented by grooves, from N 60°E, and a younger one from N 30–50°E. Thus there has been a change in direction from N 60°E to N 30–50°E and finally to N 40–50°W. There is no doubt about the sense of the different sets, nor about the age relations. Also, on the west side of Bressay the system from the north-east seems to be represented as well as the striae from the west. Both the sense and the age relations of the striae are however, difficult to interpret along this stretch of coast.

In the southern part of Lerwick and on bedrock outcrops alongside Brei Wick practically all directions of striae can be observed in the sector from N 50°W to N 20°E. Where the age relations between the striae can be observed, the striae from north-westerly directions are always the younger and those from north or north-north-east the older. However, one gets the impression that

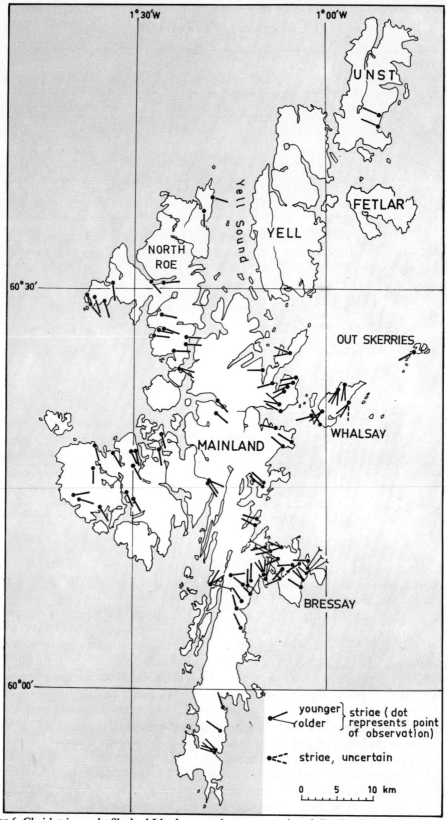

FIGURE 6. Glacial striae on the Shetland Islands: a complete representation of all striae observed during the 1964 field work

there was a continuous change in direction and, on the basis of the rather weak erosion, that the age difference between the different sets is not very great.

The striae in different parts of the Bressay–Lerwick area demonstrating ice movement from north-east or north-north-east show that before the ice moved from the central parts of Mainland to an ice cliff standing in the sea to the east it must have flowed from an ice centre to the north-east of Shetland. It is hard to believe that that centre could be anything other than the ice sheet over Scandinavia.

Additional observations and general conclusions (Fig. 6)

The pattern of glacial striae described in most of the case studies, demonstrating ice movements more or less normal to the shore around Shetland, is repeated everywhere (Fig. 6). The *Mainland* offers numerous examples in addition to those already given: the eastern coastland south of Gulber Wick; the Scalloway district; South and North Nesting and Lunnasting. The conclusion, which can be drawn after evaluation of all the striae, is that when deglaciation started over the Mainland it was covered by an ice cap with its divide in the neighbourhood of the backbone of the island. What was the relation of this ice cap to the other islands? It has already been demonstrated that in a late stage *Bressay* was covered by ice arriving from the Mainland. *Whalsay* shows a pattern of striae influenced by the retreat of an ice cliff in the sea around the island; successive changes in directions of ice flow developed during retreat, according to the depth conditions. The main feature, however, was a retreat towards the Mainland. This does not exclude the possibility that small ice masses were isolated finally on the south-western part of the island.

According to Peach and Horne (p. 791) 'there is, perhaps, no district in Shetland where the intense abrasion typical of glaciated regions is so patent as in the outskerries of Whalsay'. They were 'convinced that the ice crossed the Skerries from the north-east towards the south-west'. During our visit to *Out Skerries* we were only able to investigate Housay, the biggest of the islands. There are many beautifully striated bedrock outcrops. In some places it is also possible to determine the sense of movement with absolute certainty; i.e. the ice flowed from south-west to north-east, in the opposite direction to that proposed by Peach and Horne. It must, however, be emphasized that a more thorough investigation of sheltered localities on the other islands in the group might reveal striae from the north-east as Peach and Horne proposed. The observations presented here agree with T. Robertson's conclusion (1935) based on striae but mainly 'on the direction in which the various rocks have been transported'.

In the northernmost part of Mainland roches moutonnées with stoss sides facing to the east appear at the shoreline. Glacial grooves, up to 6 m long, and striae also demonstrate an ice movement from the east, i.e. from the island of *Yell*. This shows that the divide of the ice cap was situated east of North Roe and Yell Sound.

Peach and Horne stated that along the eastern seaboard of *Unst* the striae demonstrate an ice movement from east to west. The drift edition of the geological map 'Northern Shetland' gives only one striae locality on Unst, namely at Cata Wick; it also shows an ice movement from east to west. During a one day's visit to Unst we investigated parts of the eastern coast, but we succeeded in finding glacial grooves and striae only in the southern part of the island, including the locality at Cata Wick. Contrary to the earlier investigations we found absolutely clear evidence of a movement from west to east on several roches moutonnées. No traces of older movements from the east were observed.

We were not able to get to the westernmost, isolated island of *Foula*. The existence of striae, however, was confirmed by earlier investigations and those of the Brathay Exploration Group. Foula, then, must have at least to some extent been overridden by ice coming from the Mainland. Striae from the south-east on northern Foula, and from the south on eastern Foula,

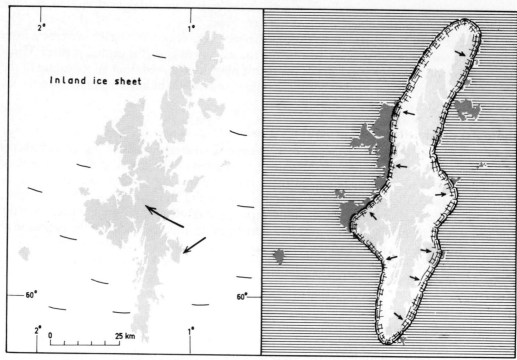

FIGURE 7. Glaciation of the Shetland Islands. During an early stage the Shetland Islands were most probably overridden by an ice sheet from the east (left map), whereas the later stage was characterized by radial flow from a local
ice cap (right map)

demonstrate local ice recession toward the south-east and the south, in accordance with the
depth conditions and the general theme of this paper.

So far, only striae belonging to a stage when ice flowed in all directions from the main ridge
of the Shetland Islands have been considered. This period can be described as an ice-cap stage.
But there are also striae demonstrating an ice movement from the east over Shetland. However,
we observed these striae only in the Bressay–Lerwick–Gulber Wick area although the striae on
Weisdale Hill probably belong to the same category. As noted earlier we did not find striae
formed by ice flowing from the east on Unst or on Out Skerries, although Peach and Horne
described a movement from the east on these islands. This does not exclude the possibility that
true striae from the east may be observed in sheltered positions at some future date.

It has been proposed already that the movement from the east meant that the ice sheet over
Scandinavia expanded to override Shetland. Other arguments have been used to justify a vast
Scandinavian ice sheet. It has been proposed (D. Flinn, 1967) that the striae radiating out from
Shetland show evidence of deflection of the local ice by 'Norwegian ice'; this explanation is in
conflict with the basic ideas advanced here to explain local differences in the directions of the
youngest striae. Observations of erratics from Scandinavia, especially the tönsbergite boulder at
Dalsetter, cannot be used as a definite proof of the overriding of Shetland by the last ice sheet;
the manner, path and time of transport of such erratics are unknown.

The general development of the glaciation of Shetland as revealed by the glacial striae may
now be summarized. At an early stage Shetland was overridden by ice coming from the east,
which most probably means Scandinavia. At a later stage an ice cap with its divide along the
backbone of the islands most probably covered the whole island group. This conclusion is

definite as far as the eastern islands, such as the Out Skerries, are concerned; on the other hand it is impossible to determine if the striae on the western islands, such as Foula, were made by a Scandinavian ice sheet or the Shetland ice cap.

Furthermore it must be regarded as quite clear that there was a successive transition from the Scandinavian ice stage to the Shetland stage; a continuous shifting of the striae gives evidence of this transition. Probably the whole Shetland ice-cap stage should be looked upon mainly as a phase of the deglaciation. During the recession of the Scandinavian ice sheet the depth conditions of the sea played a decisive role. In areas with great depths recession was accelerated because of the buoyancy effect along the ice margins and the resulting iceberg calving. This was so west of Shetland but it also happened in the Norwegian Channel with its depths of up to 400 m. The ice over Shetland was then isolated from the Scandinavian ice and formed a separate ice cap with its main divide running from northern to southern Shetland and with the flow of ice going in all directions normal to the ice cliffs. This local ice cap does not entitle us to draw any conclusion about a glaciation level lower than the highest hills on Shetland. Following the vocabulary of the Ahlmann school (H. Ahlmann, 1948, 64) it seems most probable that the ice cap over Shetland was climatologically dead but dynamically still very active. A sketch of the two main phases of the development is given in Figure 7.

DATING (Fig. 8)

As yet we do not have any direct way of dating glacial features, but some information can certainly be obtained from their degree of preservation. The good state of preservation of roches moutonnées and striae in many places on Shetland is an indication that they were formed during the Würm glaciation. A minimum age can be derived if we date the oldest organic material above the glacial features. Thus a series of C^{14} age determinations were made on those parts of lake sediment cores in which pure clay, without organic content and in many places exhibiting varves had just been superseded by more or less clayey gyttja (G. Hoppe, 1970, 1971). The results are shown in Table I. It is important to note that each date was obtained on a 3 to 6 cm long core increment, and thus may well include a sedimentation period of several hundred years. The dates should be regarded as minimum figures for the beginning of organic accumulation after deglaciation.

Four of the dates give C^{14} ages of about 10 000 years. This fits quite well with the beginning of the Preboreal. The date from Loch of Clickhimin, however, points to the Alleröd period. The core from this lake is very interesting because it includes a layer of clay at a depth of 140–150 cm, i.e. above the oldest gyttja. Above this clay the gyttja from 137–140 cm has a C^{14} age of 9620 ±750 years BP (St-1639). It is tempting to regard this sequence as reflecting the Alleröd—Younger Dryas—Preboreal periods.

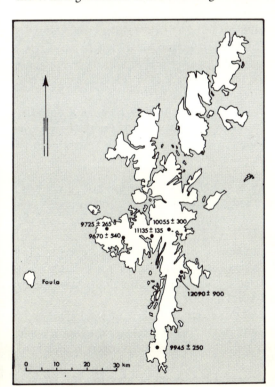

FIGURE 8. Oldest postglacial C^{14} dates in years BP

TABLE I

C^{14} *determinations*

Number	Lake	Location	Depth of sample below the surface of the lake sediments	C^{14} age in years B.P.
St-1554	Sand Water	60°17′N–1°20′W	125–128 cm	10 055 ± 300
St-1595	Stanevatstoe Loch	60°16′N–1°37′W	382–387 cm	9 725 ± 265
St-1763	Upper Loch of Brouster	60°15′N–1°36 W	444–450 cm	9 670 ± 540
St-1640	Loch of Clickhimin	60° 9′N–1°10′W	177–180 cm	12 090 ± 900
St-1559	Loch of Brow	59°56′N–1°19′W	198–201 cm	9 945 ± 250

A somewhat similar sequence was found in a peat bog at Tresta (60°15′N–1°20′W) which showed 290 cm of peat, underlain by minerogenic fine matter, probably a glaciofluvial or fluvial deposit with thin lenses of peat. From one such lens, 43 cm below the contact between peat and mineral soil and 1 cm thick, a C^{14} age of 11 135 ± 135 years was received (St-1714). The 2 cm peat just above the washed material had an age of 5865 ± 95 years (St-1607), and this date is supported by the age of a piece of wood from 264 cm below the surface, for which a value of 5145 ± 90 years (St-1608) was obtained.

From this we can conclude that the Shetland Islands were mainly free from ice by 11 000 to 12 000 years BP; probably the ice had disappeared only a short time earlier. This is in good accordance with events on the east side of the Norwegian Channel.

It is still more difficult to determine the length of the Shetland ice cap stage. On the one hand the good preservation of striae from the north-east in the Lerwick–Bressay area seems to suggest that the local ice persisted for only a short time. On the other hand the investigations of Robertson have demonstrated that the time was sufficient to permit the transport of Stavaness granite from Dury Voe west of Whalsey to the Out Skerries, a distance of at least 20 km. Thus a minimum duration of several hundred years is very probable, both because of the time the transport of boulders must have taken and because of the time necessary after this transport for the recession from the outermost skerries to the last ice remnants on central Mainland.

If the C^{14} datings of the Fugla Ness deposit are correct, then the glacierization of Shetland started not more than 35 000 years ago.

SUMMARY

The last glaciation of the Shetland Islands, the striae of which now appear in most parts of the island group, probably started less than 35 000 years ago and ended some 12 000 to 13 000 years ago. At an early stage a considerable area of Shetland was overridden by ice coming from the east, which most probably means from Scandinavia. The retreat of the Scandinavian ice sheet in the northern part of the North Sea was strongly influenced by the calving of icebergs. This mechanism was especially effective in the deep Norwegian Channel and caused the isolation of an ice cap over Shetland. This ice-cap stage, characterized by flow outwards in all directions from the backbone of the islands, lasted several hundred years *at least*.

It is most probable that the shoreline at the time of deglaciation of Shetland was only slightly lower, not more than 20–25 m lower, than it is today in spite of general sea level being some 50 m lower. If that were the case it is a good indication of an early isostatic recovery of Shetland because of the deglaciation. The recovery must have ceased at least 6000 years ago, the main evidence being submarine peat of that age which occurs off the shores of Shetland (G. Hoppe, 1965).

ACKNOWLEDGEMENTS

The author wishes to express his thanks to the Swedish Natural Science Research Council for financial support for the 1964 Shetland expedition; fellow members of the expedition: Bo Strömberg, invaluable because of his great experience in glacial striae and their interpretation, and Magnus Fries, Stig-Rune Ekman, Anders Häggblom and Christman Ehrström who were mainly engaged in sample collections for C^{14} datings; the Geological Survey of Great Britain (Scotland) for kind help in different respects; L. G. Engstrand for his help with the C^{14} datings; Birgit Hansson who has drawn the maps; and Weston Blake Jr who has critically read the manuscript and corrected the English text.

REFERENCES

AHLMANN, H. W:SON (1948) 'Glaciological research on the north Atlantic coasts', *R.G.S. Res. Ser.* 1

BIRKS, H. J. B. and M. E. RANSOM (1969) 'An interglacial peat at Fugla Ness, Shetland', *New Phytol.* 68, 777–96

CHAPELHOW, R. (1965) 'On glaciation in north Roe, Scotland', *Geogrl J.* 131, 60–70

CROLL, J. (1870) 'The boulder clay of Caithness, a product of land ice', *Geol. Mag.* 7, 209–17, 271–8

DE GEER, G. (1897) 'Stockholmstraktens geologi' in *Stockholm*, *Sveriges huvudstad* 1, Stockholm 1–27

FINLAY, T. M. (1926a) 'The old red sandstone of Shetland', 1', *Trans. R. Soc. Edinb.* 54, 553–72

FINLAY, T. M. (1926b) 'A tönsbergite boulder from the boulder clay of Shetland', *Trans. Edinb. geol. Soc.* 12, 180

FLINN, D. (1964) 'Coastal and submarine features around the Shetland islands', *Proc. Geol. Ass.* 75, 321–39

FLINN, D. (1967) 'Ice front in the North sea', *Nature, Lond.* 215, 1151–57

FLINN, D. (1970) 'The glacial till of Fair Isle, Shetland', *Geol. Mag.* 107, 273–76

FLINT, R. F. (1971) *Glacial and Quaternary geology*

HELLAND, A. (1879) 'Über die Vergletscherung der Faroer, sowie der Shetland- und Orkney-Inseln', *Z. dt. geol. Gesellsch.* 31, 716–55

HOPPE, G. (1948) 'Isrecessionen från Norrbottens kustland i belysning av de glaciala formelementen', *Geographica* 20, 1–112

HOPPE, G. (1965) 'Submarine peat in the Shetland Islands', *Geogr. Annlr* 47 A, 195–203

HOPPE, G. (1967) 'Case studies of deglaciation patterns', *Geogr. Annlr* 49A, 204–12

HOPPE, G. (1968) 'Grimsey and the maximum extent of the last glaciation of Iceland', *Geogr. Annlr* 50 A, 16–24

HOPPE, G. (1970) 'The Würm ice sheets of northern and arctic Europe', *Acta geogr. Univ. lodz.* 24, 205–15

HOPPE, G. (1971) 'Nordvästeuropas inlandsisar under den sista istiden', *Svensk Naturv.* 1971, 31–40

LJUNGNER, E. (1949) 'East–west balance of the Quaternary ice caps in Patagonia and Scandinavia', *Bull. geol. Instn. Univ. Uppsala* 33, 11–96

LÖVE, A. and D. LÖVE (1963) *North Atlantic biota and their history*

PAGE, N. R. (1972) 'On the age of the Hoxnian interglacial', *Geol. Jnl* 8, 129–42

PEACH, B. N. and J. HORNE (1879) 'The glaciation of the Shetland Isles', *Q. Jl geol. Soc. Lond.* 35, 778–811

PEACH, C. W. (1865) 'Traces of glacial drift in the Shetland Islands', *Rep. Br. Ass. Advmt Sci.* 1864, 59–61

ROBERTSON, T. (1935) 'The glaciation of Aithsting, South Nesting, Whalsay and the Outer Skerries', Manuscript in the Inst. Geol. Sciences, Highland unit unpublished records, Edinburgh.

STRÖMBERG, B. (1971) 'Isrecessionen i området kring Ålands hav', *Stockholms univ. naturgeogr. inst. forskningsrapport* 10, 1–156

RÉSUMÉ. *L'histoire glaciale des îles Shetland.* L'histoire glaciale des îles Shetland a été explorée principalement à l'aide des roches moutonnées et des stries. Les études ont été basées sur l'experience d'une pareille recherche en Scandinavie et en d'autres régions glacées, et particulièrement le resultat reçu dans les régions près du plus haut niveau de la mer. Des systèmes de stries plus vieilles, en particulier à Bressay, indiquent un «inlandsis» progressant de l'est, ce qui donne à supposer une origine scandinave. Cependant, la plupart des stries rayonnent du centre confirmant l'existence d'un glacier local en calotte pendant la fin de la glaciation. Cette calotte restait longtemps sur les îles, parce que le recul était avant tout dépendant de la fusion, tandis que le velage intense sur la mer profonde a eu une importance décisive pour la déglaciation rapide des mers voisines ou environnantes. La calotte locale de Shetland est probablement disparue depuis 12 000–13 000 ans.

FIG. 1. Les îles Shetland sont situées environ à 300 km à l'ouest de la Norvège et seul le Canal Norvègien est plus profond que 200 m

FIG. 2. Des stries dans l'ouest de Mainland central. Les gros symboles sont des observations faites par G. Hoppe et les symboles minces sont des observations faites par Geol. Survey. Les points représentent des endroits d'observation. Les courbes sous-marines sont en brasses

FIG. 3. Des stries dans Esha Ness et dans les environs. Courbes sous-marines en brasses

FIG. 4. Des stries dans Weisdale Hill. Courbes hypsométriques tous les 50 pieds

FIG. 5. Des stries dans les environs de Bressay-Lerwick-Gulber Wick. Courbes hypsométriques tous les 50 pieds. Courbes sous-marines en brasses

FIG. 6 Des stries dans les îles Shetland: une présentation totale des stries observées pendant le travail de 1964

FIG. 7. La glaciation des îles Shetland. Pendant une période antérieure les îles Shetland ont été couvertes par une glace venant de l'est (la carte gauche), alors que l'époque postérieure a été caractérisée par un flux radiaire d'un glacier calotte (la carte droite)

FIG. 8. Les C14-dates les plus vieilles en ans BP (avant 1950)

ZUSAMMENFASSUNG. *Die Glazialgeschichte der Shetland-Inseln.* Die Glazialgeschichte der Shetland-Inseln wurde im wesentlichen anhand von Rundhöckern und Gletscherschrammen untersucht. Die Untersuchungen stützen sich auf Erfahrungen bei ähnlichen Studien in Skandinavien und in anderen ehemals vergletscherten Gebieten, insbesondere auf Ergebnisse in Bereichen nahe der Höchsten Küstenlinie. Ältere Schrammensysteme, vor allem die auf Bressay, weisen auf eine Fliessrichtung des Eises von Osten hin, was eine Herkunft des Eises aus Skandinavien nahelegt. Die meisten Schrammen strahlen jedoch von den zentralen Teilen der Inselgruppe aus und beweisen so die Existenz einer lokalen Eiskappe gegen Ende der Vereisung. Diese lokale Eiskappe konnte sich über den Inseln längere Zeit halten, da sie im wesentlichen nur durch Abschmelzen aufgezehrt wurde; starkes Kalben über den tieferen Gewässern war hingegen für den schnellen Rückzug des Eises vom umgebenden Meer verantwortlich. Die Eiskappe der Shetland-Inseln verschwand wahrscheinlich vor 12 000–13 000 Jahren.

ABB. 1. Die Shetland-Inseln liegen ungefähr 300 km westlich von Norwegen, und nur in der Norwegischen Rinne übersteigt die Wassertiefe 200 m

ABB. 2. Glaziale Schrammen auf dem westlichen Zentralteil des Mainland. Kräftige Symbole zeigen Beobachtungen von G. Hoppe, dünne diejenigen des Geol. Survey. Punkte bezeichnen Beobachtungsstellen. Isobathen in Faden

ABB. 3. Glaziale Schrammen auf Esha Ness und Umgebung. Isobathen in Faden

ABB. 4. Glaziale Schrammen auf Weisdale Hill. Höhenlinien mit 50 Fuss Äquidistanz

ABB. 5. Glaziale Schrammen in der Gegend von Bressay-Lerwick-Gulber Wick. Höhenlinien mit 50 Fuss Äquidistanz; Isobathen in Faden

ABB. 6. Glaziale Schrammen auf den Shetland-Inseln: Vollständige Widergabe aller Schrammenbeobachtungen während der Geländearbeiten von 1964

ABB. 7. Vergletscherung der Shetland-Inseln. Während eines frühen Stadiums waren die Shetland-Inseln höchstwahrscheinlich von einem von Osten kommenden Eis bedeckt (linke Karte), während das spätere Stadium durch radiales Fliessen von einer lokalen Eiskappe gekennzeichnet war (rechte Karte)

ABB. 8. Die ältesten postglazialen C14—Datierungen in Jahren (BP)

V. LANDFORM CLASSIFICATION

Introduction

IN common with most geographers of his generation, David L. Linton spent many hours in practical classes instructing students in the delimitation of geographical regions as expressed upon topographical maps at scales between 1:25 000, and 1:250 000. Generations of students at King's College London and in the Universities of Edinburgh, Sheffield and Birmingham were stimulated by his cogent arguments for the recognition of a hierarchy of such regions on the map and in the field.

Early in his career as a teacher Linton was much stimulated by the methods of 'site recognition', first clearly stated by Ray Bourne in 1931. In Bourne's own words: 'A site may be defined as "an area which appears, for all practical purposes, to provide throughout its extent similar local conditions as to climate, physiography, geology, soil, and edaphic factors in general. While a site may be unique, more often the same type of site is to be met with again and again within some readily identifiable area".' In co-operation with his Edinburgh colleague, Catherine P. Snodgrass, in 1946, Linton combined the recognition of Bourne's sites with the identification of larger scale physical regions in a study of the land use of Peeblesshire and Selkirkshire. They claimed that this hierarchy of regions provides a link between general descriptive accounts of terrain and the problems of the practical man, a link which can be made continually closer as the method is more fully worked out and better applied. Peeblesshire and Selkirkshire were divided into four geomorphological regions, High Plateaux, Hills, Low Plateaux, and Hill and Valley regions, each sub-divided into a wetter and a drier version, to give eight classes of terrain. In the two counties they identified a total of twenty-six such regions.

Another important influence upon Linton's thinking was the work of Fenneman in the United States, which he emulated in 1951 in a paper on the delimitation of morphological regions. He recognized three 'major divisions' in Western Europe, the Atlantic and Central Highlands, and the Great European Plain. Within the British Isles he divided each major division into two physiographic 'provinces' and as an example of the hierarchic arrangement of regions delimited 19 'tracts' in Central Scotland, one of the 'sections' within the Irish Sea Upland province. Each tract was further divided into 'stows' and each stow was conceived as comprising an assemblage of sites.

Concurrently with his work on regions, Linton was, in the 1950s, instrumental in devising a system of geomorphological mapping on small-scale maps through the identification of breaks and changes in slope in the field. These ideas were taken up and developed more extensively by his colleagues, R. S. Waters and R. A. Savigear. At the same time Linton also inspired and devised a method of mapping the geomorphology of Britain at a scale at 1:625 000 with the intention that the British Geomorphological Research Group, of which he was Chairman, should sponsor the publication of such a map.

As in the Land Utilization Survey of Peeblesshire and Selkirkshire Linton had in mind always the practical application of the identification of small physiographic units so that in later years he came to appreciate that geomorphological mapping and the delimitation of physiographic regions could form a base for the assessment of scenery as a natural resource. In 1968 he described a cartographic evaluation of scenic resources in Scotland and compiled a numerical system for resource ratings.

It is a measure of the success of his teaching and research in these fields that the application of the regional concept in the planning of land resource development in many parts of the world is now a matter of routine, a lasting testimony of the value to be derived from elementary mapwork classes in geography laboratories.

REFERENCES

BOURNE, R. (1931) 'Regional survey and its relations to stocktaking of the agricultural and forest resources of the British Empire', *Oxf. For. Mem.* 13

LINTON, D. L. (1951) 'The delimitation of morphological regions', *London Essays in Geography* (ed. L. D. STAMP and S. W. WOOLDRIDGE) Chapter 11, 199–217

LINTON, D. L. (1968) 'The assessment of scenery as a natural resource', *Scott. geogr. Mag.* 84 (3), 219–38

LINTON, D. L. and C. P. SNODGRASS, 'Peeblesshire and Selkirkshire', *The land of Britain*. The report of the Land Utilisation Survey of Britain (ed. L. D. STAMP), parts 24–25, 381–462

WATERS, R. S. (1958) 'Morphological mapping', *Geography*, 43, 10–17

A parametric approach to landform regions

J. G. SPEIGHT

Senior Research Scientist, Division of Land Use Research, CSIRO, Canberra, Australia

Revised MS received 1 June 1973

ABSTRACT. Systematic, parametric description of landform is necessary for the application of rigorous techniques of classification and regionalization to terrain mapping in land evaluation surveys. A two-level descriptive procedure is proposed. *Landform elements* (sites, terrain components, facies) are defined as areas of land that resemble simple geometric surfaces without inflections, and that are typically described by altitude, slope, aspect, curvature, and a number of derived contextual parameters. The operational descriptive individual is commonly about 20 m in radius. *Landform patterns* (regions, recurrent landscape patterns, simple land systems, landform systems, relief units, landscapes) are complex areas of land typically comprised of a number of landform elements arranged in toposequences. These toposequences are repeated cyclically to form three-dimensional geometric patterns that may be described in terms specifying their relationship to planes of accordance, including relief and grain, and their degree of development of networks and lineations. The operational descriptive individual is commonly about 300 m in radius. Given that a region is an area homogeneous not only in attributes but also in location (contiguous, compact, and of similar size to its neighbours), then a landform element is seldom a good region, being characteristically digitate in outline and of multiple occurrence. Much greater homogeneity of location may be achieved in the delimitation of landform patterns, which may then be considered as regions in themselves, or may be agglomerated into larger regions that are somewhat less homogeneous in location or in attributes, or in both, as may be required. An outline is given of current work on the use of parametric landform description in a land resources data bank project.

THE two decades since the appearance of Linton's paper on the delimitation of morphological regions (D. L. Linton, 1951) have seen a widespread application of geomorphology in the setting-up of regions, or homogeneous mapping units, in resources surveys for use in planning agricultural development, engineering works, and military operations (G. A. Stewart, 1968). The main source of data for such work has been rapidly-increasing libraries of vertical air photos that now contain a continuous photo-coverage of a large part of the land surface of the world. On such photos, viewed as stereoscopic pairs, a geomorphologist may readily perceive contrasts in landform, and may delineate plausible boundaries between morphological regions with a stroke of his marking-pencil. The systematic description of the delineated regions is quite another matter, however, and remains the subject of debate (M. F. Thomas, 1969).

There is a great deal of confusion of terminology in this field, contributed to by the fact that landforms, vegetation, and soils are commonly considered together as characteristics of a composite kind of region, and there may be as many as six published terms for classificatory units that are essentially the same as far as landform is concerned.

Typologically, landform regions have not been satisfactorily defined. There has been an undue emphasis on dimension as a factor differentiating between types of region. Hierarchies of regional types have been set up without explicit differences between criteria appropriate to the description of different levels in a hierarchy. There has also been a tendency to let either geology or deduced landform genesis, whose intrinsic value is not denied, stand in place of objective landform description (R. L. Wright, 1972). Areas are in some cases delimited on the basis of a prior knowledge of geology which, so far as areal extent is concerned, has usually been mapped by an extrapolation based on intuitive appreciation of landform as a geological indicator. Thus landform has been the object of study of both the geologist and the geomorphologist but it has not been explicitly described by either, except in the broadest terms.

If landform mapping is thought of as based on four principles, the morphologic, the genetic, the chronologic, and the dynamic (J. A. Mabbutt and G. A. Stewart, 1963), this paper is concerned only with the morphologic, in the belief that a defensible application of any of the other three principles depends on an accurate appreciation of morphology. In Mabbutt's terms (J. A. Mabbutt, 1968) it advocates a stiffening of the 'landscape approach' to land classification through the infusion of ingredients of the 'parametric approach'. In recent surveys of the CSIRO Division of Land Research on areas in Papua New Guinea, beginning with that of Bougainville (R. M. Scott *et al.* 1967), some parameters of landform have been explicitly included in reports. A dominantly parametric approach is adopted by the CSIRO Division of Applied Geomechanics for engineering terrain evaluation (K. Grant, 1968), and parametric approaches to landform classification have been put forward by R. B. King (1970) and S. G. Möller (1972). In depositional landscapes, particularly glaciofluvial ones, the relationships between form and process may generally be so well known as to obviate the need for explicit morphological description but the methodology described here has been developed in the context of resources surveys in areas in Papua New Guinea and, to a lesser extent, in parts of Australia where fluvial denundational landscapes predominate. The descriptive technique probably reflects a preoccupation with such landscapes, and may need substantial modification to suit landscapes of different origin.

Traditionally, the landscape approach to landform mapping has been used in the reconnaissance assessment of land for non-specific purposes at low cost per unit area, achieved by extrapolation from very limited sampling. The innovations suggested here continue this tradition but require for their full development the use of a computer and a large outlay in terms of data processing systems, to be offset by subsequent economies in application to surveys of very extensive areas.

The paper seeks to show how by describing landforms from two complementary points of view, as elements and as patterns comprised of elements, a more systematic and coherent basis may be established for the recognition and delimitation of morphological regions, permitting the application of rigorous techniques of classification and regionalization that cannot be applied to established systems of morphological mapping (Wright, 1972). The term 'region' is applied loosely in this introduction to the type of morphological mapping unit employed in resources surveys though such units are usually dominated by homogeneity of landform rather than locational homogeneity: in contrast to a well-formed region a typical mapping unit may have a number of separate occurrences, and each may be quite irregular in shape. Nevertheless, degree of locational homogeneity is an implicit characteristic that does not appear to have been given due attention. Systematic description of landform is essential to any progress in the establishment of an appropriate balance between locational homogeneity and descriptive homogeneity of mapping units.

TYPOLOGY OF TERRAIN

Landforms, particularly those of fluvial denundational origin, may be viewed in relation to two models. In the first model an element of the land surface is compared to a simply-curved geometric surface without inflections and is considered in relation to other elements that are up-slope, down-slope, or to either side of it. The concept of *slope* is essential to such elements. In the second the land surface is considered as a three-dimensional cyclic or repetitive phenomenon in which simpler elements recur at quasi-regular intervals in a definable pattern. *Relief* is the concept that dominates this model. Landforms viewed in relation to these two models are called here *Landform elements* and *Landform patterns* respectively. Any piece of land may be described in terms of either model but for a given terrain, to be mapped at a given scale for a given purpose one of the two models is likely to be clearly more appropriate than the other.

Landform elements

Linton's concept of a site was an elaboration of the *facet* of S. W. Wooldridge (1932) and involved altitude, extent, slope, curvature, ruggedness, and relation to the water-table. He was conscious of the correspondence between a *site* defined in this way and the *site* as seen from the ecological point of view by R. Bourne (1931). Such elementary units of land, for practical purposes regarded as indivisible, have been called *land elements* by H. A. Haantjens (1965) and by A. B. Brink *et al.* (1966), *sites* by C. S. Christian and G. A. Stewart (1968), *terrain components* by Grant (1968) and *facies* (fatsii) by Russian workers, for example V. I. Prokayev (1962).

Slope angle and either slope profile curvature or contour curvature, or both, have been employed as landform element descriptors by R. A. G. Savigear (1956), A. Young (1964, 1972), F. R. Troeh (1964), Grant (1968), R. V. Ruhe and P. H. Walker '(1968), J. B. Dalrymple *et al.* (1968), and P. H. Walker *et al.* (1968) among others. V. L. Greysukh (1967), using arrays of altitudes (as did Troeh, 1964), constructed curves of slope versus azimuth about each given point and showed how, by counting the number of positive and negative maxima on the curve it is possible to distinguish between regular slopes, spurs, valleys, peaks, hollows, and saddles. This ingenious scheme does not, however, appear to be as relevant as the others to the distribution of processes and mantle materials on the land surface.

Walker *et al.* (1968) used two further descriptors: slope length measured from a point up-slope to a crest, and the associated height difference. These are the first parameters mentioned that go beyond local geometry towards a specification of the context in which the element is located. Context is crudely indicated in a number of soil survey texts and in the microclimatology text edited by I. A. Gol'tsberg (1967) as 'upper', 'middle', and 'lower'. (Gol'tsberg's text also employs slope and slope aspect as explicit element attributes.)

In a previous paper (J. G. Speight, 1968) I analysed detailed contour maps to characterize local geometry by slope, slope profile curvature, and contour curvature, and context by catchment area per unit of contour length (at the point of sampling), considering that this area was better defined and of greater physical significance than slope length. In subsequent unpublished work based on arrays of altitudes on a square grid pattern I have expanded the list of landform element descriptive parameters to more than twenty. These parameters are all purely geometric but have been selected for their potential significance to geomorphic processes. A summary of the current parameter list is given in the second column of Table I; the first column (which does not relate to the other columns line by line) details the primary data drawn from the array of altitudes from which the practical parameters in the second column are computed. Of the first six parameters, four have been already mentioned and altitude and slope azimuth are self explanatory. The catchment area above a short increment of contour at a point is now described not only by its area, but also by its mean length, mean height above the point, and mean slope. The term *dispersal area* is coined for the area of land *down-slope* from a short increment of contour at a point. Whilst catchment area is significant for run-on of surface water, dispersal area is significant for soil drainage. As with catchment area, its size, mean length, mean height, and mean slope are estimated. Unpublished work has shown that, for an experimental landscape, dispersal area slope and dispersal area height are the two parameters that have the highest predictive value for soil attributes of those mentioned so far.

The remaining parameters of Table I result from applying the same procedures as above to *slopes* as if they were altitudes. These parameters express concepts of local slope dominance and slope subservience. Preliminary results indicate that their use enhances the predictability of soil from landform elements.

The effect of these parameters is to provide a rather full description of a landform element

TABLE I

Landform element parameters derived from arrays of (a) altitude values and (b) slope values

Primary parameters	Practical parameters	Possible significance
(a) *Derivatives of altitude*	Altitude	Rock resistance
	Slope	Overland flow velocity; susceptibility to failure
Altitude	Slope profile curvature	Overland flow acceleration; removal of transported waste
	Slope azimuth	Irradiation (with slope); exposure
Maximum up-slope value	Contour curvature	Overland flow concentration; soil drainage impedence
Maximum down-slope value	Catchment area	Overland flow volume; supply of transported waste
Azimuth of maximum up-slope	Catchment length	Overland flow attenuation
Azimuth of maximum down-slope	Catchment height	Energy of overland flow
Azimuth of contour to the left	Catchment slope	Velocity of overland flow
Azimuth of contour to the right	Dispersal area	Soil drainage capacity
	Dispersal length	Impedence of soil drainage
	Dispersal height	Soil drainage potential
	Dispersal slope	Rate of soil drainage
(b) *Derivatives of slope*	*	
	Maximum rate of change of slope	Homogeneity/heterogeneity index
	*	
Slope	Azimuth of maximum rate of change	Relation to slope azimuth defines concavity, convexity or twist
	*	
Maximum rate of increase of slope	Slope-catchment area	
Maximum rate of decrease of slope	Slope-catchment length	Gravitational stability measures
Azimuth of maximum rate of increase	Slope-catchment contrast	
Azimuth of maximum rate of decrease	Slope-catchment gradient	
*	Slope-dispersal area	
	Slope-dispersal length	Gravitational instability measures
*	Slope-dispersal contrast	
	Slope-dispersal gradient	

* Slope-derived parameters corresponding to altitude-derived parameters but omitted because of redundancy, low reliability, or apparent uselessness.

individual (cf. Wright, 1972) of approximately two grid units radius (about 20 m radius in practice) capable of classification by the techniques of numerical taxonomy. That is, each such parcel of land may be considered as an operational taxonomic unit (O.T.U.). In trials on actual landscapes mapped with 2 m or 1·52 m (5 ft) contours it has been found that numerical classifications on the unweighted parameters of Table I assign the commonly recognized landform elements such as spur crests, hill-slopes, and foot-slopes to distinct groups and, in general, subdivide the landscape in a way that is intuitively satisfying. Figure 1 shows the elements discriminated by a particular trial classification with arbitrary names assigned to the classificatory groups. Some degree of control of the size of the delimited elements may be exercised by using broader or narrower units of classification though, since only the ultimate degree of broadening will allow crests and valleys to be combined in the one element, the grain or texture of the landscape determines a limit to such areal generalization.

The landform element characteristics discussed here are the result of an intuitive selection of those aspects of the geometry of terrain that appear relevant either to land use or to geomorphological process. It is not necessary to restrict the basis of the classification to the simple, unweighted aggregate of such characteristics, as described above, although experience in bio-

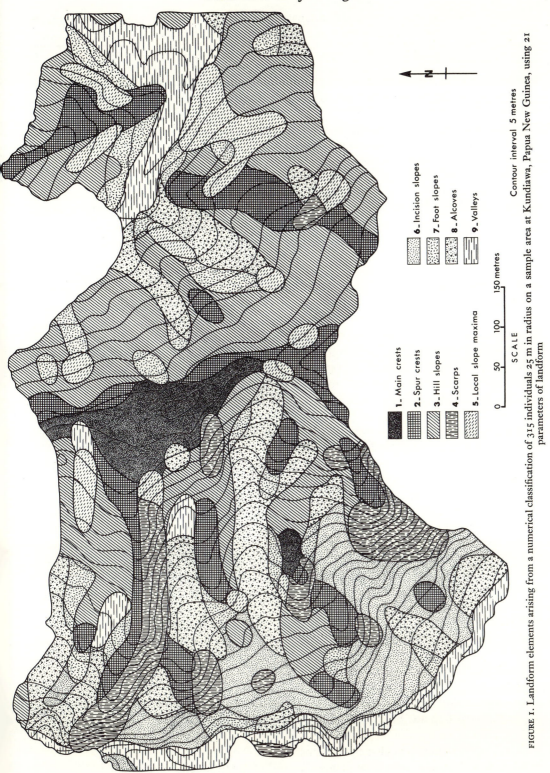

FIGURE 1. Landform elements arising from a numerical classification of 315 individuals 25 m in radius on a sample area at Kundiawa, Papua New Guinea, using 21 parameters of landform

Contour interval 5 metres

1 – Main crests
2 – Spur crests
3 – Hill slopes
4 – Scarps
5 – Local slope maxima
6 – Incision slopes
7 – Foot slopes
8 – Alcoves
9 – Valleys

SCALE

0   50   100   150 metres

logical taxonomy shows that such classifications are usually preferable to those based on a smaller number of characteristics whose importance has been traditionally accepted. A more purposive approach is to attempt to derive secondary attributes that have a more direct significance for the ultimate purposes of the classification and, if necessary, to apply numerical taxonomy to such secondary attributes. For instance, if the intention is to predict soil erosion from landform, then the erosive stress due to overland flow is a significant secondary attribute of the site that could be estimated as a function of local slope, catchment size, catchment length, catchment height, and catchment slope. On the other hand, the capability of the soil to withstand such an erosive stress probably depends largely on antecedent soil-moisture conditions which, in turn, could be estimated from the size, length, depth, and slope of the dispersal area below the site and from the slope and slope azimuth as determinants of received irradiation. Such process modelling based on a comprehensive suite of landform element attributes has not yet been attempted. Landform elements defined on the basis of such process-oriented parameters could be expected to correlate much better with soil-type than those based on the purely geometric parameters, but would have essentially similar distributions in the landscape, to the extent that landform elements similar in form are also similar in process.

Landform patterns

The mosaic of landform elements that constitutes any terrain is capable of description in terms of pattern characteristics, where 'pattern' is used in the sense of an arrangement of components. There is a consensus among all concerned with terrain evaluation that regions exist which may be recognized by the characteristic composition and relationships of their component elementary landforms. They are, as Linton expressed it, the 'morphological atoms' comprised of the 'morphological electrons and protons' represented by *sites*. Such *regions* (cf. Bourne, 1931) are variously termed *stows* (J. F. Unstead, 1933; Linton, 1951), *recurrent landscape patterns* (P. H. T. Beckett and R. Webster, 1965), *land systems* or *simple land systems* (Brink *et al.* 1966; Christian and Stewart, 1953 and 1968), *terrain patterns* (Grant, 1968) *landform systems* (Thomas, 1969), *relief units* (A. Young, 1969), and *landscapes* as employed in a restricted sense by some Russian geographers (S. V. Kalesnik, 1961; N. A. Solntsev, 1962; K. I. Gerenchuk *et al.* 1970). Most of these terms embody vegetation and soil characteristics as well as landform, but landform characteristics are essential to their definition.

A very large number of attributes may be used to delimit and to describe landform patterns. To date their description has been haphazard, possibly as a consequence of the notion that such mapping units, being basically determined by convenience in displaying the results of air-photo interpretation, are too vaguely defined to be governed by a consistent set of descriptive principles. Much reliance has been placed on the presentation of a block diagram, and on the assignment of a geological or morphogenetic 'tag', both of which are excellent aids to communication but should not be made to stand for a systematic description. Probably the most detailed parametric description of landform patterns so far published is that of R. M. Scott and M. P. Austin (1971) who described, in terms of fifteen geomorphic attributes, thirty-five patterns that had been previously mapped by non-parametric means. They were able to show, by a dendrogram and an ordination, how a numerical classification corresponded in general with the previous intuitive grouping of the patterns but reduced unintentional subjective bias and indicated a further grouping suitable for a more generalized map.

Landform pattern attributes fall into the following categories:

> Landform element composition
> A suite of toposequences formed of landform elements

Degree of development of lineations and networks
Relationships of terrain to planes of accordance
Relief and grain

As in the case of landform elements, contextual parameters may be added: geographical position, altitude, and up-slope–down-slope relationship to adjacent landform patterns.

In reading the following sections it may be helpful to refer to the examples of landform pattern descriptions in Tables II and III.

Landform elements and toposequences

Given some form of classification of landform elements, one may determine, by air-photo interpretation, or some more precise means, the elements present, their typical dimensions, and the proportion of the terrain taken up by each, and may insert mapping boundaries between regions containing elements of different kinds, or of the same kinds but in different proportions. Gerenchuk *et al.* (1970) suggest several modes of presentation of the distribution and relationships of components within a landscape (landform pattern) but their proposals are marred by the absence of criteria for systematically recognizing components of various ranks.

Landform elements are not random in their distribution but are commonly strongly ordered in *toposequences*, where a toposequence is the topographic sequence of elements passed over by a conceptual particle moving down-slope under the influence of gravity. The concept is related to the *catena* of G. Milne (1935), but does not embrace soil characteristics in its definition. As the landform element concept increases in predictive value for soils, toposequences will relate more closely to catenas.

A few landform elements do not conform well to the toposequence concept and are best described as sub-elements merely included within other elements.

As significant toposequences are identified in a given landscape it becomes evident that each has a typical planimetric length and total height which add further valuable parameters to the terrain pattern description.

Lineations and networks

There is a wide variability in the development and pattern of linear features in natural landscapes. J. R. Van Lopik and C. R. Kolb (1959) distinguish patterns with and without linear features and with and without evidence of parallelism. In the present scheme linear features are characterized by their degree of reticulation, or network development, and parallelism is characterized as strength of orientation of lineations. Topological properties of the networks (on which there is a considerable literature) are not considered relevant.

In respect of degree of reticulation, stream channels tend to be either completely reticulated, so that it is possible to traverse along the network from any channel segment to any other, or completely absent (as in swamps and doline karst); examples of poorly developed discontinuous channel networks are quite rare. Crests, on the contrary, exhibit every degree of reticulation, so that the degree of development of a crest network is an informative parameter. Given that a complete channel network is commonly present and well displayed on air photos, an estimate of channel density is called for, the simplest measure being a count of channels crossed per unit distance along an arbitrary line.

Lineations, or oriented terrain features, have traditionally been treated from three different points of view; as structural landforms, as drainage patterns, and as statistical orientation distributions. Among oriented structural landforms are included chevron ridges, strike ridges, hogbacks, homoclinal ridges, and cuestas. These features may be parametrically expressed

TABLE II

Sample description of a denudational landform pattern

Geomorphic category: low hill ridges

Altitude: from 1900 to 2300 m above sea level

Landform elements

C(1) Crest
  Length indefinite
  Slope: average 10°, maximum 15°

Occupies 4% of the landform pattern
  Width 20 m, moderately variable
  Internal relief 10 m

S(1) Steep slope
  Length 100 m, highly variable
  Slope 25°

Occupies 65% of the landform pattern
  Width indefinite
  Internal relief 10 m

S(2) Earthflow
  Length 300 m, moderately variable
  Slope 7°

Occupies 30% of the landform pattern
  Width indefinite
  Internal relief 6 m

W(1) Small stream

W(4) Through-going stream
  Width 20 m, moderately variable, not tapered
  Angular and moderately sinuous

Occupies 1% of the landform pattern

1·3 active low-water channels

Toposequences

(1) Crest—steep slope—small stream; 100 m long and 80 m high
(2) Crest—steep slope—earthflow—small stream; 400 m long and 90 m high
(3) Crest—steep slope—through-going stream; 100 m long and 90 m high

Networks

Crest network moderately well developed. Channel frequency 3·5 per km

Lineations

One weak orientation is present: direction 135°, dipping westward at 20°
Represented by several moderately-spaced vague lineaments, mainly crests, formed by resistant
  strata

Accordance relations

Grain 500 m   Relief 150 m   Accordance of summits poor
The upper surface of accordance dips at 4° towards a direction of 020°, and is parallel to the lower
  surface of accordance

---

partly in terms of lineations as outlined below and partly by assigning a characteristic orientation to slope elements. For example, in a hypothetical landform pattern a dip-slope could be described as a long gentle slope with a strong tendency to face in a particular direction, and a scarp-slope could be described as a short precipitous slope with a tendency to face in the opposite direction.

Drainage pattern types are illustrated in numerous texts, in which they are categorized as dendritic, rectangular, trellis, centrifugal, braided, meandering, disorganized, etc., but no principles of classification are given so that, though the differences are apparent, the similarities and relationships between patterns remain enigmatic. In fact, at least two distinct phenomena are involved: orientation properties of tree-like drainage networks, and *river channel patterns* in the sense of L. B. Leopold and M. G. Wolman (1957). These are best considered separately, the former as lineations (see below) and the latter, wherever stream channels are large enough to observe in detail, in the following manner. Stream channels may be characterized by: mean width, local width variability, degree of taper (especially in tidal reaches), curvedness (versus angularity) sinuosity, mean number of low-water channels in a series of cross-sections of the channel (to one decimal place), and whether the stream is active, semi-active, or relict. This scheme overcomes the difficulty that arises when a stream is found to be both braided (multiple-channel) and meandering (curved and sinuous), and thus unclassified in the scheme of Leopold and Wolman (1957).

TABLE III

Sample description of an alluvial landform pattern
Geomorphic category: Floodplain of vertical accretion
Altitude: from 0 to 10 m above sea level

Landform elements

P(1) Floodplain — Occupies 90% of the landform pattern
    Length 800 m, highly variable — Width 1200 m highly variable
    Slope less than $\frac{1}{2}°$ — Internal relief less than $\frac{1}{2}$ m
    Sub-elements:
        D(1) Ox-bow — Occupies 20% of P(1)
            Length 400 m, highly variable — Width 65 m, moderately variable
            Strongly curved

W(4) Through-going stream — Occupies 10% of the landform pattern
    Width 80 m, slightly variable, not tapered
    Curved and strongly sinuous — 1·0 active low-water channels
    Sub-elements:
        R(1) Point Bar — Occupies 10% of W (4)
            Length 300 m, highly variable — Width 40 m, highly variable
            Strongly curved

Toposequences
(1) Through-going stream—plain—external plain 800 m long and 0 m high

Networks
Crest network completely * developed — Channel frequency 1·0 per km

Lineations
None

Accordance relations
Grain indeterminate   Relief zero   Accordance of summits excellent*
The upper surface of accordance dips at less that $\frac{1}{2}°$ towards a direction of 160°, and is parallel to the
   lower surface of accordance

* Assigned conventionally.

Air-photo lineations that appear to represent faults, joints, or other geological structures have frequently been counted and measured for statistical analysis. Unfortunately, the statistics of orientation distributions are worthless in most cases: the strength of a single preferred orientation may be estimated in relation to a background of randomness but, where more than one preferred orientation is apparent or suspected, nothing may be learnt from available statistical methods. One can only present a rose diagram and let the reader draw his own conclusions. Accordingly, the orientation properties of the lineations of terrain patterns must be treated in a rather crude manner. If, on examination of a terrain pattern on an air photo, no preferred orientation is apparent, the linear patterns are random, and this is a significant attribute. If one or more preferred orientations are distinguishable, then they may each be described, in order of dominance, in the following terms: orientation direction (strike), dip angle, dip direction, strength of orientation, number of lineaments, spacing of lineaments, sharpness of lineaments, whether lineaments are dominantly crests, channels, or both, and whether the preferred orientation appears to be the result of (i) drainage consequent on a pre-existing slope, (ii) resistant strata, or (iii) faulting or jointing. Many preferred orientations, such as those due to consequent streams, carry no implication of structural planes dipping into the earth at a specific angle, but such lineations are equivalent to planes with vertical dip angles. The great advantage of characterizing preferred orientation by strike and dip, given that these parameters are readily assessed in

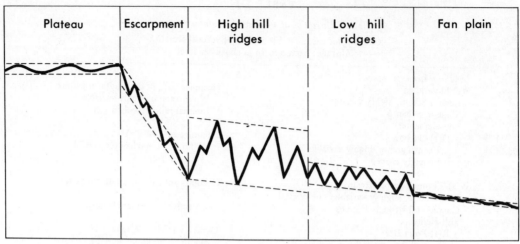

FIGURE 2. A hypothetical section showing landform patterns delimited by discordances between their respective planes of accordance

stereoscopic viewing of air photos, is that structural planes are simply described in this way, whereas the planimetric trace of the outcrop of a structural plane in crossing hilly terrain is complex and would usually have to be approximated by at least two weaker preferred orientations.

Accordance, relief, and grain

It is possible to fit to any landscape two planar or gently curved surfaces of accordance, the upper one touching the major crests or summit surfaces and the lower one, the streamline surface of G. H. Dury (1951), containing the major channels or valley floors. The vertical separation of these planes may be taken as a measure of *relief* defined in general terms as the typical height difference between a major ridge crest and an adjacent major valley floor. As the size of the region under consideration is reduced the planes of accordance must be altered so as to fit the crests and channels within the revised region boundaries. In the process adjacent areas are commonly differentiated by a discordance between their respective upper planes of accordance in the form of a step or an abrupt slope change or both. In the example of Figure 2 horizontal planes of accordance with a small separation (i.e. small relief) characterize a plateau, steeply sloping planes with moderate relief mark the escarpment, gently sloping planes of high relief the higher hills, gently sloping planes of moderate relief the foothills, and gently sloping planes of negligible relief an alluvial plain or fan.

The setting up of planes of accordance has most physical justification in areas of strong summit accordance, but whatever the degree of accordance (in itself a useful descriptive parameter) the visualization of planes of accordance assists greatly in the assessment of relief and also of grain (cf. W. F. Wood and J. B. Snell, 1960) which is defined as the typical horizontal distance between a major ridge crest and a major valley floor. Wood and Snell estimated grain and relief by selecting random points, plotting the maximum height differences observed in a set of circular sampling areas centred on each point against sample area radius, and reading the coordinates of the point on each graph at which the gradient of the curve flattened out. This seems to be equivalent to the less rigorous planes-of-accordance method, at least in the case of relief. If a typical section through the terrain is considered as a periodic curve, then relief is the amplitude of the dominant oscillation and grain is its semi-wavelength. Relief is a homologue of topo-sequence height, and grain of toposequence length, but relief and grain are generally greater than the height and length of the dominant toposequence of the landform pattern, in inverse relation to the degree of accordance of the terrain.

The orientation and slope of the planes of accordance, particularly the lower plane, have value in defining relationships between landform patterns. For instance, two patterns whose lower planes of accordance slope away from each other may together constitute a large asymmetric mountain ridge, whose existence, relief and extent, though not explicit in the description of either landform pattern, may be deduced from the attitudes of the lower planes of accordance of the two patterns taken together.

Operational landform pattern individuals

The individuals of landform pattern may be operationally defined as areas, commonly of the order of 300 m in radius, that are recognized and differentiated on the basis of their altitude, land-form element and toposequence composition, lineation and network development, accordance, relief, and grain. In contrast to landform element individuals, which may be differentiated on quantitative geometric attributes that have proved amenable to machine computation, most of the differentiae of landform patterns are hard to estimate precisely, and involve conceptual difficulties. Grain, relief, and accordance characterize the periodic, or wave-form aspects of terrain, representing wavelength, amplitude, and constancy of amplitude respectively. To deter-mine these parameters by Fourier analysis one would not expect reliable results from a wave train of less than twenty wavelengths at the very least. However, periodic phenomena in terrain rarely exceed three of four wavelengths in the one subjectively-recognized landform pattern (except for some patterns of linear dunes or sand ridges) and the suggested landform pattern individual is likely to contain no more than one wavelength. It appears that one may be placing excessive reliance on what D. Grigg (1965) calls 'a metaphysical assumption ... that there is order in the world'. Similarly, the variance of landform element composition between 'individuals' within apparently homogeneous terrain is usually large and increases as the size of the operational in-dividual is reduced. On the other hand, the size of the individual cannot be increased much beyond 300 m in radius without losing the ability to resolve many narrow landform patterns such as a single strike-ridge, an escarpment or an intermontane flood plain. Lineation and network development are more amenable to quantitative discrimination (Speight, 1968) but toposequence composition is even more difficult to specify than landform element composition because the relationship of toposequences to area is not straightforward, since several different toposequences can begin on the one crest element and end in the one channel.

MAPPING CONSIDERATIONS

Morphological regions

The above two-level development of landform typology provides a basis not only for systematic comparison of landforms from place to place but also for the setting up of morphological regions. The region concept is inescapable in communicating geographical information and must be considered further since typology alone does not define a region. In the terminology employed by Grigg (1967, 470) morphological regions may be viewed as dominantly *uniform* regions, that is, areas within which there is uniformity in selected attributes, although in some respects they may be viewed as *nodal* regions, characterized by interconnections between things within them (toposequences for example). Regionalization has been recognized to be a process similar to classification (W. Bunge, 1962, 1966; Grigg, 1965, 1967; N. A. Spence and P. J. Taylor, 1970) and therefore, given some operational definition of a geographical *individual*, open to the use of the techniques of numerical taxonomy. The stumbling-block, as pointed out by Bunge (1966), is that regions are characterized not only by homogeneity of attributes as in conventional classifications, but also by *homogeneity of location*, an implicit consideration whose significance may be missed.

For regions of finite size, homogeneity of location may be viewed as *contiguity*, as *compactness*, or as *same-sizedness*. Locational homogeneity was incorporated in a rigorous system of regionalization more than thirty years ago by M. J. Hagood, N. Danilevski, and C. E. Beum (1941). M. J. Hagood (1943) considered that the rule: 'regions shall be geographically contiguous' was a first condition in a procedure of regionalization, and accordingly grouped individuals in a region only if they were contiguous. Such a *contiguity constraint* has also been employed in other studies (e.g. D. M. Ray and B. J. L. Berry, 1966; N. A. Spence, 1968; P. J. Taylor, 1969). However, contiguity by no means ensures compactness: a spider's web, for instance, is contiguous but not compact. Bunge (1966) even denied that contiguity had any significance in regionalization, citing the somewhat special cases of small islands and peninsulas for which a strict contiguity rule might be inconvenient. Nevertheless, contiguity cannot be neglected if a region is to be a part of the earth's surface and not several separate parts.

Aspects of locational homogeneity suggested by Bunge (1966) are compactness and same-sizedness. Compactness may be achieved in the first place by reducing the size of a region, but this course has severe limitations since a given territory must then be divided into more and more regions, each of which must be individually described, and the scale of mapping must become large. In practice the number of regions into which a territory is to be divided for a particular purpose is not open to much variation, so the mean area of the regions is fixed. The compactness of a given region is then determined only by its shape, and specifically by minimizing its radius of gyration about a vertical axis through its centroid. A locationally homogeneous region will thus tend to be circular rather than either triangular, elongated, serpentine, or digitate. No two adjoining regions can be circular, however, so that the best practical approximations to circularity are many-sided polygons that are convex, or nearly so, and almost or quite equilateral.

Same-sizedness is a criterion for locational homogeneity because any one region has a claim to be as homogeneous as any other. Not only equality of area is involved: the boundary between two regions should be equidistant from the centroids of the regions so that each point is assigned to the region that is nearest to it. According to Bunge (1966) several of the definitions of locational homogeneity for a mosaic of regions produce a network of hexagons as the optimal pattern.

In summary, a scheme of morphological regionalization should be based on first, a list of specified characteristics of landform attributed to a sufficiently large array of operationally defined individuals (which may be either elements or patterns), and secondly, a set of constraints referring to contiguity, compactness, and same-sizedness of regions. It may also be desirable to introduce relational concepts other than locational homogeneity, such as altitudinal relationships between regions (not simply altitudinal similarity) for instance.

When regions are defined intuitively, as exemplified in Linton's (1951) paper, the soundness and usefulness of the regions delimited depends on the consistency with which an intuitive balance is kept between typological and locational homogeneity. If numerical methods are used, a strategy must be worked out to give due effect to each concept. Probably a step-wise technique is required, involving some degree of purely typological grouping before the first application of locational constraints.

Landform elements as regions

It has been found that the classification of landform element individuals does in fact produce homogeneous regions that are significantly larger than the individuals themselves, satisfying the criterion of contiguity to a satisfactory degree. However, they are by no means compact, being characteristically elongated and digitate, and numerous regions with the same characteristics commonly occur in proximity to one another (Fig. 1). They may also vary dramatically in size, a vast area of featureless plain being as homogeneous as a spur crest a few metres in extent.

It is not surprising to find that landform elements or sites have not (except in the 'morphological maps' developed by R. S. Waters (1958) and Savigear (1956, 1965)) been considered as regions. The regions of Bourne (1931) and the mapping units of subsequent workers in land resource surveys have all been *assemblages* of landform elements that are described but not mapped. Landform elements are mappable typological entities rather than regions and when they are mapped, the map is not likely to resemble a regional map.

Landform patterns as regions

Despite the conceptual difficulties involved in identifying a landform pattern individual, landform patterns are much more amenable to consideration as regions than are landform elements. Partly this is a matter of scale: their dimensions are frequently more suitable for display in the context of reconnaissance surveys. More importantly, being assemblages, or mosaics of the typologically inflexible landform elements they have a greater typological flexibility, so that the constraints of locational homogeneity may be more readily met.

An extreme use of the typological flexibility of landform patterns may be employed in mapping some types of plain, where, by restriction of the size of a region to match those recognized in adjacent denudational terrain, the pattern may degenerate to the point where it contains only one element (a plain) and displays no lineations, no networks, perfect near-horizontal accordance, indefinitely large grain and zero relief. Such a landform pattern may be distinguished from other similar patterns principally by its position in a toposequence that enters or passes through it.

There is of course an interaction between the conceptual size of the landform pattern individual and the estimation of landform parameters. For instance, the sloping planes of accordance delineated at one scale of mapping, by their intersection at major crests and valleys, may define the grain and relief of landform patterns mapped at a smaller scale; the grain and relief defined at the larger scale becomes microrelief within elements at the smaller scale. It is therefore essential to specify the characteristic size of the individuals that one is using. It is likely that descriptions in the terms proposed become progressively less appropriate as the size of the individual departs from an optimum (300 m radius was suggested above). Certain terrain types, however, from extremely fine-grained badlands on the one hand to extremely coarse-grained alpine terrain on the other, may be more satisfactorily described in terms of individuals whose dimensions differ markedly from the norm.

Agglomeration of landform pattern regions

The first-order landform pattern regions discussed above may prove valuable, not so much as regions whose properties are specifically required for some purpose, but rather as individuals that are homogeneous and adequately documented, both in properties and in location, to be incorporated in a variety of broader classifications and regionalizations. By a degree of purely typological classification it should be possible to produce a map equivalent (from the landform point of view) to an orthodox land system map. In contrast to the orthodox map, however, it will be convenient to vary, in the classification procedure, the weight assigned to particular terrain parameters or to apply models of terrain of any degree of complexity, tailored to any purpose.

Further regionalization (as distinct from classification) of the first-order landform pattern regions is required whenever cartographic generalization is necessary, that is, when either a smaller scale map is specified, or the visual impact of a map at a given scale is to be enhanced by displaying fewer, larger mapping units. Here there will be an opportunity for experiment with the combined application of the typological and locational aspects of regionality.

PRACTICAL APPLICATIONS

The Gladefield study

In work previously reported (Speight, 1968) an area of 3·7 km² near Gladefield homestead in the Australian Capital Territory was used to test the feasibility of mapping landform elements and landform patterns (referred to as 'land systems') parametrically. Four parameters were used to synthesize the arbitrarily selected elements 'crest', 'hill-slope', 'concave foot-slope', 'convex foot-slope', 'swale', 'plain', and 'water course' by specifying the combinations of attribute values appropriate to each type of element. A map of these elements was prepared (Speight, 1968, Fig. 3; Young, 1972, Fig. 70), and the distribution of the elements was judged to provide, both visually and quantitatively, an insight into the nature of the contrasts between three distinctive terrains occurring within the area. An attempt was also made to delimit the three terrains by use of parameters of network development and lineation, with a degree of success.

The Gold Creek study

Unpublished work (with R. M. Scott and M. P. Austin) on an experimental area of 14·3 ha at Gold Creek in the Australian Capital Territory was concerned with relating landform elements to soils. It involved expansion of the landform parameter list, development of methods to generate landform parameters from altitudes measured at the intersections of a square grid, and the use of numerical taxonomy in classification. Estimates were made of the predictive value of landform element parameters for soils, and of the predictability of soil parameters from landform elements. Relationships to vegetation and to rock outcrop were also investigated.

The Papua New Guinea data bank

Parametric description of both landform elements and landform patterns is currently being applied to the development of a land resources data bank for an area of several thousand square kilometres in Papua New Guinea. The data bank system is intended to answer questions and display maps illustrating a wide range of information on landform, vegetation, soil-mantle characteristics, and other constraints on potential land use. Basic data differs little from that typical of reconnaissance resources surveys: an air photo cover at 1:40 000 scale, and field samples from several hundred sites gained during a short field season. Both the field sampling strategy and the photo-mapping technique, however, have been influenced by the concepts outlined in this paper.

The essential, but commonly neglected, basis for extrapolation of soil mantle characteristics by the use of air photos is an adequate knowledge of the relationships between landform, vegetation, and soil. In this case a small number of dissimilar types of terrain have been intensively sampled, and the landform elements present have been determined by analysis of very detailed contour maps either prepared for the purpose by tacheometric survey or prepared previously by photogrammetry for town planning or road building purposes. When the analysis of soil–landform element relationships is complete we shall know both what kinds of landform element are significant to soils and what attributes of soil are most susceptible to prediction in the terrains that have been sampled. Meanwhile, both for the analysis of air photos and for the characteriza-tion of outlying field sites that have not been intensively surveyed, it has been necessary to set up readily recognizable but relatively crude landform elements. The relationship of these elements to those developed from the detailed maps will have to be established in order to extrapolate the soils. Elements distinguished on air photos are as follows:

Summit surface: a low-angle plane constituting the highest part of the landscape
Crest: a localized or elongated feature with marked convexity of contour or profile, constituting a local high point in the landscape
Slope: a high-angle plane without pronounced curvature
Plain: a low-angle plane with negligible internal relief, usually situated at a low level in the landscape
Channel: a linear feature constituting a local low point in the landscape and forming a watercourse.
In addition the following classes of sub-element are distinguished:
Rise: a small discernible feature standing above the surface of an element that contains it
Depression: a small discernible feature lying below the surface of an element that contains it.

Individual landform elements falling into one of these classes are characterized by their dimensions, slope, and internal relief. More than one element of each class may be identified in a given landscape, provided that differences in any of these characteristics are observed. It will be noted that, by contrast with the system of slope profile analysis of Young (1964, 1971), no consideration is given to elements characterized by curvature, other than crests. Such elements are seldom identifiable on air photos and, where they are large enough to take into account, may be approximated by a series of straight elements.

As described so far these practical landform elements are deficient in contextual attributes, but these are supplied from pattern parameters, particularly toposequences.

The actual air-photo mapping delineates landform pattern regions. These regions are thus delimited by interpretation and not by formal classification of pattern individuals. A *pro forma* has been prepared so that the photo-interpreter can fill in blank spaces for all the attribute values and descriptors necessary to specify each landform pattern as illustrated in Tables II and III. The use of the *pro forma*, whose primary purpose is to provide landform data for input to the data bank, has two significant effects on the mapping process: it keeps explicit landform attributes firmly in the mind of the photo-interpreter when he is making decisions on the drawing of boundaries and, by requiring each region to be individually described, it enhances the concept of a landform region as a unique, contiguous land area rather than an occurrence of a type of land, which is the more traditional, but less flexible viewpoint.

Aspects of locational homogeneity, other than contiguity, are only subjectively judged: whilst maintaining a high level of homogeneity in landform pattern attributes, regions are not permitted to vary greatly in size, their dimensions commonly falling between 500 m and 5 km, nor are they permitted to become either very long or very digitate.

A typological classification of landform pattern regions is to be carried out to produce agglomerated mapping units simulating land systems for comparison with those previously mapped on a part of the area. If successful, this exercise should make explicit some underlying procedures of land system classification.

Similarly, the relationship of geology to landform is to be explored by attempting to model the effect of rock characteristics on landform pattern and testing the model by comparing a map of landform pattern regions classified in terms of the model with the existing geological map. A number of other typological studies are envisaged in the general area of prediction of surface conditions, microrelief, soils, and derived land resource data from landform patterns.

ACKNOWLEDGEMENTS

I wish to thank my present and former colleagues who have offered constructive criticism of drafts of this paper: M. P. Austin, B. G. Cook, P. M. Fleming, R. W. Galloway, J. A. Mabbutt, J. R. McAlpine, B. P. Ruxton, and G. A. Stewart.

REFERENCES

BECKETT, P. H. T. and R. WEBSTER (1965) 'A classification system for terrain', *Military Engineering Experimental Establishment M.E.X.E. Rep.* 872

BOURNE, R. (1931) 'Regional survey and its relation to stock-taking of the agricultural and forest resources of the British Empire', *Oxf. For. Mem.* 13

BRINK, A. B., J. A. MABBUTT, R. WEBSTER and P. H. T. BECKETT (1966) 'Report of the working group on land classification and data storage', *Military Engineering Experimental Establishment M.E.X.E. Rep.* 940

BUNGE, W. (1962) 'Theoretical geography', *Lund Stud. Geogr. Ser. C*, 1

BUNGE, W. (1966) 'Gerrymandering, geography, and grouping', *Geogr. Rev.* 56, 256–63

CHRISTIAN, C. S. and G. A. STEWART (1953) 'General report on survey of Katherine–Darwin region, 1946', *CSIRO Aust. Land Res. Ser.* 1

CHRISTIAN, C. S. and G. A. STEWART (1968) 'Methodology of integrated surveys' in 'Aerial surveys and integrated studies', *Proc. Toulouse Conf. 1964, Natural Resource Res. Ser. UNESCO* 6, 233–80

DALRYMPLE, J. B., R. J. BLONG and A. J. CONACHER (1968) 'A hypothetical nine unit landsurface model', *Z. Geormorph.* 12, 60–76

DURY, G. H. (1951) 'Quantitative measurement of available relief and of depth of dissection', *Geol. Mag.* 88, 339–43

GERENCHUK, K. I., I. K. GORASH and A. G. TOPCHIYEV (1970) 'A method for establishing some parameters of the morphologic structure of landscapes', *Soviet Geogr.* 11, 262–71

GOL'TSBERG, I. A. (ed.) (1967) *Microclimate of the USSR*, translated for Israel Program for Scientific Translations, Jerusalem (1969)

GRANT, K. (1968) 'A terrain evaluation system for engineering', *CSIRO Aust. Div. Soil Mechanics Tech. Pap.* 2

GREYSUKH, V. L. (1967) 'The possibility of studying landforms by means of digital computers', *Soviet Geogr.* 8, 137–49

GRIGG, D. (1965) 'The logic of regional systems', *Ann. Ass. Am. Geogr.* 55, 465–91

GRIGG, D. (1967) 'Regions, models and classes' in *Models in Geography* (eds R. J. CHORLEY and P. HAGGETT) 461–509

HAANTJENS, H. A. (1965) 'Practical aspects of land system surveys in New Guinea', *J. trop. Geogr.* 21, 12–20

HAGOOD, M. J. (1943) 'Statistical methods for the delineation of regions applied to data on agriculture and population', *Social Forces* 21, 287–97

HAGOOD, M. J., N. DANILEVSKI and C. E. BEUM (1941) 'An examination of the use of factor analysis in the problem of sub-regional delineation', *Rural Sociology* 6, 216–33

KALESNIK, S. V. (1961) 'The present state of landscape studies', *Soviet Geogr.* 2, 24–34

KING, R. B. (1970) 'A parametric approach to land system classification', *Geoderma* 4, 37–46

LEOPOLD, L. B. and M. G. WOLMAN (1957) 'River channel patterns: braided, meandering and straight', *U.S. geol. Surv. Prof. paper* 282-B

LINTON, D. L. (1951) 'The delimitation of morphological regions' in *London Essays in Geography* (eds L. D. STAMP and S. W. WOOLDRIDGE) 199–217

MABBUTT, J. A. (1968) 'Review of concepts of land classification', in *Land Evaluation* (ed. G. A. STEWART) Melbourne, 11–28

MABBUTT, J. A. and G. A. STEWART (1963) 'The application of geomorphology in resources surveys in Australia and New Guinea', *Rev. Géomorph. dyn.* 14, 97–109

MILNE, G. (1935) 'Some suggested units of classification and mapping, particularly for east African soils', *Soil Res.* 4, 183–98

MÖLLER, S. G. (1972) 'A system of describing and classifying information concerning land forms', *Nat. Swedish Inst. Building Res. Doc.* D8

PROKAYEV, V. I. (1962) 'The facies as the basic and smallest unit in landscape science', *Soviet Geogr.* 3, 21–9

RAY, D. M. and B. J. L. BERRY (1966) 'Multivariate socioeconomic regionalization: a pilot study in central Canada', in *Regional statistical studies* (eds S. OSTRY and R. RYMES) Toronto, 75–130

RUHE, R. V. and P. H. WALKER (1968) 'Hillslope models and soil formation. I. Open systems', *Trans. 9th int. Cong. Soil Sci.* 4, 551–60

SAVIGEAR, R. A. G. (1956) 'Technique and terminology in the investigation of slope forms', *Premier Rapport de la Commission pour l'etude des Versants*, Union Geographique Internationale, Amsterdam, 66–75

SAVIGEAR, R. A. G. (1965) 'A technique of morphological mapping', *Ann. Ass. Am. Geogr.* 55, 514–38

SCOTT, R. M. and M. P. AUSTIN (1971) 'Numerical classification of land systems using geomorphological attributes', *Aust. Geogr. Stud.* 9, 33–40

SCOTT, R. M., P. B. HEYLIGERS, J. R. McALPINE, J. C. SAUNDERS and J. G. SPEIGHT (1967) 'Lands of Bougainville and Buka Islands, Territory of Papua and New Guinea', *CSIRO Aust. Land Res. Ser.* 20

SOLNTSEV, N. A. (1962) 'Basic problems in Soviet landscape science', *Soviet Geogr.* 3, 3–14

SPEIGHT, J. G. (1968) 'Parametric description of land form' in *Land Evaluation* (ed. G. A. STEWART) Melbourne, 239–50

SPENCE, N. A. (1968) 'A multifactor regionalization of British counties on the basis of employment data for 1961', *Regional Studies* 2, 87–104

SPENCE, N. A. and P. J. TAYLOR (1970) 'Quantitative methods in regional taxonomy', *Progress in Geogr.* 2, 1–64

STEWART, G. A. (ed.) (1968) *Land Evaluation* (Melbourne)

TAYLOR, P. J. (1969) 'The location variable in taxonomy', *Geographical Analysis* 1, 181–95

THOMAS, M. F. (1969) 'Geomorphology and land classification in tropical Africa', in *Environment and land use in Africa* (eds M. F. THOMAS and G. W. WHITTINGTON), 103–45

TROEH, F. R. (1964) 'Landform parameters correlated to soil drainage', *Proc. Soil Sci. Soc. Am.* 28, 808–12

UNSTEAD, J. F. (1933) 'A system of regional geography', *Geography* 18, 175–87

VAN LOPIK, J. R. and C. R. KOLB (1959) 'A technique for preparing desert terrain analogs', *Tech. Rept. U.S. Army Corps Engrs, Waterways Exp. Stn, Vicksburg Miss.* 3–506

WALKER, P. H., G. F. HALL and R. PROTZ (1968) 'Relation between landform parameters and soil properties', *Proc. Soil Sci. Soc. Am.* 32, 101–4

WATERS, R. S. (1958) 'Morphological mapping', *Geography* 43, 10–17

WOOD, W. F. and J. B. SNELL (1960) 'A quantitative system for classifying land forms', *U.S. Army Quartermaster Res. and Engng. Command Tech. Rep.* EP-124

WOOLDRIDGE, S. W. (1932) 'The cycle of erosion and the representation of relief', *Scott. geogr. Mag.* 48, 30–6

WRIGHT, R. L. (1972) 'Principles in a geomorphological approach to land classification', *Z. Geomorph.* 16, 351–73

YOUNG, A. (1964) 'Slope profile analysis', *Z. Geomorph. Suppl.* 5, 17–27

YOUNG, A. (1969) 'Natural resource survey in Malawi: some considerations of the regional method in environmental description', in *Environment and Land use in Africa* (eds M. F. THOMAS and G. W. WHITTINGTON), 355–84

YOUNG, A. (1971) 'Slope profile analysis: the system of best units', *Inst. Br. Geogr. Spec. Pub.* 3, 1–13

YOUNG, A. (1972) *Slopes*

RÉSUMÉ. *Une approche paramétrique des régions morphologiques (forme du terrain).* Une description systématique et paramétrique de la forme du terrain est nécessaire pour pouvoir appliquer des techniques rigoureuses de classification et de régionalisation au dressage d'une carte dans le cas de levés d'évaluation. Nous proposons une procédure descriptive à deux niveaux. *Les éléments de forme du terrain* sont définis en tant que surfaces géometriques sans inflexion qui sont décrites d'une manière typique par l'altitude, la pente, l'exposition topographique, la courbure et un certain nombre de paramètres contextuels qui en dérivent. L'unité individuelle descriptive employée normalement est d'environ 20 m de rayon. *Les arrangements de terrain* sont des surfaces complexes de la terre qui consistent typiquement d'un nombre d'éléments de forme du terrain arrangés en toposéquences. Ces toposéquences se répètent cycliquement pour former des dessins géometriques à trois dimensions qui peuvent être décrits en termes qui précisent leur relation aux plans de conformité, y inclus le relief et le grain, et par le degré du développement des réseaux et des lignages. L'unité individuelle descriptive est normalement de 300 m de rayon. Etant donné qu'une région est une aire homogène non seulement en ce qui concerne les attributs mais aussi la situation (contiguë, compacte et de dimensions semblables à ses voisines), il en émerge qu'un élément de forme du terrain est rarement une bonne région, parce qu'il est normalement digité en contours et se répète souvent. Une plus grande homogénéité de localisation peut être obtenue en délimitant les arrangements du terrain qui peuvent dès lors être considérés eux-mêmes comme des régions, ou bien peuvent être agglomérés en régions plus grandes qui sont un peu moins homogènes en ce qui concerne la localisation ou les attributs, ou tous les deux si c'est désirable. On donne un aperçu du travail en cours, travail qui porte sur l'emploi de la description paramétrique de la forme du terrain. Cette méthode sera employée dans un projet de banque d'information pour les ressources de la terre.

FIG. 1. Les éléments de formes du terrain provenant d'une classification numérique de 315 individus de rayon de 25 m sur une superficie d'essai à Kundiawa, Papua New Guinea, employant 21 paramètres de la forme du terrain. Eléments: (1) Crêtes principales (2) crêtes d'éperon (3) versants (4) côtes (5) pentes maximales locales (6) pentes d'incision (7) pieds-de-versant (8) alcôves (9) vallées

FIG. 2. Une coupe hypothétique montrant des arrangements de terrain délimités par des discordances entre leurs plans de conformité respectifs. Arrangements de terrain: plateau, côte, dorsals hauts, dorsals bas, plaine de cônes alluviaux

ZUSAMMENFASSUNG. *Eine Methode der Definition von Oberflächenformenregionen mit Hilfe von Parametern.* Systematische Beschreibung von Landformen mit Hilfe von Parametern ist notwendig für die Anwendung von genauen Methoden der Klassifizierung und Regionalisierung bei der Kartierung von Terrain in Untersuchungen über Land Resourcen. Ein beschreibendes Verfahren in zwei Stufen wird vorgeschlagen. *Oberflächenformenelemente* (Site, Terrain Komponent, Fazies) werden als Gebiete, die einer einfachen geometrischen Fläche ähneln, definiert und mit Hilfe von Höhenlage, Hangneigung, Orientierung, Krümmung und einer Anzahl abgeleiteter vom Zusammenhang abhängiger Parameter beschrieben. Die operative Beschreibungseinheit (operational descriptive individual) misst meist etwa 20 m in Radius. *Oberflächenformenmuster* (Regionen, wiederkehrende Oberflächenformenmuster, einfache Land Systeme, Land Formen Systeme, Reliefeinheiten, Landschaften) sind komplexe Gebiete, die typischerweise eine Anzahl von Oberflächenformenelementen umfassen, die in einer topographischen Abfolge (Topofolge) angeordnet sind. Diese Topofolgen werden zyklisch wiederholt und bilden geometrische Muster, die hinsichtlich ihrer Beziehung zu Akkordanzebenen, einschliesslich Relief, Körnung, und des Grades der Entwicklung von Netzwerken und Lineationen, beschrieben werden können. Die operative Beschreibungseinheit hat einen Radius von meist ungefähr 300 m. Ist eine Region nicht allein in ihren Eigeschaften sondern auch in ihrer Lage ein homogenes Gebiet (angrenzend, kompakt und von ähnlicher Grösse wie seine Nachbarn), dann ist ein Oberflächenformenelement selten eine gute Region, denn es ist kennzeichnenderweise fingerförmig im Umriss und häufig im Vorkommen.

Wesentlich grössere Homogenität kann bei der Abgrenzung von Oberflächenformenmuster erreicht werden, die dann selbst als Regionen betrachtet oder zu grösseren Regionen zusamengefasst werden können, die etwas weniger homogen in ihrer Lage oder ihren Eigenschaften sind oder beides, je nach Erfordernis. Eine Übersicht über derzeitige Arbeit über den Gebrauch von parametrischen Beschreibungen der Oberflächenformen in einem Land Resourcen Datenbank Projekt wird gegeben.

ABB. 1. Oberflächenformenelemente eines ausgewählten Untersuchungsgebiets bei Kundiawa, Papua New Guinea. Die Elemente gehen aus einer numerischen Klassifikation von 315 induviduellen Gebieten, 25 m im Radius, unter der Benutzung von 21 Oberflächenformenparameter, hervor. Elemente: (1) Hauptkamm (2) Spornkamm (3) Hang (4) Steilabfall (5) lokales Hangneigungsmaximum (6) Einschneidungshang (7) Fusshang (8) Nischen (9) Täler.

ABB. 2. Ein hypothetisches Profil von Oberfächenformenmuster, welche durch Diskordanzen zwischen ihren zugehörigen Akkordanzebenen begrenzt sind. Oberflächenformenmuster: Plateau, Steilabfall, hoher Rücken, niedriger Rücken, Fächerebene

Detailed geomorphological mapping and land evaluation in Highland Scotland

ROGER S. CROFTS

Research Assistant in Geography, University College London

Revised M.S. received 1 August 1973

ABSTRACT. Three approaches to land classification are defined—genetic, parametric and landscape. Detailed geomorphological mapping has certain methodological characteristics in common with a landscape approach based on pattern recognition, but it differs from other landscape approaches because of the larger scale of mapping and data evaluation. Landforms are mapped at approximately 1 : 10 000 scale on the basis of their genesis, morphology, structure and chronology. This geomorphological inventory forms the basis for applications in pure and applied geomorphology as exemplified from the Scottish Highlands. Preliminary analysis of deglaciation landforms and mass-wasting phenomena in time and space can be accomplished by the extraction of data from the map. Fragile and unstable areas subject to rapid physical change can be identified and used as a basis for the evaluation of road and recreational developments. The original map requires simplification to facilitate both comparison and extrapolation to other areas and to be of optimum value to the lay user. Landform elements—*facets* and *sites*—are identified at the largest map scales. These are grouped into patterns and combined with other information pertaining to the physical environment to delimit *land potential zones*. These zones form a basis for land-use planning.

OVER a period of two decades D. L. Linton made several important contributions in the field of landform mapping and landscape classification. He first evaluated the utility of continental-scale mapping and classification pioneered in North America (N. M. Fenneman, 1928) to Britain and western Europe (Linton, 1951). He found that the greater complexity of relief and smaller size of area under consideration merited larger-scale mapping such as he carried out in central Scotland. Using the geographical work of J. F. Unstead (1933) and S. W. Wooldridge (1932) and the ecological studies of R. Bourne (1931) he elaborated a hierarchy of landform classification units at a scale appropriate to Britain: site, stow, tract, pays, section, province, continental subdivision: which have formed a basis for subsequent terrain analysis and land evaluation.

The recognition of the value of a detailed inventory of landform led to the Land Form Survey (British Geomorphological Research Group, 1959–62; Linton in R. S. Waters, 1958) and resulted in the development of morphological mapping and slope profile analysis (Waters, 1958; R. A. G. Savigear, 1965). Realizing the importance of mapping landforms he was influential in the instigation of a programme of geomorphological mapping to cover England, Scotland and Wales at 1 : 625 000 scale (Linton, 1967). Finally, with the growing interest in environmental management and the stimulation of the Countryside (Scotland) Act 1967, he developed a simple classification of terrain attributes based on absolute altitude and relative relief as one possible base for the evaluation of scenic beauty (Linton, 1968).

Implicit in Linton's work is the distinction between the mapping of individual features in the landscape, for example landforms, and the designation of areas with homogeneous attributes. The literature on land classification and evaluation (G. A. Stewart, 1968; UNESCO, 1968) does not discuss geomorphological mapping as a possible preliminary method in the evaluation of land and relatively few geomorphologists (Subcommission, 1963; M. F. Thomas, 1969; J. Tricart *et al.*, 1971) have argued the validity of the method for this purpose. It is the contention of the author that this should be a prime aim of geomorphological mapping. The extensive literature on land classification seeks to distinguish different methods of classifying land and

TABLE I

Major landform and landscape mapping techniques

	Regional physiographic analysis	Terrain analysis	Land research	Parametric approach
Objective	Landscape evolution; delimitation of physiographic regions	Classification and prediction of terrain at a distance	Classify and evaluate land for economic development	Division and classification of land on basis of selected attribute values
Methods	Field and air-photo collection of morphological, chronological and genetic data	Identification of terrain patterns and their components from field and air-photo survey	Recognition of large-scale terrain patterns, land systems, and small-scale components, land units, from air-photo analysis and field sampling	Quantitative definition of land attributes from maps, aerial sensors and field survey
Scale of analysis	Continental	Local and regional	Regional and subcontinental	Local and regional
Practical application	Teaching and research; secondary data for planning	Basis for small-scale land evaluation and feasibility for military use	Basis for investigations on specific and general regional economic development	Basis for specific and general regional economic development
Sponsors	Education institutions	National governments (defence)	National governments	National governments, educational institutions
Discipline	Geography	Earth sciences	Interdisciplinary	Geomorphology
Selected references	Fenneman, 1928 Gvozdetskiy, 1962 Hammond, 1954 Herbertson, 1905 Pecsi, 1970	Beckett and Webster, 1965, 1972 Brink et al., 1966 Dowling and Beaven, 1969	Bawden and Stobbs, 1963 Christian and Stewart, 1968 Grant, 1968 Ignat'yev, 1968	Scott and Austin, 1971 Speight, 1968

	Geomorphological mapping	Qualitative morphometry	Quantitative morphometry	Morphological mapping and slope profile analysis
Objective	Classification of landforms by morphogenesis	Subjective slope descriptors (in relation to man)	Quantitative description and simulation of relief	Measurement and definition of two-dimensional slope characteristics
Methods	Field and air-photo survey of land-form genesis, morphology, morphometry and age	Measurement of relief and slope from maps, air photos and field survey	Statistical analysis of relief and slope from maps and field survey	Field measurement of slope shape
Scale of analysis	Local and regional	All scales	All scales	Local and regional
Practical application	Basis for research and small-scale land evaluation	Teaching method; secondary for comparison with other physical factors	Teaching and research; secondary developments in land use and water management	Teaching and research; secondary comparison with other methods
Sponsors	Educational institutions (occasionally governments)	Educational institutions	Educational institutions	Educational institutions
Discipline	Geomorphology	Geography	Geology/ Geomorphology/ Hydrology	Geomorphology
Selected references	Demek, 1973 Klimaszewski, 1956 Tricart 1965 Tricart et al., 1971 Verstappen, 1970	Glock, 1932 Smith, 1935 Wooldridge, 1932	Horton, 1945 Strahler, 1956 Turner and Miles, 1968 Veatch, 1935	Bridges and Doornkamp, 1963 Pitty, 1969 Savigear, 1965 Waters, 1958

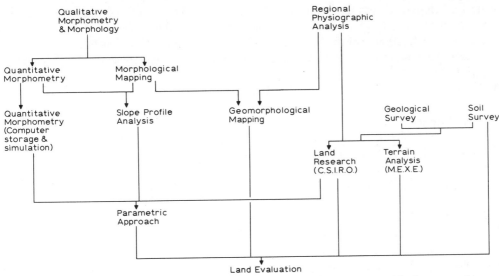

FIGURE I. Schematic representation of the chronological development of landform and landscape mapping

landscape for evaluating potential for development (C. S. Christian and G. A. Stewart, 1968; J. A. Mabbutt, 1968; Thomas, 1969). It is not appropriate here to cover this ground again but rather to compare these types with those whose prime concern is with the mapping of landforms.

GEOMORPHOLOGICAL MAPPING IN RELATION TO LAND CLASSIFICATION TECHNIQUES

Basically the techniques of geomorphological mapping, qualitative and quantitative morphometry, morphological mapping and slope profile analysis (Fig. 1 and Table I) are all examples of feature-based mapping and landscape analysis in which either the total landscape is described—geomorphological mapping—or specific aspects of landform and landscape are described—morphology and morphometry. Only in the case of geomorphological mapping does genesis of landform and landscape become an integral part of the method. In contrast the other group of techniques—regional physiographic analysis, land research, terrain analysis and to a certain extent parametric methods (Fig. 1 and Table I)—resort to aerial scanning of land to identify landform and landscape patterns. Essentially these latter techniques are integrated studies considering geological, geomorphological, pedological, biogeographical and hydrological factors. However, there is a distinction between two basic styles—terrain analysis and land research (Christian and Stewart, 1968). The former is rightly termed a physiographic approach (Thomas, 1969) based on the recognition of recurring patterns of surface materials, morphology and drainage—*facets*—identifiable at 1:50 000 scale on aerial photographs for military purposes (P. H. T. Beckett and R. Webster, 1965 and 1972). The land-research approach is essentially a geomorphological approach based on the recognition of large-scale landscape patterns—*land systems*—which can be subdivided into smaller components or *land units*.

Mabbutt (1968) has made a somewhat different division into genetic, landscape and parametric approaches to land classification. The former is equated with regional physiographic analysis, but as it is actually based on macro morphology it is a physiographic rather than a genetic approach. The landscape approach—land research and terrain analysis—is developed from the concepts of land types of Bowman (1916), Bourne (1931), and Linton (1951) (cf. G. R.

Heath, 1956). The parametric approach, elaborated in this volume, p. 213, by J. G. Speight, is a logical development from the latter together with the experience gained from morphological and morphometric analysis. It attempts to quantify land attributes to give a more objective evaluation of land. The value of the landscape and parametric approaches has been demonstrated in the appraisal of land resources in developing countries both for general economic development (A. B. Brink *et al.*, 1966) and for specific developments in the field of road engineering evaluation (J. W. F. Dowling and P. J. Beaven, 1969) and military ballistics (R. G. Pope, J. M. Hawkes and R. Davies, 1971).

The feature-based mapping techniques as such are omitted from Mabbutt's (1968) discussion, presumably in the belief that they do not contribute to land classification and evaluation. As the morphological and morphometric techniques are based largely on morphology without reference to surface materials, drainage, soils or other parameters, their utility as descriptive and predictive tools for land evaluation is limited. Numerous geomorphologists (M. Klimaszewski, 1956; Thomas, 1969; Tricart, 1965, 1971a and b; K. Walton, 1968) have realized the potential of geomorphological mapping as a partial inventory of natural resources and therefore a tool for land evaluation and land-resource management.

The method of geomorphological mapping has been defined by the Subcommission on Geomorphological Mapping of the International Geographical Union (1963 and 1968). The ultimate aim is 'to supply the national economy with geomorphological maps as a detailed picture of the relief being an important element of the geographical environment and to enable the rational use of this environment' (Klimaszewski 1963, 7). This is accomplished in three ways:

(1) By depicting the surface relief in terms of morphometry, morphology, genesis and age and incorporating the impact of structure and lithology.
(2) By assessing the rates of operation of geomorphological processes, particularly those which are pertinent to economic development.
(3) By producing a map which synthesizes the field data and portrays it in a form suitable for use by other earth scientists and particularly planners.

After the initial survey four possible lines of development can be followed. First, the data can be synthesized to depict the elements listed under item 1 above to produce a geomorphological map such as those found in a number of national atlases (Poland, Belgium, Hungary, Denmark). These are assessments of a component of the natural environment of whole countries and stand alongside national maps of geology, hydrology and soil. Secondly, data may be extracted for geomorphological research purposes. Thirdly, data important in the evaluation of resources for economic development can be extracted. Fourthly, the landforms depicted on the map can be classified as a prelude to land evaluation. This fourth line of development forms a link between the feature-based and areal designation types of approach and leads on to the assessment of aspects of the physical environment for land evaluation.

THE BASIS OF MAPPING

For the work described in this paper it was necessary to decide which of the two approaches should be adopted.

The MEXE terrain analysis method has been applied in part of Argyll and compared with soil association mapping (Beckett and Webster, 1969; J. S. Bibby, 1969). In the area '. . . the pattern of change was so rapid that there were very few units about which, if mapped, one could make precise enough statements about some attribute of interest to farmers and engineers . . .' (Beckett, personal communication, 1972). Webster further found that although *Land Systems* and

Land Elements could be defined, there was difficulty in defining a *Facet* and was forced to modify the definition to that of a *Patterned Facet* (Bibby, personal communication, 1972).

As a consequence it was decided to use a geomorphological mapping technique. There are a number of possible methods ranging from A. L. Perel'man's (1967) complex geochemical one to the genetic approach advocated by most geomorphologists (J. P. Bakker, 1963; T. Nakano, 1961; D. St-Onge, 1964; Thomas, 1969). The latter stresses causal relationships of processes in time and space and uses the important distinction between erosional and depositional types (J. A. Mabbutt and G. A. Stewart, 1963) which has long been a basis of landform classification (J. D. Falconer, 1915). It also embodies other descriptive tools—chronology, morphology, morphometry and to a certain extent structure and lithology. This latter method of geomorphological mapping was favoured because of the genetic basis and also because it has been employed successfully at very large scales in similar terrain in Europe.

SCALES OF MAPPING

Scales of mapping employed for geomorphological maps have ranged from about 1:5000 in complex coastal situations to 1:1 M on the proposed Geomorphological Map of Europe and even larger in national atlases. Landscape type maps show a similar variation from about 1:50 000 to the 1:500 000 scales employed for the smaller-scale land system maps. The scale of mapping is determined by a number of factors, principally the objectives of the survey, the requirements of the user or sponsor, the type of terrain, the current and potential land use of the area, the size of area and the time available. These considerations also determine the method and variety of approach. For example, for surveys of large areas the use of a land research approach which identifies land systems at scales between 1:250 000 and 1:500 000 is most appropriate, whereas at a scale of 1:50 000 the terrain analysis approach would be preferred. The regularity of terrain over much of Australia favours the land systems approach, on the other hand the greater morphogenetic complexity of western Europe lends itself to the geomorphological mapping approach.

Geomorphological mapping has been carried out in Britain at numerous scales. The proposed Geomorphological Map of Great Britain is being compiled at 1:250 000 for eventual publication at 1:625 000 (Linton, 1967). A number of large-scale maps covering small areas have been constructed as academic exercises (R. H. Ryder, 1968). Recently, large-scale geomorphological maps have been made as integral parts of resource surveys in Scotland (W. Ritchie, J. S. Smith and A. Mather, 1966; R. S. Crofts, 1969). Small-scale geomorphological and landform mapping in Britain has been of great value in resource assessment, as in Linton's (1968) scenic evaluation of Scotland. The areas covered in the present study were small and had to be mapped in the field and from aerial photographs at large or detailed scales (see note 1) owing to the irregularity of the terrain, dominated as it is by an abundance and diversity of glacial landforms.

The 1:10 560 map scale recommended by the IGU Subcommission was chosen and the readily available Ordnance Survey sheets at this and in some areas the 1:10 000 scale were used as base maps. The whole of Highland Scotland is covered by either early twentieth-century uncontoured editions, or the post-second world war contoured provisional edition or, for limited areas recently photogrammetrically-compiled sheets. Air-photograph cover at approximately the same scale is available for the whole area. In sandy coastal areas the 1:10 560 scale is inadequate to depict the important spatial details required for purposes of evaluation. For such areas, as recommended by G. F. Gellert in the Subcommission report (1963), a three-fold enlargement to the 1:3520 scale was made (Fig. 2).

METHOD OF MAPPING

Mapping was carried out in the field, supported by air photograph interpretation as suggested

FIGURE 2. Areas of detailed geomorphological mapping in the Highlands and Islands of Scotland carried out by members of the Department of Geography, University of Aberdeen (R. S. Crofts, A. Mather, W. Ritchie and J. S. Smith)

by Beckett and Webster (1965) and H. Th. Verstappen (1970). A standard legend based on the Subcommission's recommendations (1968) was used (Figs 3 and 6). Certain problems arise in depicting field data on the maps as landforms vary in size and hence some forms are over-emphasized, compare for example the planation surfaces and glacial forms shown on Figure 3. The inclusion of non-genetic morphological and morphometric information can mask the landforms, whilst the addition of generic data concerning the nature of surface materials cause further confusion. Some of these problems may be overcome on the final version of the map by using colours and replacing symbols by areal shading.

A section of a field map redrawn at a reduced scale but without a morphological base is shown on Figure 3. The complexity of the terrain as mapped is evident and simplification and synthesis are required. This can be achieved by the separation of the morphogenetic-chronologic elements of use in pure and applied geomorphological studies (Verstappen, 1970) and the classification of the data for planning purposes as is discussed in detail below.

APPLICATIONS OF DETAILED MORPHOLOGICAL MAPPING

Academic applications

A map of landforms and landform units may serve as the starting point for enquiries in certain fields of pure geomorphology such as those of the genesis and chronological development of the total landscape or its component parts and the spatial distributions and rates of operation of current processes. The spatial relationship between landforms having the same or a similar genesis and their relationship with the overall shape of the land surface at various scales can be readily assessed from the field map.

Mapping in this study has been focused upon recently deglaciated areas in which a 'freshness' and variety of form are characteristic (Fig. 2). In Feughside four stages in deglaciation can be inferred from the distribution of meltwater channel and esker-kame systems. First, a period of active movement of the ice sheet which overwhelmed the whole Feugh basin as evidenced by the deposition of till and erosion of sub-horizontal meltwater channels at high elevations. Ice stagnation followed when the surface of the ice had downwasted below the level of the basin rim and the supply of ice was cut off. Consequently overflow channels were incised across the lowest cols of the basin and depositional forms became aligned with the directions of flow of the meltwater. After further downwasting the topography within the basin exercised a strong control, resulting in the down-slope flow of meltwater and consequent deposition across the basin floor. Finally, a number of small ice remnants melted to form hillside kettle holes and subaqueous deposits were laid down. Although this postulated scheme of development requires to be tested in the field, it illustrates the use to which detailed geomorphological maps may be put in the construction of genetic hypotheses.

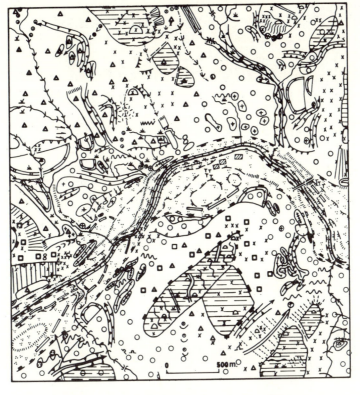

MORPHOMETRY			
		▭ ▭ Abandoned channel	⌇⌇⌇ Hummocky till
▽▽▽ Rock cliff 3 - 10 m.		⊑ Rock gorge	⌣ Hummocky fluvio-glacial
▼▼▼ Rock cliff 10 - 50 m.		°⊙⊚ Seepage	(++++) Esker
⌒ Summit		⊘ Lake	⊕ Kame
Planation surface with backslope		)))))) Catenary valley	Kame terrace

MATERIALS			
		>>> V-shaped valley	⌑ Kettle hole
△ ▵ Till		))) Flat-floored valley	Shoulder glaciated valley
○ ○ Fluvio-glacial		Alluvial fan	Cliffed shoulder glaciated valley
▢ ▫ Boulder field		River banks undercut <1 m, 1 - 3 m	Ice moulded surface
Alluvium		Stable 1-3 m, 3 - 10 m, >10 m	Ice grooved surface
Marsh		Terrace edges 1-3m, 3-10 m, >10 m.	Meltwater channel
Peat		Degraded	Meltwater channel rock, drift

		SLOPES	ANTHROPIC
x x Rock			
⊛ ⊛ Weathered rock		//|\\ Scree	⊨ Bridge
Shattered rock		⊦⊦⊦⊦ Creep	⊤⊤⊤ Made ground

FLUVIAL			
		§§§ Solifluction	⌐ Gravel pit
Permanent stream		GLACIAL	Building
Sporadic stream		Till ridges	+++++ Embankment.

FIGURE 3. A sample area of the detailed geomorphological map, upper Dee valley, Aberdeenshire

TABLE II

The distribution of mass-wasting phenomena by slope gradient, altitude and slope orientation

		Slope gradient				Altitude (metres)					Slope orientation								Total
		0–5°	5–13°	13–31°	>31°	152–305	305–456	456–610	610–760	>760	N	NE	E	SE	S	SW	W	NW	
Rock fall and scree	a	0·06	4·87	11·90		0·75	5·43	6·87	3·43	0·35	1·65	1·46	1·70	4·50	3·05	1·28	1·30	1·89	16·83
	b	0·36	28·95	70·69		4·46	32·24	40·84	20·38	2·08	9·80	8·67	10·12	26·72	18·12	7·64	7·70	11·23	100·00
	c																			2·42
Debris fall	a				0·19			0·12	0·05	0·02			0·03	0·05	0·01	0·01	0·05	0·04	0·19
	b				100·00			63·16	26·32	10·52			15·79	26·32	5·26	5·26	26·32	21·05	100·00
	c																			0·03
Debris slide	a		0·15	0·81		0·01	0·31	0·47	0·13	0·04	0·15	0·12	0·18	0·12	0·13	0·09	0·09	0·08	0·96
	b		15·50	84·50		1·04	32·57	48·18	13·84	4·37	15·50	11·97	19·15	12·69	13·53	9·16	9·16	8·84	100·00
	c																			0·14
Debris flow	a	0·11	0·41	0·36	0·30		0·07	0·46	0·62	0·03	0·32	0·09	0·04	0·11	0·18	0·21	0·21	0·02	1·18
	b	8·93	34·64	30·90	25·53		5·96	38·72	52·77	2·55	27·23	7·23	3·40	9·36	15·16	17·96	17·96	1·70	100·00
	c																			0·1
Creep	a	0·14	2·19	14·84	6·24	0·24	4·82	13·83	4·48	0·04	2·43	3·40	2·10	3·22	4·25	3·02	2·15	2·84	23·41
	b	0·60	9·38	63·38	26·64	1·02	20·59	59·10	19·12	0·17	10·38	14·55	8·97	13·74	18·14	12·90	9·19	12·13	100·00
	c																			3·37
Peat erosion	a	1·29	1·06	0·30	0·04		0·25	1·71	0·73		0·60	0·39	0·27	0·25	0·29	0·46	0·23	0·20	2·69
	b	47·99	39·36	11·16	1·49		9·30	63·62	27·08		22·14	14·69	10·04	9·31	10·71	17·11	8·56	7·44	100·00
	c																			0·39

a, area in km²; b, per cent area of each category; c, per cent of total mapped area

The analysis of mass-wasting is an actively pursued field of research in geomorphology but few workers have mapped the distribution of the different forms involved. They were mapped in detail in the Dee valley, marginal to the Cairngorm and Lochnagar massifs approximately between the 220 and 760 m contours, an area of 695 km². C. F. S. Sharpe's scheme (1938) was used as a basis for the recognition of slip and flowage features. The aerial extent of each form was calculated from the field maps using a 100 m² grid which allowed estimates to the nearest 10 m². Slope gradient, orientation and altitude, all of which may have an influence on the distribution of the forms, were also measured. The classes of slope gradients employed—0 to 5°, 5 to 13°, 13 to 31° and <31°—were based on the work of Waters (1958) and L. F. Curtis, J. C. Doorn-kamp and K. J. Gregory (1965). Altitudes were grouped into 152 m intervals and the direction orientations of slopes into eight sectors centred upon each of the cardinal compass points (grid north orientation).

The data for six mass-movement types are shown in Table II. Surprisingly mass-wasting forms only cover 6·51 per cent of the total area, reflecting a relative stability of slopes between 220 and 760 m. Creep and solifluction (3·37 per cent) and rock fall/scree (2·42 per cent) are the two main types and cover 45·25 km². The majority of mass-wasting types, particularly slide phenomena, are evident on the steeper slopes, but flow types also occur on low angle slopes, Maximum activity is between 456 and 610 m. The preferred orientations of all slope activity, apart from peat erosion, is south-east and south, reflecting perhaps the importance of exposure to solar heat and a greater number of freeze/thaw cycles per annum. Follow-up work on the physical properties of the materials and the hydrological characteristics of the slopes could shed further light on the significance of slope orientation.

This study of the spatial patterns of landforms resulting from glaciation and mass-wasting has a predictive value in so far as it has shown that certain forms occur in particular positions in the landscape such that, for example, contemporary solifluction and soil creep are found on slopes between the high and intermediate plateaux and on the lower oversteepened valley walls. Peat erosion is virtually restricted to the edges of the intermediate plateaux, whilst screes, slips and slides are concentrated on the upper sectors of oversteepened valley walls.

Applications in resource evaluation

Perhaps the most important application of mapping is in the assessment of potential for development in that it reveals any geomorphological advantages or constraints in the area (Christian and Stewart, 1960; Subcommission on Geomorphological Mapping, 1963). The importance of providing specialized and specific maps of the physical environment based on the simplification and synthesis of the original data has often been demonstrated by geomorphologists (Klimaszew-ski, 1956, 1968; Mabbutt and Stewart, 1963) and especially by soil scientists (L. J. Bartelli *et al.*, 1966). The demand from planners for such information to help them in the compilation of structure and land use plans is increasing (R. W. Kiefer, 1965). Some (Bakker, 1963) believe that in attempting to make his views intelligible to the non-specialists, oversimplification could undermine belief in the geomorphologists' powers to contribute to the solution of problems. However, if he first produces a detailed map of physical characteristics and then derives from this a simplified version which the planner can use in the construction of general development plans, reference back to the more elaborate map may be made when and if required.

After the initial testing in Feughside and the Dee valley (Crofts, 1969), the technique has been utilized in projects of the University of Aberdeen concerned with the physical basis of resource evaluation and economic development.

An obvious first application of these maps is as inventories of particular resources such as building aggregate to evaluate the type and amount of aggregate within a given distance of the

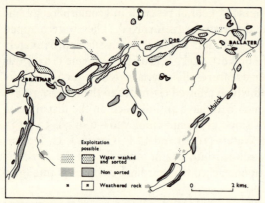

FIGURE 4. Distribution of aggregates of sand/gravel size and their exploitation potential, upper Dee valley, Aberdeenshire

point of demand suitable for use by the construction industry. Such evaluations have to be based on knowledge of formation of glaciofluvial landforms, and the requirements of the construction industry(A. G. McLellan, 1967). Figure 4 is an evaluation of this particular resource based on the geomorphological maps of the upper Dee valley. Three types of aggregate are available, water-washed and sorted glaciofluvial deposits in esker and kame systems, non- or poorly-sorted types of glacial and slope deposits and the 'sorted' gravels resulting from *in situ* chemical weathering of bedrock. The classification of the feasibility of exploitation is based on the volume and homogeniety of the deposits. This could be elaborated by information on the physical properties of the materials according to the preferences of the user.

Secondly, the recognition of currently active geomorphological processes during mapping highlights environmentally fragile situations, which being potential constraints upon development require to be pinpointed and brought to the attention of the planner. A simple example of this is the oft-proposed road link between the Dee valley above Braemar and the Spey valley at Kincraig via the Geldie and Feshie valleys. A geomorphological appraisal of the feasibility of the project where necessary, should permit a reconciliation between conservation and development interests. In general the proposed road follows a natural route crossing a col in the Dee–Spey

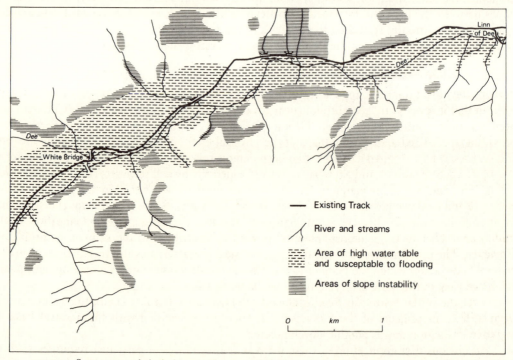

FIGURE 5. Some geomorphological constraints on road development in the upper Dee valley, Aberdeenshire

watershed at the low elbow of river capture (Linton, 1949). Part of the proposed road follows the existing gravel track shown on Figure 5. Two geomorphological factors which act as technical and hence economic restraints have been incorporated in the map. Any road construction scheme necessitating the removal of supporting debris at or near the base of the valley side slope will inevitably cause failure on the slopes above, particularly if the slope is already actively creeping or flowing. Expensive retaining structures would be required or, alternatively, the road must be realigned. The latter alternative would also be difficult as the road would have to be constructed across the Dee flood-plain, and whilst for the most of its length the road could be above the annual average flood level, not even an embankment could protect it from abnormal flood conditions resulting from intense rain storms known to occur at frequent intervals. During such events the flood-plain is inundated to a considerable depth and debris from gullying on the valley slopes is transported across the foot-slopes and deposited on the valley floor (Crofts, 1969). It is on the basis of such observations and other physical, economic and social facts that the planner and road engineer have to determine a least-cost route.

Geomorphological survey can also be employed as an aid in the evaluation of the multiple uses of the countryside. Two examples will be given.

Proposals for the development of daytime and overnight recreational facilities have been formulated for the Muir of Dinnet area, Dee valley. The siting of certain facilities as outlined in a detailed plan were at variance with some geomorphological constraints. Camping areas were to be sited on an alluvial fan astride a stream of highly variable discharge and the main picnic and camping area gave immediate access to an area of permanently high water table (the infilled outer margin of a large but shallow kettle hole lake). A geomorphological map at a scale of 1:2640 was constructed in the field and from this a drainage map was abstracted, these together with other geomorphological and ecological data collected during the field survey were used in the formulation of a revised plan by the planners.

Similarly, the rational planning of the use of coastal areas demands a knowledge not only of the physical processes operating or likely to operate there naturally but also the likely effects of man's influence upon them. An inventory of the physical characteristics of the sandy coasts of the Scottish Highlands and Islands and the processes operating is being compiled under the sponsorship of the Countryside Commission for Scotland. The areas covered so far are shown on Figure 2 (Crofts and Mather, 1972; Crofts and Ritchie, 1973; Mather and Crofts, 1972; Ritchie, 1971; Ritchie and Mather, 1969, 1970 a and b). Stretches of sandy coast are infrequent and their uncontrolled use has led to a degradation of the resource through the intensification of pre-existing instability within the system and the creation of new areas of instability (Mather Ritchie, and Crofts, 1973).

The relevance of detailed geomorphological mapping is best shown by reference to Mellon Udrigle, Wester Ross (Fig. 6) (Crofts and Mather, 1972, 42–6). The area consists of three sectors—a steep sandy intertidal beach, a very wide backshore with abundant sand accumulation and a machair (or dune pasture) at a senile stage of development and in a highly degraded state. The backshore sand reservoir would appear to be adequate to regenerate the eroding machair but this is not happening. A comparative analysis of the texture and shape of sand grains in the backshore and machair suggests that the reservoir on the backshore is currently being derived from the erosion of the machair as well as from the beach, and that very little sand is returned to the machair. The instability of the machair, caused by a combination of wind erosion, overgrazing and intensive recreational activity, would perhaps have been naturally repaired but for the present dominance of offshore winds which transport the sand seawards. The preferred method of restabilization of the machair would be to allow a steady-state situation to develop, but this does not appear to be possible with the present wind pattern. Management strategies

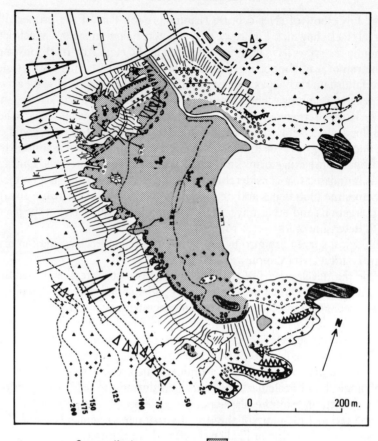

-----	Contours (feet)		Bare sand
7°	Slope angle		Cobbles
⩔⩔	Rock slope >15° sharp or convex upper break		Boulders
⩛⩛	Drift slope >15° sharp or convex upper break		Marsh
⩔⩔⩔	Live marine cliffs, rock, drift	⊢	Terracettes
⩔⩔⩛	Fossil marine cliffs, rock, drift		Debris slide
▭ ▭	Ridge crest	T T	Talus
	River terrace edge rock, drift		Wind undercut face
	Raised shoreline edge rock, drift		Coastal edge prograding
H.W.M.	H.W.M.O.S.T. rock, drift		Coastal edge stable
L.W.M.	L.W.M.O.S.T. rock drift		Embryo dune
⟶	Drainage channel	‖‖‖	Machair
◎	Tidal pool		Building
+ .	Rock	═══	Road fenced and metalled
▲ ▲	Till	═══	Unfenced road
	Intertidal rock platform	=====	Track

FIGURE 6. Detailed geomorphological map of the coast at Mellon Udrigle, Wester Ross

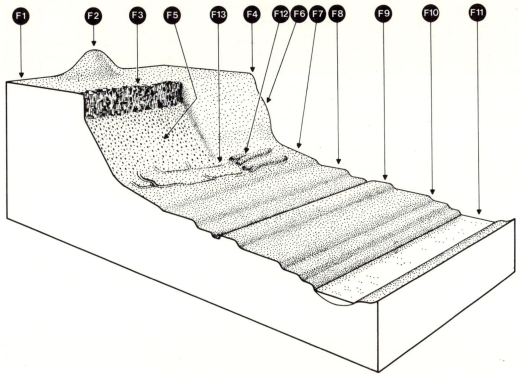

FIGURE 7. Block diagram of typical arrangement of facets comprising a mature glaciated valley recurrent landscape pattern:

F1 Planation surface	F8 Relict glacial terraces
F2 Hill summit	F9 Relict river terraces
F3 Rock cliff	F10 Flood-plain terrace
F4 Hill crest	F11 River and stream bottomlands
F5 Unstable hill-slopes	F12 Glaciofluvial depositional features
F6 Stable hill-slopes	F13 Glaciofluvial erosional features

and tactics involving the prohibition of access, planting and artificial surface stabilization, are needed in order to restore the resource to a stable state.

LANDFORM CLASSIFICATION

A detailed geomorphological map needs to be simplified if it is to be of direct use to non-specialists. This may be achieved by employing some type of classificatory system, the type and level of which is determined by user requirements. A series of such classificatory maps at various levels of abstraction is preferable.

The features shown on a geomorphological map occur at three levels in an hierarchically conceived arrangement of landform components as described by Thomas (1969)—namely the *facet*, the *unit landform* and the *landform complex*. The wide range in size of the components creates problems of scale at the classificatory stage. It is necessary for the sake of simplicity to adopt a basic taxonomic unit capable of identification both at the 1 : 10 000 scale and when reduced to smaller scales for presentation to planners. The basic unit of area adopted is the *facet*, described by Wooldridge (1932) as 'the physiographic atom' and defined here as a unit of land surface having homogeniety of origin and overall similarity of surface form, material and age. So defined facets are comparable with Thomas' (1969) facet and resemble the facets defined by Beckett and Webster (1965), the facets of V. I. Prokayev (1962) and the unit of J. B. Dalrymple, R. J. Blong and A. J. Conacher (1968) but it has no equivalent in Linton's (1951) hierarchy. However, the facets of Becket and Webster and of Thomas are identifiable at the 1:50 000 scale, whereas those defined here may be identified at the 1:10 000 scale. Webster was unable to recognize such units at the 1:50 000 scale in Argyll (Bibby pers. comm., 1972) and was forced to adopt the term *patterned facet* consisting of 'a few elements . . . linked together by simple inter-relationships in a recurrent pattern within the unit' (Bibby, 1969). It is here argued that *facets* as defined above, being in Thomas' terminology a 'genetically single feature', can be meaningfully identified in the Scottish Highlands at the 1:10 000 scale. Recently the Nature Conservancy (1973) has applied the method successfully to Beinn Eighe. Examples of *facets* delimited in Highland Scotland are portrayed on the block diagram in Figure 7. They are similar to the *catena* of G. Milne (1935) and the catenary pattern of land units of H. A. Haantjens (1965) in that topographic situation has a controlling effect on the distribution and recognition of facets.

Further precision can be gained by the subdivision of facets into *sites*. These are equivalent to Linton's site and are the fundamental unit of relief, 'the indivisible flat or slope' (Linton, 1951, p. 215), the physiographic electron. A flood-plain facet, for example, comprises three sites—the channel floor of low gradient, relatively coarse alluvium and a high water table; the flat surface of the flood-plain based on coarse alluvium with a cover of fine overbank deposits and moderate drainage; and the intervening relatively steep banks of the channel with coarse alluvium and good drainage. Each site is identified primarily by its morphology (cf. Thomas, 1969), and within it there is uniformity of lithology and drainage.

The other practically useful level of generalization from the detailed geomorphological map is at the approximately 1:250 000 scale (Fig. 7). The objective is to group facets, employing the terminology initially used by Beckett and Webster (1965), into characteristic associations, *recurrent landscape patterns* (RLP). These are the *tracts* of Unstead (1933) and the *land systems* of Beckett and Webster (1972) and the CSIRO (Christian and Stewart, 1953) although the latter are of greater areal extent. Other RLPs such as mountain massifs, dissected upland plains and highland valleys can be identified in the Scottish Highlands.

The map of facets (Fig. 8) has been compiled directly from the detailed geomorphological map by collapsing the data into genetically-compatible aerially-contiguous units. Its simplicity compared to the geomorphological map (Fig. 3) is self-evident. The facets have not been divided into sites as this would overcomplicate the document.

LAND CLASSIFICATION AND POTENTIAL

Geomorphological data, either in the form of a detailed geomorphological map or a map of facets, are not the only criteria which may be used to classify land for potential development.

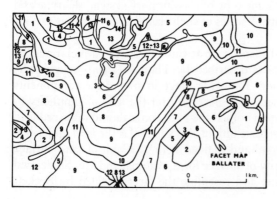

FIGURE 8. Facet map of the Ballater area. Key as for Fig. 7 with addition of F14 Glacial depositional features

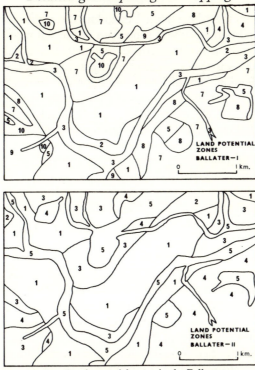

FIGURE 9. Land potential zones in the Ballater area

LAND POTENTIAL ZONES—BALLATER—I

LAND POTENTIAL	DETERMINING CRITERIA
1. High and moderate density housing; light, low-factories; all road types	Low gradient, well-drained surfaces, mixed sediments, moderate bearing capacity, with negligible settlement
2. Minor roads; mixed agriculture	Low gradient, poorly drained surfaces, mixed sediments, low bearing capacity with moderate settlement
3. No potential other than water-based recreation and water supply	Low gradient permanently waterlogged areas of mixed sediments, high settlement (includes watercourses)
4. Grazing; forestry	Intermediate gradient, seasonally high water tables, mixed sediments, low bearing capacity, with moderate settlement
5. Hill grazing; forestry	High gradient, unstable slopes
6. Hill grazing; forestry; recreation	High gradient stable slopes, in rock, drift or drift over rock, low bearing capacity
7. Low density housing; minor roads; mixed agriculture; forestry	Intermediate gradient slopes with coherent drift over rock, moderate bearing capacity with low settlement
8. High and moderate density housing; factories; all road types; mixed agriculture; forestry	Low gradient slopes with coherent drift over rock, high bearing capacity with low settlement
9. Low density housing; minor roads; mixed agriculture; forestry	Irregular surface of coherent mixed sediments, moderate bearing capacity with low settlement
10. Unsuitable for all but recreational uses	Irregular rock surfaces, high bearing capacity with negligible settlement

LAND POTENTIAL ZONES—BALLATER—II

1. High development potential—housing, industry, communications, agriculture
2. Moderate development potential—housing, industry, communications
3. Low development potential—housing, industry, communications, except for agriculture and forestry
4. Areas suitable for pasture and forestry
5. Areas unsuitable for development

Many other physical factors must be analysed including the engineering properties of the substrate, its hydrological condition and those properties of the soil which have a bearing upon its agricultural capability. To these must be added economic and social factors. The aim of the survey of the physical factors is first to delimit *land potential zones*, such as those depicted on the geomorphological evaluation maps of J. Pokorny and M. Tycznska (1963), the land suitability maps of engineers (Kiefer, 1965) and geologists (E. Dobrovolny and H. R. Schmoll, 1968) and the land capability maps of soil surveyors (Bibby and D. Mackney, 1969; Mackney and C. P. Burnham, 1966; A. A. Klingebiel and P. H. Montgomery, 1961). Secondly, on the basis of the survey data, to recommend the optimum potential land use.

Such a scheme was developed and tested in the Ballater area of the Dee valley where residential, recreational and industrial development was proposed. Data at the facet and site levels plus information on the engineering properties of the substrate have been combined to produce *land potential zones* (Figs 9 I and II). Two levels of classification are presented based on the most feasible use—residential, industrial, communications, cultivation, forestry, pasture and recreation.

The physical criteria used in the evaluation are gradient, hydrology, slope stability and the character of the substrate. Each criterion was first mapped individually, then areas with similar characteristics were grouped and the preferred use was designated. In many cases one physical factor determines the potential land use. For example, many of the gradients in facets 2, 4, 5, 6, 12 and 13 (Fig. 8) make them unsuitable for the erection of buildings. On the other hand the lack of relative relief and low gradients in facets 8, 9, 10 and 11 make them potentially suitable for building. Other criteria are then brought into the evaluation, for example, facets 10 and 11 are subject to flooding and abandoned river channels on the other terraces are poorly drained. In all these areas the substrate may include a large proportion of fine grained materials which are prone to shrinkage when loaded. Irregularity of relief on facet 12, and slope instability on facets 3 and 5 militate against certain uses and therefore limit the potential of the site. However, potential is relative, the paucity or absence of land with a low inclination may make it necessary to designate more steeply sloping areas for certain uses normally located on flatter areas.

Once the area has been divided on the basis of physical constraints upon, and advantages for, development the result is compared with the possible uses. Obviously certain areas have high potential and are suitable for a whole range of uses. This classification cannot go as far as determining the preferred individual use in this situation as these involve non-physical factors. Therefore Figure 9 I shows a number of possible uses for each area and the determining criteria. This data can be further summarized (Fig. 9 II) to give a general impression of the potential of an area. A comparison between the facet (Fig. 8) and land potential maps (Fig. 9) shows the great similarity in area divisions. This reflects the distinctive physical character of the landform units of the area. Undoubtedly this situation would not hold in many areas but it does illustrate the importance of land classification based on facets.

Maps of this type can be the physical base for evaluating planning applications and for the compilation of structure plans. Such a large scale of presentation is required in the Scottish Highlands as the majority of planning applications are for small areas.

CONCLUSIONS

This study has attempted to demonstrate the applicability of detailed geomorphological mapping to two types of landscape analysis, first in preliminary studies of the spatial and temporal aspects of the geomorphology of an area, and secondly in the classification of landform as a prerequisite to land evaluation as a result of which a partial inventory of the natural resources of an area is produced. Detailed and large-scale geomorphological mapping covering the whole country, similar to the national geological and soil surveys and comparable with the work of the French and Polish geomorphologists, would be a worthwhile task but difficult to justify in economic terms. But detailed geomorphological mapping could be carried out as an integral part of resource inventory and evaluation surveys based on consumer-contractor principles.

The scales of survey discussed—1:10 560 and 1:3520—are larger than those used by many geomorphologists and land surveyors but are dictated by the nature of the terrain, the necessity for a detailed depiction of landform, the intricacy of current land use and the small scale of developments to which the surveys are related. The field map is a basic document from which data can be abstracted, simplified and classified. It is a tool for the solution of applied physical geographical problems and also a basis for academic research.

ACKNOWLEDGEMENTS

Particular thanks for guidance and encouragement are due to Professor K. Walton and other colleagues at Aberdeen University where the work was carried out. Field-work costs were defrayed by grants from the Countryside Commission for Scotland and the Scottish Tourist Board. The figures were drawn by the cartographic staffs in the Geography Departments at Aberdeen University and University College London. The constructive criticisms of an earlier draft by Professor E. H. Brown, Dr R. U. Cooke and the referees are gratefully acknowledged.

NOTE

1. The terms large- and small-scale are used in two contexts, first when referring to map scale and mapping in which context large-scale covers a small area in detail, secondly in referring to size of territory when large scale is continental or sub-continental in dimensions. Mapping scales are approximately grouped into small scale, > 1:250 000; medium scale, 1:100 000 to 1:250 000; large-scale, 1:10 000 to 1:100 000; detailed, < 1:10 000.

REFERENCES

BAKKER, J. P. (1963) 'Different types of geomorphological map', *Geog. Studies* (Warsaw) 46, 13–23

BARTELLI, L. J., A. A. KLINGEBIEL, J. V. BAIRD and M. R. HEDDLESON (1966) *Soil surveys and land use planning*, Soil Science Soc. Amer.

BAWDEN, M. G. and A. R. STOBBS (1963) *The land resources of eastern Bechuanaland*, Directorate of Overseas Surveys

BECKETT, P. H. T. and R. WEBSTER (1965) 'A classification system for terrain', *Military Engineering Experimental Establishment M.E.X.E.* 872

BECKETT, P. H. T. and R. WEBSTER (1969) 'A review of studies on terrain, by the Oxford-MEXE-Cambridge group 1960–69', *Military Engineering Experimental Establishment M.E.X.E. Rep.* 1123

BECKETT, P. H. T. and R. WEBSTER (1972) 'The development of a system of terrain evaluation over large areas', *R. Engrs' J.* 85, 243–58

BIBBY, J. S. (1969) 'Land classification and the South Argyll Mainland Study', unpub. MS

BIBBY, J. S. and D. MACKNEY (1969) 'Land use capability classification', *Tech. Monogr. Soil Surv. G.B.* 1

BOURNE, R. (1931) 'Regional survey and its relation to stock-taking of the agricultural and forest resources of the British Empire', *Oxf. For. Mem.* 13

BOWMAN, I. (1916) 'The Andes of southern Peru', *Am. Geog. Soc. Publ.* 2

BRIDGES, E. M. and J. C. DOORNKAMP (1963) 'Morphological mapping and the study of soil patterns', *Geography* 48, 175–81

BRINK, A. B., J. A. MABBUTT, R. WEBSTER and P. H. T. BECKETT (1966) 'Report of the working group on land classification and data storage', *Military Enginerring Experimental Establishment M.E.X.E. Rep.* 940

BRITISH GEOMORPHOLOGICAL RESEARCH GROUP (1959–62) *Land Form Survey* Reports 1–5

CHRISTIAN, C. S. and G. A. STEWART (1953) 'General report on survey of Katherine-Darwin region, 1946', *Land Res. Ser. CSIRO Aust.* 1

CHRISTIAN, C. S. and G. A. STEWART (1960) 'Land research in northern Australia', *Aust. Geogr.* 7, 217–31

CHRISTIAN, C. S. and G. A. STEWART (1968) 'Methodology of integrated surveys', in 'Aerial surveys and integrated studies', *Proc. Toulouse Conf. 1964, Natural Resource Res. Ser. UNESCO* 6, 233–80

CROFTS, R. S. (1969) 'Terrain analysis', in *Royal Grampian Country*, Dept. of Geogr., Univ. Aberdeen, pp. 4–8

CROFTS, R. S. and A. MATHER (1972) *Beaches of Wester Ross*, Dept. of Geogr. Univ. Aberdeen

CROFTS, R. S. and W. RITCHIE (1973) *Beaches of Mainland Argyll*, Dept. of Geogr., Univ. Aberdeen

CURTIS, L. F., J. C. DOORNKAMP and K. J. GREGORY (1965) 'The description of relief in field studies in soils', *J. Soil Science* 16, 16–30

DALRYMPLE, J. B., R. J. BLONG and A. J. CONACHER (1968) 'A hypothetical nine unit landsurface model', *Z. Geomorph.* 12, 60–76

DEMEK, J. (ed.) (1973) *Manual of detailed geomorphological mapping*, Int. Geogr. Un. comm. Geomorph. Survey and Mapping

DOBROVOLNY, E. and H. R. SCHMOLL (1968) 'Geology as applied to urban planning: an example from the Greteer Anchorage area borough, Alaska', *23 Int. Geol. Cong.* 12, 39–56

DOWLING, J. W. F. and P. J. BEAVEN (1969) 'Terrain evaluation for road engineers in developing countries', *J. Inst. Highw. Eng.* (June), 5–15

FALCONER, J. D. (1915) 'Land forms and landscapes', *Scott. geogr. Mag.* 31, 244–53

FENNEMAN, N. M. (1928) 'Physiographic subdivisions of the United States', *Ann. Ass. Am. Geogr.* 18, 261–363

GLOCK, W. S. (1932) 'Available relief as a factor of control in the profile of a landform', *J. Geol.* 40, 74–83

GRANT, K. (1968) 'A terrain evaluation system for engineering', *Tech. Pap. Soil Mech. Sect., CSIRO Aust.* 2

GVOZDETSKIY, N. A. (1962) 'An attempt to classify the landscapes of the USSR', *Soviet Geogr.* 3, 30–9

HAANTJENS, H. A. (1965) 'Practical aspects of land system surveys in New Guinea', *J. trop. Geogr.* 21, 12–20

HAMMOND, E. H. (1954) 'Small-scale continental landform maps', *Ann. Ass. Am. Geogr.* 44, 33–42

HEATH, G. R. (1956) 'A comparison of two basic theories of land classification', *Photogramm. Engng* 22, 144–68

HERBERTSON, A. J. (1905) 'The major natural regions: an essay in systematic geography', *Geogrl J.* 20, 300–12

HORTON, R. E. (1945) 'Erosional development of streams and their drainage basins', *Bull. geol. Soc. Am.* 56, 275–370

IGNAT'YEV, G. M. (1968) 'Landscape methods abroad', *Soviet Geogr.* 9, 857–63

KIEFER, R. W. (1965) 'Land evaluation for land use planning', *Build Sci.* 1, 109–26

KLIMASZEWSKI, M. (1956) 'The principles of the geomorphological map of Poland', *Przegl. geogr.* 28, 32–40

KLIMASZEWSKI, M. (1963) 'Landform list and signs used in the detailed geomorphological map', *Geogr. Studies* Warsaw 46, 139–77

KLIMASZEWSKI, M. (1968) 'Problems of the detailed geomorphological map', *Folia Geogr., Geogr.-Phys.* 2

KLINGEBIEL, A. A. and P. H. MONTGOMERY (1961) 'Land capability classification', *U.S. Dept. Agric. Soil Conserv. Serv., Agric. Handbook* 210

LINTON, D. L. (1949) 'Some Scottish river captures re-examined-I', *Scott. geogr. Mag.* 65, 123–32

LINTON, D. L. (1951) 'The delimitation of morphological regions', in *London Essays in Geography* (ed. L. D. STAMP and S. W. WOOLDRIDGE), 199–217

LINTON, D. L. (1967) 'A geomorphological map of Great Britain at 1:625 000 scale', *Zpr. geogr. Úst ČSAV* 6, 16

LINTON, D. L. (1968) 'The assessment of scenery as a natural resource', *Scott. geogr. Mag.* 84, 218–38

McLELLAN, A. G. (1967) *The distribution of sand and gravel deposits in west central Scotland*, Dept. of Geogr., Univ. Glasgow

MABBUTT, J. A. (1968) 'Review of concepts of land classification', in *Land Evaluation* (ed. G. A. STEWART), Melbourne, 11–28

MABBUTT, J. A. and G. A. STEWART (1963) 'The application of geomorphology in integrated resources surveys in Australia and New Guinea', *Rev. Géomorph. dyn.* 14, 97–109

MACKNEY, D. and C. P. BURNHAM (1966) 'The soils of the Church Stretton district of Shropshire', *Mem. Soil Surv. G.B.*

MATHER, A. and R. S. CROFTS (1972) *Beaches of west Inverness-shire and north Argyll*, Dept. of Geogr., Univ. Aberdeen

MATHER, A., W. RITCHIE and R. S. CROFTS (1973) 'Highland sand lands', *Geogr. Mag.* 45, 863–7

MILNE, G. (1935) 'Some suggested units of classification and mapping, particularly for east African soils', *Soil Res.* 4, 183–98

NAKANO, T. (1961) 'Landform classification—its principle and its application', *J. Geogr.*, Tokyo 70, 53–64

NATURE CONSERVANCY (1973) *Report of survey of Beinn Eighe National Nature Reserve*

PECSI, M. (1970) *Geomorphological regions of Hungary*

PEREL'MAN, A. L. (1967) *Geochemistry of Epigenesis*

PITTY, A. F. (1969) *A scheme of hillslope analysis*, Dept. of Geogr., Univ. Hull

POKORNY, J. and M. TYCZNSKA (1963) 'Method of evaluation of relief for land planning purposes', *Geogr. Studies* Warsaw 46, 95–100

POPE, R. G., J. M. HAWKES and R. DAVIES (1971) 'Elementary soil-vehicle mechanics', *R. Engrs' J.* 85, 84–103

PROKAYEV, V. I. (1962) 'The facies as the basic and smallest unit in landscape science', *Soviet Geogr.* 3, 21–9

RITCHIE, W. (1971) *Beaches of Barra and the Uists*, Dept. of Geogr., Univ. Aberdeen

RITCHIE, W. and A. MATHER (1969) *Beaches of Sutherland*, Dept. of Geogr., Univ. Aberdeen

RITCHIE, W. and A. MATHER (1970a) *Beaches of Lewis and Harris*, Dept. of Geogr., Univ. Aberdeen

RITCHIE, W. and A. MATHER (1970b) *Beaches of Caithness*, Dept. of Geogr., Univ. Aberdeen

RITCHIE, W., J. S. SMITH and A. MATHER (1966) *Terrain analysis of the northern coastal zone of the Cromarty Firth*, Dept. of Geogr., Univ. Aberdeen

RYDER, R. H. (1968) *Geomorphological mapping of the island of Rhum*, Unpub. M.Sc. Univ. Glasgow

SAVIGEAR, R. A. G. (1965) 'A technique of morphological mapping', *Ann. Ass. Am. Geogr.* 55, 514–39

SCOTT, R. M. and M. P. AUSTIN (1971) 'Numerical classification of land systems using geomorphological attributes', *Aust. Geogr. Stud.* 9, 33–40

SHARPE, C. F. S. (1938) *Landslides and related phenomena*

SMITH, G.-H. (1935) 'The relative relief of Ohio', *Geogr. Rev.* 25, 272–84

SPEIGHT, J. G. (1968) 'Parametric description of land form', in *Land Evaluation* (ed. G. A. STEWART) Melbourne, 239–50

ST-ONGE, D. (1964) 'Geomorphological map legends, their problems and their value in optimum land utilization', *Geogrl. Bull.*, Ottawa, 22, 5–12

STEWART, G. A. (ed.) (1968) *Land evaluation*, Melbourne

STRAHLER, A. H. (1956) 'Quantitative slope analysis', *Bull. Geol. Soc. Am.* 67, 571–96

SUBCOMMISSION ON GEOMORPHOLOGICAL MAPPING (1963) 'Problems of geomorphological mapping', *Geogr. Studies* Warsaw, 46

SUBCOMMISSION ON GEOMORPHOLOGICAL MAPPING (1968) 'The unified key to the detailed geomorphological map of the world in 1:25 000 to 1:50 000 scale', *Folia Geogr., Geogr.-Phys.* 2

THOMAS, M. F. (1969) 'Geomorphology and land classification in tropical Africa', in *Environment and land use in Africa* (eds M. F. THOMAS and G. W. WHITTINGTON), 103-45

TRICART, J. (1965) 'La cartographie géomorphologie detaillée', *Principes et méthodes de la géomorphologie* Paris, 182-215

TRICART, J. (1971a) 'Carte géomorphologique et description du milieu natural', *Mém. et Doc., Centre de Doc. Cartogr. C.N.R.S.* 12, 165-80

TRICART, J. (1971b) Cartographie géomorphologique et classement des terres pour la conservation', *Mém. et Doc., Centre de Doc. Cartogr. C.N.R.S.* 12, 215-22

TRICART, J. *et al.* (1971) 'Cartographie Géomorphologique, travaux de la R.C.P. 77', *Mém. et Doc., Centre de Doc. Cartogr. C.N.R.S.* 12

TURNER, A. K. and R. D. MILES (1968) 'Terrain analysis by computer', *Proc. Indiana Acad. Sci.* 77 (1967), 256-70

UNESCO (1968) 'Aerial Surveys and Integrated Studies', *Proc. Toulouse Conf. 1964, Natural Resource Res. Ser.* 6

UNSTEAD, J. F. (1933) 'A system of regional geography', *Geography* 18, 175-87

VEATCH, J. O. (1935) 'Graphic and quantitative comparisons of land types', *J. Am. Soc. Agron.* 27, 505-10

VERSTAPPEN, H. TH. (1970) 'Introduction to the ITC system of geomorphological survey', *Geogr. Tijdschr.*, N.S. 4, 85-91

WALTON, K. (1968) 'The approach of the physical geographer to the countryside', *Scott. geogr. Mag.* 84, 212-18

WATERS, R. S. (1958) 'Morphological mapping', *Geography* 43, 10-17

WOOLDRIDGE, S. W. (1932) 'The cycle of erosion and the representation of relief', *Scott. geogr. Mag.* 48, 30-6

RÉSUMÉ. *Cartographie géomorphologique de détail et évaluation du terrain des hautes terres écossaises.* On peut préparer des cartes morphologiques et analyser le terrain selon trois critères: genèse, caràctere, ou paysage. La préparation de cartes géomorphologiques de détail emploie la recognaissance des éléments du paysage, mais cette méthode differe des autres techniques de 'paysage' parce que la préparation des cartes et l'évaluation des renseignements fonctionnent à une échelle plus grande. Les formes de terrain sont cartographiées à l'echelle 1:10000 selon leur genèse, morphologie, structure, et chronologie. On emploie cette méthode morphologique dans une étude de géomorphologie pure et appliquée des hautes terres écossaises. On peut entreprendre une analyse des éléments de morphologie déglaciaire des phénomènes liès aux mouvements de masse par l'extraction des renseignements de cette carte. On peut identifier des zones fragiles et instables où des changements physiques sont rapides. Ces renseignements de base permettent une évaluation du potentiel physique pour la construction des routes et l'installation des equipements de loisir. Il faut simplifier cette carte pour faciliter la comparaison avec d'autres régions et pour offrir une meilleure valeur aux non-specialistes. Modèles de paysage—facettes et sites—sont identifiés aux echelles les plus grandes. Ces combinaisons, avec d'autres renseignements concernent le milieu physique, permettent une délimitation de *Zones de Potentiel Physique*. Ces zones pourraient servir comme point de depart pour l'aménagement du territoire.

FIG. 1. Schéma chronologique de la préparation des cartes de géomorphologie et de paysage

FIG. 2. Couverture par cartes géomorphologiques de détail dans les hautes terres et les iles d'Ecosse

FIG. 3. Partie d'une carte géomorphologique de détail: haute vallée du Dee, Aberdeenshire

FIG. 4. Repartition d'aggrégats de dimensions de sable/gravier et potentiel pour leur extraction: haute vallée du Dee, Aberdeenshire

FIG. 5. Quelques contraintes géomorphologiques sur la construction des routes: haute vallée du Dee, Aberdeenshire

FIG. 6. Carte géomorphologique de détail: littoral du Mellon Udrigle, Wester Ross

FIG. 7. Diagramme de la répartition typique de facettes dans une vallée glaciaire:

F1 Surface d'aplanissement	F8 Terrasses glaciaires reliques
F2 Sommet	F9 Terrasses fluviales reliques
F3 Falaise rocheuse	F10 Terrasse de plaine inondable
F4 Crête	F11 Plaines inondables de deposition
F5 Pentes instables	F12 Phénomènes de déposition glaciofluviale
F6 Pentes stables	F13 Phénomènes d'érosion glaciofluviale
F7 Piemont	

FIG. 8. Carte de facettes autour de Ballater
Comme la Figure 7, avec l'addition de F 14 Phénomènes de déposition glaciaire.

FIG. 9. Zones de potentiel physique autour de Ballater

ZONES DE POTENTIEL PHYSIQUE—BALLATER—I

POTENTIAL PHYSIQUE	CRITERES DETERMINANTS
1. Zones résidentielles à haute ou moyenne densité, usines légères et basses, toutes espèces de routes	Pente modèrée, surfaces bien drainées, dépôts mélangés, moyenne surface d'appui, presque sans habitation
2. Routes secondaires, agriculture	Pente modérée, surfaces mal drainées, dépôts mélangés, mauvaise surface d'appui, avec quelques habitations

3. Sans potentiel sauf loisirs et approvisione-
ment des eaux

Pente modérée, dépôts mélanges et engagés dans l'eau, zone d'habitation le long des cours d'eau

4. Pâturages, forêts

Pente moyenne, dépôts mélangés et engagés dans l'eau en hiver, mauvaise surface d'appui, quelques habitations

5. Pâturages maigres, forêts

Pente rapide et instable

6. Pâturage maigres, forêts, loisirs

Pente rapide, stable, et rocheuse, dépôts superficiels ou dépôts couvrant la roche-mère, mauvaise surface d'appui

7. Zones résidentielles à faible densité, routes secondaires, agriculture, forêts

Pente moyenne avec dépôts superficiels couvrant la roche, moyenne surface d'appui, peu d'habitation

8. Zones résidentielles à haute ou moyenne densité, usines, toutes espèces de routes, agriculture, forêts

Pente modérée, dépôts superficiels couvrant la roche-mère, bonne surface d'appui, peu d'habitation

9. Zones résidentielles à faible densité, routes secondaires, agriculture, forêts

Surface irrégulière de dépôts mélangés, moyenne surface d'appui, peu d'habitation

10. Inutile sauf pour loisirs

Surfaces rocheuses et irrégulières, bonne surface d'appui, presque sans habitation

ZONES DE POTENTIEL PHYSIQUE—BALLATER—II

1. Grand potential de développement—résidence, industrie, transports, agriculture
2. Moyen potentiel de développement—résidence, industrie, transports
3. Faible potentiel de développement—résidence, industrie, transports
4. Zones convenables aux pâturages et forêts
5. Zones peu convenables au développement

ZUSAMMENFASSUNG. *Detaillierte geomorphologische Aufzeichnung und Landbewertung in Schottischen Hochland.* Drei Methoden der Aufzeichnung von Landformen und Geländeanalyse werden bestimurt—genetisch, parametrisch und landschaftlich. Detaillierte geomorphologische Aufzeichnung fällt in die Landschaftsmethode, welche der Wiedererkennung von Landschaftsmustern beruht. Sie unterscheidlt sich jedoch von anderen Landschaftsmethoden, weil der Mabstal der Aufzeichnung und Datenauswerbung viel grosser ist. Landschaftsformen werden im Masstab 1:10 000 auf der Basis von Genesis, Morphologie, Struktur und Chronologie aufgezeichnet. Dieses geomorphologische Inventar bildet den Hintergrund für Anwendungen in reiner und angewandter geomorphologie ausgeführt hier am Beispiel des Schottischen Hochlands. Vorlänfige Analyse der entgletscherten Landformen und Erscheinungen von Massenbewegung in Zeit und Raum Können zum Beispiel mit dem Auszug von Daten von der Landkarte erreicht werdcn. Veränderliche und unstabile gebiete, welche zu rapiden physischen Veränderungen neigen, Können identifiziert werden und als Basis für die Bewertung von Entwicklungsplänen für Strassenban und Erholungsgebiete benutzt werden. Die ursprünglichen Aufzeichnungen müssen vereinfacht werden um Verglisch und Extrapolierung mit anderen Gebieten zu ermöglichen und um von grösseren Wert für den Laien zu sein. Landschafts muster—Facetten und Lagen—werden von Karten im grössten Masstab identifiziert. Resultierende groupierungen zusammen mit anderer Information die physische Umgebung Betreffend ermöglichen die Abgrenzung von Landpotential zone. Diese zonen bilden die Grundlage für die Flächennutzungs plannung.

ABB. 1. Schematische Darstellung der chronologischen Entwicklung von Landformen und Landschaftsaufzeichnung
ABB. 2. Gebiete detaillierter geomorphologischer Aufzeichnung im Hochland und den Inseln Schottlands
ABB. 3. Muster einer detaillierter geomorphologischen landkarte, gebit im oberen Deetal, Aberdeenshire
ABB. 4. Verteilung von Kiesanhäufungen und ihr Ausnutzungspotential, oberes Deetal, Aberdeenshire
ABB. 5. Einige geomorphologische Beschränkungen für den Strassenbau im oberen Deetal, Aberdeenshire
ABB. 6. Detaillierte geomorphologische Karte der Küste bei Mellon Udrigle, Wester Ross
ABB. 7. Graphische Darstellung einer typischen Anordnung von Facetten, die wiederkehrenden Landschaftsmuster eines geteiflen gletschertales umfassen:

F1 Planierte Oberfläche   F8 Übergebliebene Geltscherterrassen
F2 Gipfel   F9 Übergebliebene Flussterrassen
F3 Felsenkliff   F10 Überschwemmungsebene
F4 Bergkranz   F11 Fluss-und Strombochflutbetten
F5 Unstabile Berghänge   F12 Fluvioglaciale Ablagemerkmale
F6 Stabile Berghänge   F13 Fluvioglaciale Erosionsmerkmale
F7 Bergfuss

ABB. 8. Facettenkarte der Ballater Gegend Erläuterungen wie in Abbildung 7 mit zusätzlich: F14 Gletscher Ablagemerkmale
ABB. 9. Landpotential zonen in der Ballater Gegend

LANDPOTENTIAL ZONEN—BALLATER—I

LANDPOTENTIAL	BESTIMMENDE MERKMALE
1. Hohe und mässige Wohndichte, leichte Fabriken, alle Strassentypen	Geringe Neigung, gut entwässerte Oberflächen, gemischte Ablagerungen, mässige Tragfähigkeit mit unbedeutender Setzungen
2. Nebenstrassen, genischte Landwirtschaft	Geringe Neigung, schlecht entwässerte Oberflächen, gemischte Ablagerungen, geringe Tragfähigkeit mit mässiger Setzungen
3. Kein Potential Ausser auf Wasser beruhender Erholung und Wasserversorgung	Geringe Neigung, ständig durchtränkte gebiete gemischter Ablagerungen, starke Setzungen (Wasseradern eingeschlossen)
4. Weide, Forstwirtschaft	Mittlete Neigung, hoher Wasserspiegel jahreszeitlich bedingt, genische Ablaglerungen, geringe Tragfähigkeit mit mässingen Setzungen
5. Bergweide, Forstwirtschaft	Steile Neigung, unstabile Hänge
6. Bergweide, Forstwirtschaft, Erholung	Steile Neigung, stabile Felschänge, Drift oder Drift überFelsen, geringe Tragfähigkeit
7. Geringe Wohndichte Nebenstrassen, gemischte Landwirtschaft, Forstwirtschaft	Mittlere Neigung, Abhänge mit zusammenhängen dem Drift über Felsen, mässige Tragfähigkeit mit geringen Setzungen
8. Hohe und mässige Wohndichte, Fabriken, alle Strassentypen, gemischte Landwirtschaft, Forstwirtschaft	Geringe Neigung, Abhange mit zusammenhängendem Drift über Felsen, hohe Tragfähigkeit mit geringen Setzungen
9. Geringe Wohndichte, Nebenstrassen, gemischte Landwirtschaft, Forstwirtschaft	Unregelmässige Oberfläche aus zusammen hängenden gemischten Ablagerungen, mässige Tragfähigkeit mit geringen Setzungen
10. Ungeeignet fur alle Nutzung ausser Erholung	Unregelmässige Felsen-oberflächen, hohe Tragfähigkeit mit unbedeistenden Setzungen

LANDPOTENTIAL ZONEN—BALLATER—II

1. Hohes Entwicklungspotential—Wohnen, Industrie, Kommunikationen, Landwirtschaft
2. Mässiges, Entwicklungspotential—Wohnen, Industrie, Kommunikationen
3. Geringes Entwicklungspotential—Wohnen, Industrie, Kommunikationen. ausser land—und Forstwirtschaft
4. Geeignete Gebiete für Weide und Forstwirtschaft
5. Für Entwicklung ungeeignete gebiete

The British Geomorphological Research Group

The British Geomorphological Research Group was founded in January 1961 to encourage research in geomorphology, to undertake large-scale projects of research or compilation in which the co-operation of many geomorphologists is involved, and to hold field-meetings and symposia. In January 1970 the Group was constituted a formal Study Group of the Institute of British Geographers. Under the chairmanship of Professor Eric H. Brown, it has a membership of over 400; its publications include a series of Technical Bulletins, a Register of Current Research in Geomorphology, and a Bibliography of British Geomorphology.

Details of membership can be obtained from the Honorary Treasurer and Membership Secretary, Dr J. R. Hails, Institute of Oceanographic Sciences, Beadon Road, Taunton, Somerset. Lists of publications for sale may be obtained from the Hon. Secretary of the Group, Dr E. Derbyshire, Department of Geography, The University of Keele, Keele, Staffs., ST5 5BG.

The following publications are available postage free, if payment sent with order, from: Geo Abstracts, University of East Anglia, Norwich, NOR 88C, England.

Technical Bulletins: (£0.40 each)
1. Field methods of water hardness determination, IAN DOUGLAS, 1969
2. Techniques for the tracing of subterranean drainage, DAVID P. DREW and DAVID I. SMITH, 1969
3. The determination of the infiltration capacity of field soils using the cylinder infiltrometer, RODNEY C. HILLS, 1970
4. The use of the Woodhead sea bed drifter, ADA PHILLIPS, 1970
5. A method for the direct measurement of erosion on rock surfaces, C. HIGH and F. K. HANNA, 1970
6. Techniques of till fabric analysis, J. T. ANDREWS, 1970
7. Field method for hillslope description, LUNA B. LEOPOLD and THOMAS DUNNE
8. The measurement of soil frost-heave in the field, PETER A. JAMES
9. A system for the field measurement of soil water movement, BRIAN J. KNAPP
10. An instrument system for shore process studies, ROBERT M. KIRK

Current Register of research 1970–71 (£0.60)
Current Register of research 1972–73 (£1.50)
A Bibliography of British Geomorphology (ed. K. M. CLAYTON), Philip 1964 (£0.93)

The Institute of British Geographers

Details of Membership are available from the Administrative Assistant, Institute of British Geographers, 1 Kensington Gore, London, SW7 2AR (Tel. 01–584 6371).

Papers or monographs intended for publication must be sent in the first place to the Hon. Editor, Dr B. Robson, Department of Geography, Cambridge University, Downing Place, Cambridge. Papers for reading at the Annual Conference (even if they are subsequently to be considered for publication) should, however, be sent to the Hon. Secretary, Professor R. Lawton, Department of Geography, University of Liverpool.

Requests for copies of publications should be made to the Administrative Assistant, who also has available 'Notes for the guidance of authors submitting papers for publication by the Institute'.

Publications in Print

These publications may be obtained from any bookseller at the prices quoted above. Members of the Institute may buy copies of Transactions at two-thirds of these prices and copies of the Special Publications at a discount of 20 per cent. Application for these should be made to the Administrative Assistant, Institute of British Geographers, 1 Kensington Gore, London, SW7 2AR.